W9-BFX-712

SUFFOLK UNIVERSITY LIBRARY

DISCARDED BY
SUFFOLK UNIVERSITY
SAWYER LIBRARY

**INTRODUCTION
TO
SUPERCONDUCTIVITY**

The late F. K. Richtmyer was Consulting Editor of the series from its inception in 1929 to his death in 1939. Lee A. DuBridge was Consulting Editor from 1939 to 1946; and G. P. Harnwell from 1947 to 1954. Leonard I. Schiff served as consultant from 1954 until his death in 1971.

**McGRAW-HILL
BOOK COMPANY**

New York
St. Louis
San Francisco
Auckland
Düsseldorf
Johannesburg
Kuala Lumpur
London
Mexico
Montreal
New Delhi
Panama
Paris
São Paulo
Singapore
Sydney
Tokyo
Toronto

MICHAEL TINKHAM
*Gordon McKay Professor of Applied Physics
and Professor of Physics
Harvard University*

Introduction to Superconductivity

This book was set in Times New Roman.
The editors were Bradford Bayne and Michael Gardner;
the production supervisor was Dennis J. Conroy.
The drawings were done by J & R Services, Inc.
Kingsport Press, Inc., was printer and binder.

Library of Congress Cataloging in Publication Data

Tinkham, Michael.
  Introduction to superconductivity.

  (International series in pure and applied physics)
  Bibliography: p.
  Includes index.
  1. Superconductivity.   I.   Title.
QC612.S8T49   537.6′23   74-32166
ISBN 0-07-064877-8

QC 612
S 8
T49

94788

**INTRODUCTION
TO
SUPERCONDUCTIVITY**

Copyright © 1975 by McGraw-Hill, Inc. All rights reserved.
Printed in the United States of America. No part of this publication may be reproduced,
stored in a retrieval system, or transmitted, in any form or by any means,
electronic, mechanical, photocopying, recording, or otherwise,
without the prior written permission of the publisher.
1 2 3 4 5 6 7 8 9 0 KP KP 7 9 8 7 6 5

To MARY,
JEFFREY, and CHRISTOPHER

CONTENTS

PREFACE

This book has evolved from a set of lecture notes originally written for a graduate course at Harvard University during the fall term of 1969. They were subsequently rewritten during a sabbatical leave at the Cavendish Laboratory in 1971–1972 and during a repeat of the course in 1973.

The objective of the lectures, and of this book, is to provide an up-to-date introduction to the intriguing subject of superconductivity and some of its potential applications. The emphasis is on the rich array of phenomena and how they may be understood in the simplest possible way. Consequently, the use of thermal Green functions has been completely avoided, despite their fashionability and undeniable power in the hands of skilled theorists. Rather, the power of phenomenological theory in giving insight is emphasized, and microscopic theory is often narrowly directed to the task of computing the coefficients in phenomenological equations. It is hoped that this emphasis will make the treatment more palatable to the experimentalist, and also complement the more generous coverage of the formal theoretical aspects of the subject in most books presently available. Finally, the author was motivated by the hope that if the theoretical techniques were kept as elementary as possible, the work might have more value to undergraduates and technologists with incomplete backgrounds in theoretical physics.

In a sense this book forms an updated and greatly expanded version of the Les Houches lectures of the author, written in 1961. However, so much development of the subject has occurred in the intervening years that these notes were really rewritten (twice) from start to finish. In the process, the author has drawn frequently on the excellent book of de Gennes, "Superconductivity in Metals and Alloys," and on the two-volume treatise "Superconductivity" edited by Parks. There is little in the book which has not been published previously in some form, but some topics—particularly fluctuation effects—have developed too recently to have appeared in previous books.

No attempt has been made to give an exhaustive or definitive treatment. Such a treatment required the two volume Parks treatise mentioned above. Rather, the author has chosen to introduce the reader to a selection of topics which reflect his own focus on the electrodynamic properties of superconductors, which, after all, give the subject its unique interest. The time limitation of a semester lecture course provided unrelenting discipline in limiting the number of topics and the depth of treatment.

The book starts with an introductory survey which lays out the ground to be covered in the book, and gives some of the milestones in the historical development of the subject. The reader is advised to treat this as an overview only, intended to introduce concepts and language, with the detailed explanations to be developed in subsequent chapters. He definitely should not puzzle over issues which are only sketchily introduced at this point.

The second chapter is devoted to "basic BCS," the microscopic theory developed by Bardeen, Cooper, and Schrieffer to explain the superconducting state. This theory is placed at the beginning because no serious discussion of superconductivity is possible without concepts derived from the theory. Unfortunately, this chapter has by far the most forbidding formal nature of any part of the book, but this should not be allowed to discourage the reader. Little use of the mathematical details will be made in the following chapters, and so this chapter can be skimmed for the general ideas (which are summarized in the concluding section), and referred to later if more detailed understanding of some particular point is required.

With Chap. 3, we move into the phenomenological level of treatment, which characterizes the rest of the book. First, the implications of the nonlocal electrodynamics in determining the effective penetration depth of a magnetic field into bulk and thin-film superconductors are explored, the thorough discussion of the latter topic reflecting a historical interest of the author. A simplified discussion is then given of the intermediate state, in which superconducting and normal material coexist in the presence of a magnetic field.

Chapter 4 develops the Ginzburg-Landau theory from the same phenomenological point of view used by the original authors. The theory is then applied to an extensive catalog of classic problems: domain-wall energy, critical current density, fluxoid quantization, critical fields of films and foils, the upper critical field H_{c2}, the Abrikosov vortex state, and the surface nucleation field H_{c3}. The concepts treated here underlie the subjects treated in the following chapters, in addition to illustrating the power of the Ginzburg-Landau approach.

In Chap. 5, the magnetic properties of type II superconductors are developed in some detail. After the equilibrium flux density has been worked out, attention is focused on the creep and flow of the flux under the influence of transport currents. In this way, insight is obtained into the considerations which limit potential applications of type II superconductors in high-field magnets. The chapter concludes with a discussion of the factors governing the design of superconducting magnets to cope with time-varying fields, including the use of twisted multicore composite conductors to minimize ac losses while maintaining thermal stability.

Chapter 6 is devoted to the Josephson effect and macroscopic quantum phenomena. These subjects represent some of the purest and most fundamental aspects of superconductivity, yet also provide the basis for sensitive instruments which have revolutionized electromagnetic measurements. Both aspects are reflected in the treatment given; in particular, the detailed discussion of practical SQUID magnetometers is the first to appear in a textbook.

Although for years it was thought that the effects of thermodynamic fluctuations were unobservably small in superconductors, the advent of the superconducting detectors just mentioned has made it possible to observe such effects both above and below T_c. Chapter 7 surveys these phenomena in both electrical conductivity and diamagnetism. For example, it is shown how fluctuation effects put a limit (though an astronomical one) on the lifetime of "persistent" currents below T_c, and how they also give rise to "precursors" of superconductivity above T_c. Because this subject has flowered since the date of the Parks treatise, this book is the first containing a thorough discussion of this interesting and informative new aspect of superconductivity.

The final chapter is devoted to introductory discussions of three topics: the Bogoliubov method, gapless superconductivity, and time-dependent Ginzburg-Landau theory. These topics go beyond the elementary Ginzburg-Landau phenomenology, and bring in more microscopic considerations. Yet the basic concepts and conclusions have been drawn inevitably into the discussions of the topics treated earlier; moreover, taken together, they lay the groundwork for work going on at the present frontiers of research. Hence, it seems fitting to close the book with a peek at these topics, where the last word is by no means in.

Finally, the author is pleased to thank the reviewers of the manuscript for constructive suggestions; the detailed reading of the final manuscript by Dr. Richard Harris is especially appreciated. The comments of students who have used the notes also were particularly helpful. The speedy and accurate typing of Miss Patricia McCarthy in preparing the final manuscript was an invaluable incentive to continued progress. More generally, the author wants to thank his numerous students, colleagues, and collaborators, especially in Berkeley, Orsay, Harvard, and the other Cambridge, for making his exploration of superconductivity the pleasure it has been. Although it would be impossible to list them all here, I cannot close this Preface without explicitly acknowledging numerous seminal discussions over the years with M. R. Beasley, J. Clarke, P. G. de Gennes, R. A. Ferrell, and R. E. Glover, III. If this book serves to initiate others into the fascination I have found in this subject, it will have well served its intended purpose.

MICHAEL TINKHAM

**INTRODUCTION
TO
SUPERCONDUCTIVITY**

1
INTRODUCTORY SURVEY

Superconductivity was discovered in 1911 by H. Kamerlingh Onnes in Leiden, just three years after he had first liquefied helium, giving him the refrigeration technique required to reach temperatures of a few degrees Kelvin. For decades, a fundamental understanding of this phenomenon eluded the many scientists working in the field, but in recent years a remarkably complete and satisfactory picture has emerged. It is the purpose of this book to introduce the reader to this picture.

We start this chapter by reviewing the basic observed electrodynamic phenomena and their early phenomenological description. We then briefly sketch the evolution of the more recent concepts which are central to our present understanding. It is hoped that this quasi-historical review of the development of the subject, though terse, will provide helpful guideposts to orient the reader as he proceeds through the subsequent chapters, in which the ideas are developed in more detail.

1-1 THE BASIC PHENOMENA

What Kamerlingh Onnes[1] observed was that the electrical resistance of various metals such as mercury, lead, and tin disappeared completely in a small temperature range at a critical temperature T_c, characteristic of the material. The complete disappearance of resistance is most sensitively demonstrated by experiments with persistent currents in superconducting rings, as shown schematically in Fig. 1-1. Once set up, such currents have been observed to flow without measurable decrease for a year, and a lower bound of some 10^5 years for their characteristic decay time has been established using nuclear resonance to detect any slight decrease in the field produced by the circulating current. In fact we shall see that under many circumstances we expect absolutely no change in field or current to occur in times less than $10^{10^{10}}$ years! Thus, *perfect conductivity* is the first traditional hallmark of superconductivity. It is also the prerequisite for most potential applications, such as high-current transmission lines or high-field magnets.

The next hallmark to be discovered was *perfect diamagnetism*, found in 1933 by Meissner and Ochsenfeld.[2] They found not only that a magnetic field is excluded from a superconductor (see Fig. 1-2), as might appear to be explained by perfect conductivity, but also that a field is *expelled* from an originally normal sample (if it is of high quality) as it is cooled through T_c. This certainly could *not* be explained by perfect conductivity, which would tend to trap flux *in*. The existence of a reversible *Meissner effect* implies that superconductivity will be destroyed by a critical magnetic field H_c which is related thermodynamically to the free-energy difference between normal and superconducting states in zero field, the so-called condensation energy of the superconducting state. More precisely, H_c is determined by equating the energy $H^2/8\pi$ per unit volume, associated with holding the field out against the magnetic pressure, with the condensation energy. That is,

$$\frac{H_c^2(T)}{8\pi} = f_n(T) - f_s(T) \qquad (1\text{-}1)$$

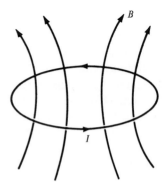

FIGURE 1-1
Schematic diagram of persistent current experiment.

[1] H. Kamerlingh Onnes, *Leiden Comm.* 120b, 122b, 124c (1911).
[2] W. Meissner and R. Ochsenfeld, *Naturwissenschaften* **21**, 787 (1933).

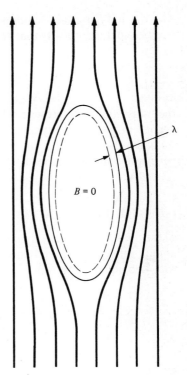

FIGURE 1-2
Schematic diagram of exclusion of
magnetic flux from interior of massive
superconductor. λ is the penetration
depth, typically only 500 Å.

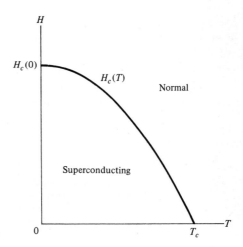

FIGURE 1-3
Temperature dependence of critical field.

where f_n and f_s are the Helmholtz free energies per unit volume in the two phases. It was found empirically that $H_c(T)$ is quite well approximated by a parabolic law

$$H_c(T) \approx H_c(0)\left[1 - \left(\frac{T}{T_c}\right)^2\right] \qquad (1\text{-}2)$$

as illustrated in Fig. 1-3. While the transition in zero field at T_c is of second order, the transition in the presence of a field is of first order, there being a discontinuous change in the thermodynamic state of the system and an associated latent heat. Actually, the diamagnetism is perfect only for bulk samples, since the field does penetrate a finite distance λ, typically approximately 500 Å (angstroms).

1-2 THE LONDON EQUATIONS

These two basic electrodynamic properties, which give superconductivity its unique interest, were well described in 1935 by the brothers F. and H. London[1] with two equations governing the microscopic electric and magnetic fields

$$\mathbf{E} = \frac{\partial}{\partial t}(\Lambda \mathbf{J}_s) \qquad (1\text{-}3)$$

and

$$\mathbf{h} = -c \operatorname{curl}(\Lambda \mathbf{J}_s) \qquad (1\text{-}4)$$

where

$$\Lambda = \frac{4\pi\lambda_L^2}{c^2} = \frac{m}{n_s e^2} \qquad (1\text{-}5)$$

is a phenomenological parameter. It was expected that n_s, the "number density of superconducting electrons," would vary continuously from zero at T_c to a limiting value of the order of n, the density of conduction electrons, at $T \ll T_c$. In (1-4) we introduce our notational convention of using $\mathbf{h}$ to denote the value of the flux density on a microscopic scale, reserving $\mathbf{B}$ to denote a macroscopic average value. Although notational symmetry suggests using $\mathbf{e}$ for the microscopic local value of $\mathbf{E}$ in the same way, to avoid constant confusion with the charge e of the electron we shall do so only in the few cases* where it is really useful. These notational conventions are discussed further in the Appendix.

The first of these equations describes perfect conductivity, since any electric field accelerates the superconducting electrons rather than simply sustaining their

---

[1] F. and H. London, *Proc. Roy. Soc. (London)* **A149**, 71 (1935).

* The fundamental basis for our notational asymmetry in treating $\mathbf{E}$ and $\mathbf{B}$ is in the Maxwell equations

$$\operatorname{curl}\mathbf{h} = 4\pi\mathbf{J}/c \qquad \operatorname{curl}\mathbf{e} = -\dot{\mathbf{h}}/c$$

Superconductors in equilibrium can have nonzero $\mathbf{J}_s$ as described by the London equations, causing $\mathbf{h}$ to vary on the scale of λ. But in equilibrium, or even steady-state, $\dot{\mathbf{h}} = 0$, so that $\mathbf{e}$ is zero or at least constant. Thus the use of both $\mathbf{e}$ and $\mathbf{E}$ offers no advantage. The distinction is valuable only in discussing time-dependent phenomena such as motion of flux-bearing vortices in type II superconductors.

velocity against resistance as in a normal conductor. The second London equation (1-4), when combined with the Maxwell equation

$$\text{curl } \mathbf{h} = \frac{4\pi \mathbf{J}}{c} \qquad (1\text{-}6)$$

leads to the equation

$$\nabla^2 \mathbf{h} = \frac{\mathbf{h}}{\lambda_L^2} \qquad (1\text{-}7)$$

This implies that a magnetic field is exponentially screened from the interior of a sample with penetration depth λ_L, i.e., the Meissner effect. Thus, the parameter λ_L in (1-5) is to be operationally defined as a penetration depth.

A simple, but unsound, "derivation" of (1-3) can be given by computing the response to a uniform electric field of a perfect normal conductor, i.e., a free-electron gas with mean free path $\ell = \infty$. In that case, $d(m\mathbf{v})/dt = e\mathbf{E}$ and $\mathbf{J} = ne\mathbf{v}$, from which (1-3) follows. But this computation is not valid for the spatially nonuniform fields in the penetration depth, for which (1-3) and (1-4) are most useful. The fault is that the response of an electron gas to electric fields is nonlocal, so that the current at a point is determined by the electric field averaged over a region of radius $\sim \ell$ about that point. Consequently, only fields that are uniform over a region of this size give a full response; in particular, the conductivity becomes *infinite* as $\ell \to \infty$ *only* for fields filling all space. Since we are dealing here with an interface between a region with field and one with none, it is clear that even for $\ell = \infty$, the effective conductivity remains finite. For the case of an alternating current, this is the familiar extreme anomalous limit of the skin effect, in which there is a finite surface resistance even as $\ell \to \infty$. If instead one considered a transient situation with a magnetic field suddenly applied to a normal metal with $\ell = \infty$, the field would penetrate to a depth which increased with time as $t^{1/3}$. Thus normal conductivity even with infinite electronic mean free path cannot explain a permanent flux exclusion. Moreover, neither can it explain persistent currents in a ring unless the electron scattering at the surface is assumed to be completely specular.

A much more profound motivation for the London equations is the quantum one, emphasizing use of the vector potential $\mathbf{A}$, given by F. London[1] himself. Noting that the canonical momentum $\mathbf{p} = m\mathbf{v} + e\mathbf{A}/c$, and arguing that in the absence of an applied field one would expect a ground state with zero net momentum (as shown in a theorem[2] of Bloch), one is led to the relation for the local average velocity in the presence of the field

$$\langle \mathbf{v}_s \rangle = \frac{-e\mathbf{A}}{mc}$$

This will hold if one postulates that for some reason the wavefunction of the superconducting electrons is "rigid" and retains its ground state form with

---

[1] F. London, "Superfluids," vol. I, Wiley, New York, 1950.

[2] This theorem is apparently unpublished, though famous. See p. 143 in the preceding reference.

$\langle \mathbf{p} \rangle = 0$. Denoting the number density of electrons participating in this rigid ground state by n_s, we then have

$$\mathbf{J}_s = n_s e \langle \mathbf{v}_s \rangle = \frac{-n_s e^2 \mathbf{A}}{mc} = \frac{-\mathbf{A}}{\Lambda c} \qquad (1\text{-}8)$$

Taking the time derivative of both sides yields (1-3) and taking the curl leads to (1-4). Thus (1-8) contains both London equations in a compact and suggestive form. Since (1-8) is evidently not gauge-invariant, it will only be correct for a particular gauge choice. This choice, known as the London gauge, is specified by requiring that div $\mathbf{A} = 0$ (so that div $\mathbf{J} = 0$), that the normal component of $\mathbf{A}$ over the surface be related to any supercurrent through the surface by (1-8), and that $\mathbf{A} \to 0$ in the interior of bulk samples.

This argument of London leaves open the actual value of n_s, but a natural upper limit is provided by the total density of conduction electrons n. If that is inserted in (1-5), we obtain

$$\lambda_L(0) = \left(\frac{mc^2}{4\pi n e^2} \right)^{1/2} \qquad (1\text{-}9)$$

The notation is chosen to indicate that this should be a limiting value as $T \to 0$, while n_s is expected to decrease continuously to zero as $T \to T_c$. Careful measurements of the RF penetration depths of normal and superconducting samples have shown that the superconducting penetration depths λ are always larger than $\lambda_L(0)$, even after an extrapolation of the data to $T = 0$. This excess penetration depth can be interpreted qualitatively in the London framework as indicating the incomplete rigidity of the wavefunction, so that $n_s < n$, but a quantitative interpretation requires introduction of an additional concept: the coherence length ξ_0.

1-3 THE PIPPARD NONLOCAL ELECTRODYNAMICS

Pippard[1] introduced the coherence length while proposing a nonlocal generalization of the London equation (1-8). This was done in analogy to Chambers' nonlocal generalization[2] of Ohm's law from $\mathbf{J}(\mathbf{r}) = \sigma \mathbf{E}(\mathbf{r})$ to

$$\mathbf{J}(\mathbf{r}) = \frac{3\sigma}{4\pi} \int \frac{\mathbf{R}[\mathbf{R} \cdot \mathbf{E}(\mathbf{r}')] e^{-R/\ell}}{R^4} \, d\mathbf{r}'$$

where $\mathbf{R} = \mathbf{r} - \mathbf{r}'$; this formula takes account of the fact that the current at a point $\mathbf{r}$ depends on $\mathbf{E}(\mathbf{r}')$ throughout a volume of radius $\sim \ell$ about $\mathbf{r}$. Pippard argued that the superconducting wavefunction should have a similar characteristic dimension ξ_0 which could be estimated by an uncertainty-principle argument, as follows:

[1] A. B. Pippard, *Proc. Roy. Soc.* (*London*) **A216**, 547 (1953).
[2] This approach of Chambers is discussed, for example, in J. M. Ziman, "Principles of the Theory of Solids," p. 242, Cambridge, New York, 1964.

Only electrons within $\sim kT_c$ of the Fermi energy can play a major role in a phenomenon which sets in at T_c, and these electrons have a momentum range $\Delta p \approx kT_c/v_F$, where v_F is the Fermi velocity. Thus $\Delta x \gtrsim \hbar/\Delta p \approx \hbar v_F/kT_c$, leading to the definition of a characteristic length

$$\xi_0 = a\frac{\hbar v_F}{kT_c} \qquad (1\text{-}10)$$

where a is a numerical constant of order unity, to be determined. For typical elemental superconductors such as tin and aluminum, $\xi_0 \gg \lambda_L(0)$. If ξ_0 represents the smallest size of wavepackets the superconducting charge carriers can form, then one would expect a weakened supercurrent response to a vector potential $\mathbf{A}(\mathbf{r})$ which did not maintain its full value over a volume of radius $\sim \xi_0$ about the point of interest. Thus ξ_0 plays a role analogous to the mean free path in the nonlocal electrodynamics of normal metals. Of course, if the ordinary mean free path were less than ξ_0, one might expect a still further reduction in the response to an applied field.

Collecting these ideas into a concrete form, Pippard proposed replacement of (1-8) by

$$\mathbf{J}_s(\mathbf{r}) = -\frac{3}{4\pi\xi_0\Lambda c}\int\frac{\mathbf{R}[\mathbf{R}\cdot\mathbf{A}(\mathbf{r}')]}{R^4}e^{-R/\xi}\,d\mathbf{r}' \qquad (1\text{-}11)$$

where again $\mathbf{R} = \mathbf{r} - \mathbf{r}'$, and the coherence length ξ in the presence of scattering was assumed to be related to that of pure material ξ_0 by

$$\frac{1}{\xi} = \frac{1}{\xi_0} + \frac{1}{\ell} \qquad (1\text{-}12)$$

Using (1-11), he computed the penetration depth for various values of ξ_0 and λ_L and compared the results with experimental data. He found[1] that he could fit the data on both tin and aluminum by the choice of a single parameter $a = 0.15$ in (1-10). [We shall see later that the microscopic theory of Bardeen, Cooper, and Schrieffer[2] (BCS) confirms this form, with a numerical constant 0.18.] For both metals, λ is considerably larger than λ_L because $\mathbf{A}(\mathbf{r})$ decreases sharply over a distance $\lambda \ll \xi_0$, giving a weakened supercurrent response, and hence increased field penetration. Moreover, the increase of λ with decreasing mean free path predicted by (1-11) and (1-12) was consistent with data on a series of tin-indium alloys with varying mean free path. Thus, apart from extremely minor quantitative changes, Pippard's nonlocal electrodynamic equation (1-11) not only fits the experimental data to the accuracy with which it is available, but also anticipates completely the form of electrodynamics found several years later from the microscopic theory.

[1] T. E. Faber and A. B. Pippard, *Proc. Roy. Soc.* **A231**, 336 (1955).
[2] J. Bardeen, L. N. Cooper, and J. R. Schrieffer, *Phys. Rev.* **108**, 1175 (1957).

1-4   THE ENERGY GAP AND THE BCS THEORY

The next major step in the evolution of our understanding of superconductors was the establishment of the existence of an energy gap between the ground state and the quasi-particle excitations of the system. This concept had been suggested very early by Daunt and Mendelssohn[1] to explain the observed absence of thermoelectric effects, and it had been postulated theoretically by various workers. However, the first quantitative experimental evidence arose from precise measurements of the electronic specific heat of superconductors by Corak et al.[2] These measurements showed that the electronic heat well below T_c was dominated by an exponential dependence, so that

$$C_{es} \approx \gamma T_c a e^{-bT_c/T} \qquad (1\text{-}13)$$

where the normal-state electronic specific heat is $C_{en} = \gamma T$, and $a \approx 10$ and $b \approx 1.5$ are numerical constants. Such an exponential dependence implies a minimum excitation energy per particle of $\sim 1.5kT_c$.

At about the same time, measurements of electromagnetic absorption in the region of $h\nu \sim kT_c$ were first carried out. Using millimeter-microwave techniques, it was possible[3] to reach this region in aluminum, which has a low $T_c \approx 1.2°\text{K}$ (kelvin) and hence a small gap, but it was not possible to carry the measurements to temperatures much below T_c. Working from the far infrared side as well as the microwave side, Glover and Tinkham[4] were able to make a more complete study of thin lead films at temperatures far below $T_c \approx 7.2°\text{K}$. These measurements and similar ones on tin films could be interpreted quite convincingly in terms of an energy gap of three to four times kT_c. This result was consistent with the calorimetric one if excitations always were produced in pairs, as would be expected if they obeyed Fermi statistics. The spectroscopic measurement gives the minimum total energy E_g required to create the pair of excitations; the thermal one measures the energy $E_g/2$ per statistically independent particle.

At this point, Bardeen, Cooper, and Schrieffer[5] (BCS) produced their epoch-making pairing theory of superconductivity, which forms the subject of the next chapter. In the BCS theory, it was shown that even a weak attractive interaction between electrons, such as that caused in second order by the electron-phonon interaction, causes an instability of the ordinary Fermi-sea ground state of the electron gas with respect to formation of bound pairs of electrons occupying states with equal and opposite momentum and spin. These so-called Cooper pairs have a spatial extension of order ξ_0, and, crudely speaking, comprise the superconducting charge carriers anticipated in the phenomenological theories.

[1] J. G. Daunt and K. Mendelssohn, *Proc. Roy. Soc. (London)* **A185**, 225 (1946).
[2] W. S. Corak, B. B. Goodman, C. B. Satterthwaite, and A. Wexler, *Phys. Rev.* **96**, 1442 (1954); **102**, 656 (1956).
[3] M. A. Biondi, M. P. Garfunkel, and A. O. McCoubrey, *Phys. Rev.* **101**, 1427 (1956).
[4] R. E. Glover and M. Tinkham, *Phys. Rev.* **104**, 844 (1956); **108**, 243 (1957).
[5] J. Bardeen, L. N. Cooper, and J. R. Schrieffer, *Phys. Rev.* **108**, 1175 (1957).

One of the key predictions of this theory was that a minimum energy $E_g = 2\Delta(T)$ should be required to break a pair, creating two quasi-particle excitations. This $\Delta(T)$ was predicted to increase from zero at T_c to a limiting value such that

$$E_g(0) = 2\Delta(0) = 3.528kT_c \qquad (1\text{-}14)$$

for $T \ll T_c$. Not only did this result agree with the measured gap widths, but the BCS prediction for the shape of the absorption edge above $hv = E_g$ was also in quantitative agreement with the data of Glover and Tinkham. This agreement provided one of the most decisive early verifications of the microscopic theory.

1-5   THE GINZBURG-LANDAU THEORY

Although a considerable body of work followed the appearance of the BCS theory, serving to substantiate its predictions for various processes such as nuclear relaxation and ultrasonic attenuation in which the energy gap and excitation spectrum play a key role, the most exciting developments of the ensuing decade came in another direction. This direction is epitomized by the Ginzburg-Landau (GL) theory of superconductivity, which concentrates entirely on the superconducting electrons rather than on excitations. Already in 1950, 7 years before BCS, Ginzburg and Landau[1] had introduced a complex pseudowave function ψ as an order parameter for the superconducting electrons such that the local density of superconducting electrons (as defined in the London equations) was given by

$$n_s = |\psi(x)|^2 \qquad (1\text{-}15)$$

Then, using a variational principle and working from an assumed expansion of the free energy in powers of ψ and $\nabla\psi$, they derived a differential equation for ψ

$$\frac{1}{2m^*}\left(\frac{\hbar}{i}\nabla - \frac{e^*}{c}\mathbf{A}\right)^2 \psi + \beta|\psi|^2\psi = -\alpha(T)\psi \qquad (1\text{-}16)$$

which is very analogous to the Schrödinger equation for a free particle, but with a nonlinear term. The corresponding equation for the supercurrent

$$\mathbf{J}_s = \frac{e^*\hbar}{i2m^*}(\psi^*\,\nabla\psi - \psi\,\nabla\psi^*) - \frac{e^{*2}}{m^*c}|\psi|^2\mathbf{A} \qquad (1\text{-}17)$$

was also the same as the usual quantum-mechanical one for particles of charge e^* and mass m^*. With this formalism they were able to treat two features which were beyond the scope of the London theory, namely,

1   Nonlinear effects in fields strong enough to change n_s (or $|\psi|^2$)
2   Spatial variation of n_s

---

[1] V. L. Ginzburg and L. D. Landau, *Zh. Eksperim. i Teor. Fiz.* **20**, 1064 (1950).

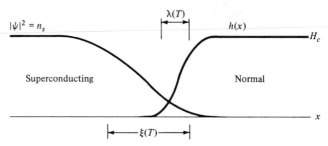

FIGURE 1-4
Interface between superconducting and normal domains in the intermediate state.

A major early triumph of the theory was in handling the so-called intermediate state of superconductors, in which superconducting and normal domains coexist in the presence of $H \approx H_c$. The interface between two such domains is shown schematically in Fig. 1-4.

When first proposed, the theory appeared rather phenomenological, and its importance was not generally appreciated, especially in the Western literature. However, in 1959 Gor'kov[1] was able to show that the GL theory was in fact a limiting form of the microscopic theory of BCS (suitably generalized to deal with spatially varying situations), valid near T_c, in which ψ is directly proportional to the gap parameter Δ. More physically, ψ can be thought of as the wavefunction of the center-of-mass motion of the Cooper pairs. The GL theory is now universally accepted as a masterstroke of physical intuition which embodies in a most simple way the macroscopic quantum-mechanical nature of the superconducting state crucial for understanding its unique electrodynamic properties.

The Ginzburg-Landau theory introduces a characteristic length, now called the temperature-dependent coherence length,

$$\xi(T) = \frac{\hbar}{|2m^*\alpha(T)|^{1/2}} \qquad (1\text{-}18)$$

which characterizes the distance over which $\psi(\mathbf{r})$ can vary without undue energy increase. In a pure superconductor far from T_c, $\xi(T) \approx \xi_0$, the Pippard coherence length; near T_c, however, $\xi(T)$ diverges as $(T_c - T)^{-1/2}$, since α vanishes as $(T - T_c)$. Thus these two "coherence lengths" are related but distinct.

The ratio of the two characteristic lengths defines the GL parameter

$$\kappa = \frac{\lambda}{\xi} \qquad (1\text{-}19)$$

[1] L. P. Gor'kov, *Zh. Eksperim. i Teor. Fiz.* **36**, 1918 (1959) [*Soviet Phys.—JETP* **9**, 1364 (1959)].

Since λ and ξ diverge in the same way at T_c, this dimensionless ratio is approximately independent of temperature. For typical pure superconductors $\lambda \approx 500$ Å and $\xi \approx 3000$ Å, so $\kappa \ll 1$. In this case, one can see that there is a positive surface energy associated with a domain wall between normal and superconducting material in the intermediate state. [The qualitative argument is simply that one pays an energetic cost $\sim \xi H_c^2/8\pi$ for the variation of ψ from its superconducting value to zero, while reducing the diamagnetic energy only by $\sim \lambda H_c^2/8\pi$ (see Fig. 1-4).] This positive surface energy stabilizes a domain pattern with a scale of subdivision intermediate between the microscopic length ξ and the macroscopic sample size.

1-6   TYPE II SUPERCONDUCTORS

In 1957 (the same year as BCS) Abrikosov[1] published a remarkably significant paper, almost overlooked at the time, in which he investigated what would happen in GL theory if κ were large instead of small, i.e., if $\xi < \lambda$, rather than the reverse. Reversing the argument indicated above, this should lead to a *negative* surface energy, so that the process of subdivision into domains should proceed until it is limited by the microscopic length ξ. Because this behavior is so radically different from the classic behavior described earlier, he called these "type II superconductors" to distinguish them from the earlier "type I" variety. He showed that the exact breakpoint was at $\kappa = 1/\sqrt{2}$. For materials with $\kappa > 1/\sqrt{2}$ he found that, instead of a discontinuous breakdown of superconductivity in a first-order transition at H_c, there was a continuous increase in flux penetration starting at a first critical field H_{c1} and reaching $B = H$ at a second critical field H_{c2}, as shown schematically in Fig. 1-5. Because of the partial flux penetration, the diamagnetic energy of holding the field out is less, so H_{c2} (which turns out to be given by $\sqrt{2}\kappa H_c$) can be much greater than the thermodynamic critical field H_c. This property has made possible high-field superconducting solenoids.

Another result of Abrikosov's analysis was that, in the so-called mixed state, or Schubnikov phase, between H_{c1} and H_{c2}, the flux should penetrate not in laminar domains but in a regular array of flux tubes, each carrying a quantum of flux

$$\Phi_0 = \frac{hc}{2e} = 2.07 \times 10^{-7} \text{ gauss-cm}^2 \qquad (1\text{-}20)$$

Within each unit cell of the array there is a vortex of supercurrent concentrating the flux toward the center. Although Abrikosov predicted a square array, it was later shown that he had made a numerical error, and that a triangular array should have a lower free energy. This vortex array has been demonstrated experimentally by a magnetic decoration technique coupled with electron microscopy.[2]

[1] A. A. Abrikosov, *Zh. Eksperim. i Teor. Fiz.* **32**, 1442 (1957) [*Soviet Phys.—JETP*, **5**, 1174 (1957)].

[2] U. Essmann and H. Träuble, *Phys. Letters* **24A**, 526 (1967).

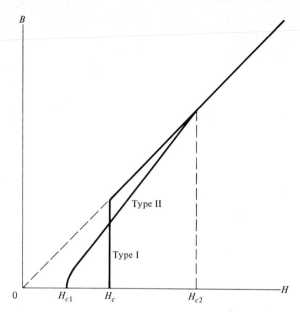

FIGURE 1-5
Comparison of flux penetration behavior of type I and type II superconductors with the same thermodynamic critical field H_c. $H_{c2} = \sqrt{2}\kappa H_c$. The ratio of B/H_{c2} from this plot also gives the approximate variation of R/R_n, where R is the electrical resistance for the case of negligible pinning, and R_n is the normal-state resistance.

Of course random inhomogeneities in the crystal lead to "pinning" of vortices at favorable locations, so that in some cases one only finds a liquidlike pattern of flux tubes.

   We have already noted that type II superconductors are not perfectly diamagnetic, and since $|\psi|^2$ turns out to go to zero in the centers of the vortices, we are not surprised to find that there is no energy gap in the cores. Thus, we are led to ask whether the final hallmark—perfect conductivity—is also lost. The answer is a bit equivocal. In the presence of a transport current, the flux tubes experience a Lorentz force $\mathbf{J} \times \boldsymbol{\Phi}_0/c$ (analogous to the macroscopic force density $\mathbf{J} \times \mathbf{B}/c$) tending to make them move sideways, in which case a longitudinal "resistive" voltage is induced. In an ideal homogeneous material, the flux motion is resisted only by a viscous drag, and the type II superconductor shows a resistance comparable to that in the normal state, but reduced by a factor $\sim B/H_{c2}$. In real materials, however, there is always some inhomogeneity to pin the flux, so that there is no resistance until a finite current is reached, such that the Lorentz force is capable of exceeding the pinning force. In magnet materials, the pinning is made strong enough to give large critical currents. Actually, even with pinning, the resistance is at best only astronomically small, not exactly zero, because thermally activated fluctuations can occasionally overcome the pinning, leading to a small resistance due to flux "creep."

1-7 PHASE, JOSEPHSON TUNNELING, FLUXOID QUANTIZATION, AND PERSISTENT CURRENTS: THE ESSENCE OF SUPERCONDUCTIVITY

Faced with all these fallen hallmarks, we might well ask what really is the essential universal characteristic of the superconducting state. The answer is the existence of the many-particle condensate wavefunction $\psi(\mathbf{r})$, which has amplitude and phase and which maintains phase coherence over macroscopic distances. This condensate is analogous to, but not identical with, the familiar Bose-Einstein condensate, with Cooper pairs of electrons replacing the single bosons which condense in superfluid helium, for example.

Since the phase and particle number are conjugate variables, reflecting complementary aspects of the wave-particle dualism, there is an uncertainty relation

$$\Delta N \, \Delta \varphi \gtrsim 1 \qquad (1\text{-}21)$$

which limits the precision with which N and φ can be simultaneously known. However, since $N \sim 10^{22}$, both N and φ can be known to within small fractional uncertainties, and the phase may be treated as a semiclassical variable.

The physical significance of the phase degree of freedom was first emphasized in the work of Josephson,[1] who predicted that pairs should be able to tunnel between two superconductors even at zero voltage difference, giving a supercurrent density

$$J = J_0 \sin \left(\varphi_1 - \varphi_2 \right) \qquad (1\text{-}22)$$

where J_0 is a constant and φ_i is the phase of ψ in the ith superconductor at the tunnel junction. Although originally received with some skepticism, this prediction has been extremely thoroughly verified. Subsequently, Josephson junctions have been utilized in ultrasensitive voltmeters and magnetometers, and in making the most accurate available measurements of the ratio of fundamental constants h/e.

The most basic implication of the existence of a phase factor in $\psi(\mathbf{r}) \equiv |\psi(\mathbf{r})| e^{i\varphi(\mathbf{r})}$, however, is operative in the simple case of a superconducting ring. In that case, the single-valuedness of ψ requires that $\varphi(\mathbf{r})$ return to itself modulo 2π on going once around the circuit. Just as the corresponding condition in an atom leads to the quantization of orbital angular momentum in integral multiples of $\hbar$,

[1] B. D. Josephson, *Phys. Letters* **1**, 251 (1962).

here this condition requires that the *fluxoid* Φ' take on only values $n\Phi_0 = nhc/2e$. The fluxoid is a quantity introduced by F. London, which can be written

$$\Phi' = \Phi + \frac{m^*c}{e^{*2}}\oint \frac{\mathbf{J}_s \cdot d\mathbf{s}}{|\psi|^2} \qquad (1\text{-}23)$$

where $\Phi = \oint \mathbf{A} \cdot d\mathbf{s}$ is the ordinary magnetic *flux* through the integration loop. If the ring has walls thick compared to λ, the path of integration can be taken inside the skin depth where $J_s \to 0$; then (1-23) implies that $\Phi = \Phi'$, so that the flux itself has the quantized value $n\Phi_0$. This property was demonstrated experimentally[1] in 1961. When J_s is not small, as in the vortices in a type II superconductor, both terms may be equally important, and the value of the flux itself is unrestricted; only the fluxoid, directly related to the phase integral, always has precise quantum values.

This concept forms the basis for understanding the quantum nature of persistent currents in a ring. The current cannot decrease by infinitesimal amounts, but only in quantum jumps in which the fluxoid quantum number decreases by one or more units. Such a quantum jump could occur readily if only a single electron were involved, as in an atomic transition. However, in the superconductor such a quantum jump for the macroscopic wavefunction ψ requires a collective transition of all the pairs involved. The extremely long life of persistent currents (despite the fact that they are in principle only metastable) results from the extreme improbability of such a simultaneous quantum jump by $\sim 10^{20}$ particles. Until a quantum jump occurs, there is no decay whatsoever of the persistent current.

1-8 FLUCTUATION EFFECTS

The foregoing remarks about persistent currents bring out the importance of thermodynamic fluctuations in causing finite, though astronomically small, resistance below T_c. The other side of the coin is that *above* T_c fluctuations cause some vestiges of superconductivity to remain. These were first observed by Glover,[2] who found that the conductivity of amorphous films of superconductors diverges as $(T - T_c)^{-1}$ as one approaches T_c from above. This "Curie-Weiss" form of temperature dependence with appropriate coefficient was also predicted theoretically at about the same time. Somewhat later, the corresponding effect was also observed[3] in the diamagnetic susceptibility of pure bulk samples. In this case the basic divergence is as $(T - T_c)^{-1/2}$.

---

[1] B. S. Deaver and W. M. Fairbank, *Phys. Rev. Letters* **7**, 43 (1961); R. Doll and M. Näbauer, *Phys. Rev. Letters* **7**, 51 (1961).

[2] R. E. Glover, III, *Phys. Letters* **25A**, 542 (1967).

[3] J. P. Gollub, M. R. Beasley, R. S. Newbower, and M. Tinkham, *Phys. Rev. Letters* **22**, 1288 (1969).

These measurements and the associated theory show that in principle the effects of the superconducting interaction persist to arbitrarily high temperatures, but that a fairly strong cutoff sets in at about $2T_c$. Thus, there is not only some resistance below T_c but also some superconductivity above T_c, although the apparently abrupt switchover observed by Kamerlingh Onnes is still a good working approximation for most purposes. It is worth noting that the superconducting transition is basically much sharper than other second-order phase transitions such as those in magnetic materials because the coherence length ξ_0 is so much larger than the interatomic distance. Thus each electron interacts with many others, whereas in magnetic materials only a few near neighbors are strongly coupled, leading to much more prominent fluctuation effects.

With this quick overview behind us, let us now proceed to put some flesh on the bare bones of this outline.

2

THE BCS THEORY

In this book we shall emphasize the phenomenological rather than the microscopic theory of superconductivity. Nonetheless, it seems appropriate to start our systematic presentation with a treatment of "basic BCS," since this theory,[1] presented in 1957, completely revolutionized the quality of our understanding of superconductivity. Moreover, it provides the framework for calculating the parameters of the phenomenological theory as well as the a priori theoretical justification for its existence, form, and success.

There is a pedagogical disadvantage to starting with this survey of the BCS theory in that it appears to be one of the more forbidding chapters of the book, largely because the technique of second quantization is used. The reader who doubts his preparation in this area should be assured that the treatment has been kept as simple as possible, and that enough explanation of the method is given to allow a novice to follow the presentation. It is also worth emphasizing that the phenomenological theory, to which the rest of the book is largely devoted, was generally developed to explain experimental observations independently of, and prior to, its "deduction" from microscopic theory. The phenomenological theory

[1] J. Bardeen, L. N. Cooper, and J. R. Schrieffer, *Phys. Rev.* **108**, 1175 (1957).

can be studied, appreciated, and used in practice with only a limited, "cultural" understanding of the microscopic theory. Thus, although we have reversed the order and put the microscopic theory first for logical simplicity in presentation, the reader is urged not to allow himself to get bogged down in this chapter. If this threatens, he should skim quickly ahead to the summary at the end of the chapter, and in reading later chapters simply refer back to this chapter when necessary to clarify specific points.

2-1   COOPER PAIRS

The basic idea that even a weak attraction can bind pairs of electrons into a bound state was presented by Cooper[1] in 1956. What he showed was that the Fermi sea of electrons is unstable against the formation of at least one bound pair, regardless of how weak the interaction is, so long as it is attractive. This result is a consequence of the Fermi statistics and of the existence of the Fermi-sea background, since it is well known that binding does not ordinarily occur in the two-body problem in three dimensions until the strength of the potential exceeds a finite threshold value.

   To see how this binding comes about, we consider a simple model of two electrons added to a Fermi sea at $T = 0$, with the stipulation that the extra electrons interact with each other but not with those in the sea, except via the exclusion principle. Thus we seek a two-particle wavefunction. By the general arguments of Bloch, we expect the lowest energy state to have zero total momentum, so that the two electrons must have equal and opposite momenta. This suggests building up an orbital wavefunction of the sort

$$\psi_0(\mathbf{r}_1, \mathbf{r}_2) = \sum_{\mathbf{k}} g_{\mathbf{k}} e^{i\mathbf{k} \cdot \mathbf{r}_1} e^{-i\mathbf{k} \cdot \mathbf{r}_2}$$

Taking account of the antisymmetry of the total wavefunction with respect to exchange of the two electrons, ψ_0 is converted either to a sum of products of $\cos \mathbf{k} \cdot (\mathbf{r}_1 - \mathbf{r}_2)$ with the antisymmetric singlet spin function $(\alpha_1 \beta_2 - \beta_1 \alpha_2)$ or to a sum of products of $\sin \mathbf{k} \cdot (\mathbf{r}_1 - \mathbf{r}_2)$ with one of the symmetric triplet spin functions $(\alpha_1 \alpha_2, \alpha_1 \beta_2 + \beta_1 \alpha_2, \beta_1 \beta_2)$. (In these expressions, α_1 refers to the "up" spin state of particle 1, whereas β_1 refers to its "down" state.) Anticipating an attractive interaction, we expect the singlet coupling to have lower energy, because the cosinusoidal dependence of its orbital wavefunction on $(\mathbf{r}_1 - \mathbf{r}_2)$ gives a larger probability amplitude for the electrons to be near each other. Thus we consider a two-electron singlet wavefunction of the form

$$\psi_0(\mathbf{r}_1 - \mathbf{r}_2) = \left[\sum_{k > k_F} g_{\mathbf{k}} \cos \mathbf{k} \cdot (\mathbf{r}_1 - \mathbf{r}_2) \right] (\alpha_1 \beta_2 - \beta_1 \alpha_2) \qquad (2\text{-}1)$$

[1] L. N. Cooper, *Phys. Rev.* **104**, 1189 (1956).

By inserting (2-1) into the Schrödinger equation of the problem, one can show that the weighting coefficients g_k and the energy eigenvalue E are to be determined by solving

$$(E - 2\epsilon_k)g_k = \sum_{k' > k_F} V_{kk'} g_{k'} \qquad (2\text{-}2)$$

In this, the ϵ_k are unperturbed plane-wave energies, and the $V_{kk'}$ are the matrix elements of the interaction potential

$$V_{kk'} = \Omega^{-1} \int V(\mathbf{r}) e^{i(\mathbf{k'} - \mathbf{k}) \cdot \mathbf{r}} \, d\mathbf{r} \qquad (2\text{-}3)$$

where $\mathbf{r}$ is the distance between the two electrons and Ω is the normalization volume. This $V_{kk'}$ characterizes the strength of the potential for scattering a pair of electrons with momenta $(\mathbf{k'}, -\mathbf{k'})$ to momenta $(\mathbf{k}, -\mathbf{k})$. If a set of g_k satisfying (2-2) with $E < 2E_F$ can be found, then a bound-pair state exists.

Since it is hard to analyze this situation for general $V_{kk'}$, Cooper introduced the very serviceable approximation that all $V_{kk'} = -V$ for $\mathbf{k}$ states out to a cutoff energy $\hbar\omega_c$ away from E_F, and that $V_{kk'} = 0$ beyond $\hbar\omega_c$. Then the right side of (2-2) is a constant, independent of $\mathbf{k}$, and we have

$$g_k = V \frac{\sum g_{k'}}{2\epsilon_k - E} \qquad (2\text{-}4)$$

Summing both sides and canceling $\sum g_k$, we obtain

$$\frac{1}{V} = \sum_{k > k_F} (2\epsilon_k - E)^{-1} \qquad (2\text{-}5)$$

Replacing the summation by an integration, with $N(0)$ denoting the density of states at the Fermi level for electrons of one spin orientation, this becomes

$$\frac{1}{V} = N(0) \int_{E_F}^{E_F + \hbar\omega_c} \frac{dE}{2\epsilon - E} = \tfrac{1}{2} N(0) \ln \frac{2E_F - E + 2\hbar\omega_c}{2E_F - E}$$

For the usual weak-coupling case, where $N(0)V \ll 1$, the energy E can then be written as

$$E \approx 2E_F - 2\hbar\omega_c e^{-2/N(0)V} \qquad (2\text{-}6)$$

Thus, indeed, there *is* a bound state with negative energy with respect to the Fermi surface made up entirely of electrons with $k > k_F$, that is, with kinetic energy in excess of E_F. The contribution to the energy of the attractive potential outweighs this excess kinetic energy, leading to binding regardless of how small V is. Note that the form of the binding energy is not analytic at $V = 0$; that is, it cannot be expanded in powers of V. As a result, it cannot be obtained by perturbation theory, a fact that greatly delayed the genesis of the theory.

Returning to the wavefunction, we see that the dependence on the relative coordinate $\mathbf{r} = \mathbf{r}_1 - \mathbf{r}_2$ is proportional to

$$\sum_{k > k_F} \frac{\cos \mathbf{k} \cdot \mathbf{r}}{2\xi_k + E'}$$

where we have gone over to energies measured from the Fermi energy, so that

$$\xi_k = \epsilon_k - E_F \qquad \text{and} \qquad E' = 2E_F - E > 0 \qquad (2\text{-}7)$$

(Because of the sign change, E' is now the *binding* energy relative to $2E_F$.) Since g_k depends only on ξ_k, this solution has spherical symmetry; hence it is an S state as well as a singlet spin state. Note that the weighting factor $(2\xi_k + E')^{-1}$ has its maximum value $1/E'$ when $\xi_k = 0$, that is, for electrons at the Fermi level, and that it falls off with higher values of ξ_k. Thus, the electron states within a range of energy $\sim E'$ above E_F are those most strongly involved in forming the bound state. Since $E' \ll \hbar\omega_c$ for $N(0)V < 1$, this shows that the detailed behavior of $V_{kk'}$ out around $\hbar\omega_c$ will not have any great effect on the result. This fact gives us some justification for making such a crude approximation to $V_{kk'}$. A second consequence of this small range of energy states is that, by the uncertainty principle argument of Pippard cited earlier, it implies that the size of the bound pair is not less than $\sim \hbar v_F/E'$. Since kT_c turns out to be of the order of E', this implies that the size of the Cooper pair state is $\sim \xi_0 = a\hbar v_F/kT_c$, much larger than the interparticle distance. Thus, the pairs are highly overlapping.

2-2 ORIGIN OF THE ATTRACTIVE INTERACTION

We now must examine the origin of the negative $V_{kk'}$ needed for superconductivity. If we take the bare Coulomb interaction $V(\mathbf{r}) = e^2/r$, and carry out the computation of $V(\mathbf{q})$

$$V(\mathbf{q}) = V(\mathbf{k} - \mathbf{k}') = V_{kk'} = \Omega^{-1} \int V(\mathbf{r}) e^{i\mathbf{q} \cdot \mathbf{r}} \, d\mathbf{r}$$

we find

$$V(\mathbf{q}) = \frac{4\pi e^2}{\Omega q^2} = \frac{4\pi e^2}{q^2} \qquad (2\text{-}8)$$

where the last equality holds for unit normalization volume Ω. Evidently this $V(\mathbf{q})$ is always positive.

Now, if we take the dielectric function $\varepsilon(\mathbf{q}, \omega)$ of the medium into account, $V(\mathbf{q})$ is reduced by a factor $\varepsilon^{-1}(\mathbf{q}, \omega)$. The most obvious ingredient in $\varepsilon(\mathbf{q}, \omega)$ is the screening effect[1] of the conduction electrons. This introduces a screening length

---

[1] See, for example, J. M. Ziman, "Principles of the Theory of Solids," chap. 5, Cambridge Univ. Press, 1964.

$1/k_s \approx 1\text{Å}$. In the Fermi-Thomas approximation, ε is given by $\varepsilon = 1 + k_s^2/q^2$, so that

$$V(\mathbf{q}) = \frac{4\pi e^2}{q^2 + k_s^2} \qquad (2\text{-}9)$$

Thus the electronic screening has eliminated the divergence at $\mathbf{q} = 0$, but it still leaves a positive $V_{\mathbf{kk'}}$. Hence, no superconductivity would result.

Negative terms come in only when one takes the motion of the ion cores into account. The physical idea is that the first electron polarizes the medium by attracting positive ions; these excess positive ions, in turn, attract the second electron, giving an effective attractive interaction between the electrons. If this attraction is strong enough to override the repulsive screened Coulomb interaction, it gives rise to a net attractive interaction, and superconductivity results. Historically, the importance of the electron-lattice interaction in explaining superconductivity was first suggested by Fröhlich[1] in 1950. This suggestion was confirmed experimentally by the discovery[2] of the "isotope effect," i.e., the proportionality of T_c and H_c to $M^{-1/2}$ for isotopes of the same element.

Since these lattice deformations are resisted by the same stiffness that makes a solid elastic, it is clear that the characteristic vibrational, or phonon, frequencies will play a role. (For the electronically screened Coulomb interaction, the characteristic frequency is the plasma frequency, which is so high that we can assume instantaneous response.) From momentum conservation, we can see that if an electron is scattered from $\mathbf{k}$ to $\mathbf{k'}$, the relevant phonon must carry the momentum $\mathbf{q} = \mathbf{k} - \mathbf{k'}$, and the characteristic frequency must then be the phonon frequency $\omega_{\mathbf{q}}$. As a result, it is plausible that the phonon contribution to the screening function be proportional to $(\omega^2 - \omega_{\mathbf{q}}^2)^{-1}$. Evidently this resonance denominator gives a *negative* sign if $\omega < \omega_{\mathbf{q}}$, corresponding to the physical argument above; for higher frequencies, i.e., electron energy differences larger than $\hbar\omega_{\mathbf{q}}$, the interaction becomes repulsive. Thus the cutoff energy $\hbar\omega_c$ of Cooper's attractive matrix element $-V$ is expected to be of the order of the Debye energy $\hbar\omega_D = k\Theta_D$, which characterizes the cutoff of the phonon spectrum.

Careful analyses of the best way to treat the coupled electron-phonon system have been given by Fröhlich[3] and by Bardeen and Pines,[4] but detailed calculations relating to specific materials are only beginning to give quantitative results. The first attempt to test the theoretical criterion for superconductivity systematically throughout the periodic table was a calculation by Pines.[5] He used the "jellium" model, in which the solid is approximated by a fluid of electrons and

[1] H. Fröhlich, *Phys. Rev.* **79**, 845 (1950).

[2] E. Maxwell, *Phys. Rev.* **78**, 477 (1950); C. A. Reynolds, B. Serin, W. H. Wright, and L. B. Nesbitt, *Phys. Rev.* **78**, 487 (1950).

[3] H. Fröhlich, *Proc. Roy. Soc. (London)* **A215**, 291 (1952).

[4] J. Bardeen and D. Pines, *Phys. Rev.* **99**, 1140 (1955).

[5] D. Pines, *Phys. Rev.* **109**, 280 (1958).

point ions, with complete neglect of crystal structure and Brillouin zone effects as well as of the finite ion-core size. As shown in the book of de Gennes,[1] for example, the jellium model in a certain approximation leads to

$$V(\mathbf{q}, \omega) = \frac{4\pi e^2}{q^2 + k_s^2} + \frac{4\pi e^2}{q^2 + k_s^2} \frac{\omega_\mathbf{q}^2}{\omega^2 - \omega_\mathbf{q}^2} \qquad (2\text{-}10)$$

The first term is the screened Coulomb repulsion, whereas the second term is the phonon-mediated interaction, which is attractive for $\omega < \omega_\mathbf{q}$. Unfortunately (2-10) is too simplified to be of much use as a criterion for superconductivity, since it reduces to zero for $\omega = 0$, and it is always negative for $\omega < \omega_\mathbf{q}$, regardless of material parameters. It does, however, illustrate that the phonon-mediated interaction is of the same order of magnitude as the direct one, so that the concept of achieving a net, negative interaction-matrix element in this way is not unreasonable.

2-3   THE BCS GROUND STATE

Having seen that the Fermi sea is unstable against the formation of a bound Cooper pair when the net interaction is attractive, it is clear that in that case we must expect pairs to condense until an equilibrium point is reached. This will occur when the state of the system is so greatly changed from the Fermi sea (because of the large number of bound pairs) that the binding energy for an additional pair has gone to zero. Evidently it would not be easy to handle such a complicated state unless an ingenious mathematical form could be found. Such a form was provided by the BCS wavefunction.

When we write down wavefunctions for more than two electrons, the scheme of handling the antisymmetry used above for a single Cooper pair becomes quite awkward, and it is convenient to replace it by a scheme of $N \times N$ "Slater determinants" to specify N-electron antisymmetrized product functions. The Slater determinants, in turn, are more compactly expressed using the language of second quantization, in which the occupied states (including spin index) are specified by the use of "creation operators" such as $c_{\mathbf{k}\uparrow}^*$, which creates an electron of momentum $\mathbf{k}$ and spin up. It is also necessary to introduce annihilation operators $c_{\mathbf{k}\uparrow}$ which empty the corresponding state. In this notation, the singlet wavefunction discussed above is written

$$|\psi_0\rangle = \sum_{k > k_F} g_\mathbf{k} c_{\mathbf{k}\uparrow}^* c_{-\mathbf{k}\downarrow}^* |F\rangle \qquad (2\text{-}11)$$

where $|F\rangle$ represents the Fermi sea with all states filled up to k_F. This form makes it obvious that pairs of time-reversed states are always occupied together, a

[1] P. G. de Gennes, "Superconductivity in Metals and Alloys," p. 102, W. A. Benjamin, New York, 1966.

feature that Anderson[1] showed is maintained in the case of dirty superconductors, where a generalized pairing scheme is needed since $\mathbf{k}$ is no longer a good quantum number. One may verify that (2-11) is equivalent to the form (2-1) given above for singlet pairing by summing the two 2×2 Slater determinants with the (equal) coefficients $g_{\mathbf{k}}$ and $g_{-\mathbf{k}}$.

Since electrons obey Fermi statistics, the creation and annihilation operators introduced above obey the characteristic anticommutation relations of fermion operators

$$[c_{\mathbf{k}\sigma}, c^*_{\mathbf{k}'\sigma'}]_+ \equiv c_{\mathbf{k}\sigma} c^*_{\mathbf{k}'\sigma'} + c^*_{\mathbf{k}'\sigma'} c_{\mathbf{k}\sigma} = \delta_{\mathbf{k}\mathbf{k}'} \delta_{\sigma\sigma'}$$
$$[c_{\mathbf{k}\sigma}, c_{\mathbf{k}'\sigma'}]_+ = [c^*_{\mathbf{k}\sigma}, c^*_{\mathbf{k}'\sigma'}]_+ = 0 \qquad (2\text{-}12)$$

where σ refers to the spin index. The particle number operator $n_{\mathbf{k}\sigma}$ is defined by

$$n_{\mathbf{k}\sigma} = c^*_{\mathbf{k}\sigma} c_{\mathbf{k}\sigma} \qquad (2\text{-}13)$$

which has an eigenvalue of unity when operating on an occupied state, and gives zero when operating on an empty state. For our purposes, only elementary manipulations using these rules will be required in carrying out applications of this formalism. We use it simply as a compact notation for dealing with many-electron wavefunctions and operators which act on them.

We approach the BCS wavefunction by observing that the most general N-electron wavefunction expressed in terms of momentum eigenfunctions and with the Cooper pairing built in is

$$|\psi_N\rangle = \sum g(\mathbf{k}_i, \ldots, \mathbf{k}_l) c^*_{\mathbf{k}_i\uparrow} c^*_{-\mathbf{k}_i\downarrow} \cdots c^*_{\mathbf{k}_l\uparrow} c^*_{-\mathbf{k}_l\downarrow} |\phi_0\rangle$$

where $|\phi_0\rangle$ is the vacuum state with no particles present, $\mathbf{k}_i$ and $\mathbf{k}_l$ designate the first and last of the M $\mathbf{k}$-values in the band which are occupied in a given term in the sum, and g specifies the weight with which the product of this set of $N/2$ pairs of creation operators appears. Since there are

$$\frac{M!}{[M - (N/2)]! \, (N/2)!} \approx 10^{(10^{20})}$$

ways of choosing the $N/2$ states for pair occupancy, there will be that many terms in the sum and that many of the $g(\mathbf{k}, \ldots)$ to determine. This is obviously hopeless. What BCS did was to argue that with so many particles involved it would be a good approximation to use a Hartree self-consistent field or "mean-field" approach, in which the occupancy of each state $\mathbf{k}$ is taken to depend only on the *average* occupancy of other states. In its simplest form, this relaxes the constraint on the total number of particles being N, since occupancies are treated only statistically. However, because the number of particles is huge, no serious error is made by working with a system in which only $\bar{N}$ is fixed. Essentially we work in a grand canonical ensemble.

[1] P. W. Anderson, *J. Phys. Chem. Solids* **11**, 26 (1959).

What BCS took as their form for the ground state was

$$|\psi_G\rangle = \prod_{k=k_1,\,\ldots,\,k_M} (u_k + v_k c_{k\uparrow}^* c_{-k\downarrow}^*)|\phi_0\rangle \qquad (2\text{-}14)$$

where $|u_k|^2 + |v_k|^2 = 1$. This form implies that the probability of the pair $(k\uparrow, -k\downarrow)$ being occupied is $|v_k|^2$, while the probability that it is unoccupied is $|u_k|^2 = 1 - |v_k|^2$. (For simplicity, we can consider u_k and v_k all real, but it will later prove interesting to let them differ by a phase factor $e^{i\varphi}$, where φ will turn out to be the phase of the macroscopic condensate wavefunction.) Evidently this $|\psi_G\rangle$ can be expressed as a sum

$$|\psi_G\rangle = \sum_N \lambda_N |\psi_N\rangle \qquad (2\text{-}15)$$

where each term represents the part of the expansion of the product form (2-14) containing $N/2$ pairs. [These $|\psi_N\rangle$ are special cases of the general form mentioned above, in which $g(k, \ldots)$ is given by $\prod_k u_k \prod_{k'} v_{k'}$, where k runs over the $(M - N/2)$ unoccupied pair states and k' runs over the $N/2$ occupied pair states.] If all the u_k and v_k are finite, there is a finite probability of any N from 0 to $2M$. However, the values of $|\lambda_N|^2$ are very sharply peaked about the average value

$$\bar{N} = \sum_k 2|v_k|^2 \qquad (2\text{-}16)$$

As an illustration of the formal manipulation of these second quantized forms, let us run through the mechanics of how $\bar{N}$ is calculated, although (2-16) is obviously correct in view of the physical significance of v_k mentioned above. To start,

$$\bar{N} = \langle N_{op}\rangle = \left\langle \sum_{k,\sigma} n_{k\sigma}\right\rangle = \langle\psi_G| \sum_k (c_{k\uparrow}^* c_{k\uparrow} + c_{k\downarrow}^* c_{k\downarrow})|\psi_G\rangle$$

$$= 2\sum_k \langle\psi_G| c_{k\uparrow}^* c_{k\uparrow}|\psi_G\rangle$$

since the electrons all occur in pairs with antiparallel spin. Putting in $|\psi_G\rangle$ explicitly, this becomes

$$\bar{N} = 2\sum_k \langle\phi_0| (u_k^* + v_k^* c_{-k\downarrow} c_{k\uparrow}) c_{k\uparrow}^* c_{k\uparrow}(u_k + v_k c_{k\uparrow}^* c_{-k\downarrow}^*)$$

$$\times \prod_{l\neq k} (u_l^* + v_l^* c_{-l\downarrow} c_{l\uparrow})(u_l + v_l c_{l\uparrow}^* c_{-l\downarrow}^*)|\phi_0\rangle$$

In writing this we have used the property that $\langle A\phi|\psi\rangle = \langle\phi|A^\dagger|\psi\rangle$, and that the adjoint of a product of operators is the product of the adjoints in reverse order. Also, we have been able to rearrange the order of factors in the products to group together all those concerning a given pair state k or l, because, by (2-12), commutation of even numbers of dissimilar Fermi operators introduces no sign change. As we proceed to evaluate this expression, we may think of $|\phi_0\rangle$ as being the product of vacuum states for each k value. This enables the factor relating to each pair to be evaluated separately. Multiplying out the factor for $l \neq k$, we have

$$|u_l|^2 + u_l^* v_l c_{l\uparrow}^* c_{-l\downarrow}^* + v_l^* u_l c_{-l\downarrow} c_{l\uparrow} + |v_l|^2 c_{-l\downarrow} c_{l\uparrow} c_{l\uparrow}^* c_{-l\downarrow}^*$$

When we take the $\langle \phi_0 | \quad | \phi_0 \rangle$ matrix element, the middle two terms give zero, since they change the occupancy of the lth pair. The last term creates and then annihilates the pair, leading to a factor of unity. [More carefully, using the commutation relations (2-12), the operators in the last term can be transformed by successive binary interchanges to $-c_{l\uparrow} c_{-l\downarrow} c_{l\uparrow}^* c_{-l\downarrow}^*$, $+(c_{l\uparrow} c_{l\uparrow}^*)(c_{-l\downarrow} c_{-l\downarrow}^*)$, and $+(1 - c_{l\uparrow}^* c_{l\uparrow})(1 - c_{-l\downarrow}^* c_{-l\downarrow})$, both factors of which give unity when operating on $| \phi_0 \rangle$.] Thus, each factor for $l \neq k$ simply reduces to $|u_l|^2 + |v_l|^2 = 1$. When the same procedure is followed in the product with $l = k$, the cross terms in $u_k v_k$ still drop out, since the extra factor $c_{k\uparrow}^* c_{k\uparrow}$ leaves particle conservation unaffected. Moreover, because $c_{k\uparrow} | \phi_0 \rangle$ gives zero, the $|u_k|^2$ term also drops out, leaving simply $|v_k|^2$. Thus, we recover (2-16) as anticipated.

To estimate the sharpness of the peak at $\bar{N}$, one needs to evaluate

$$\langle (N - \bar{N})^2 \rangle = \langle N^2 - 2N\bar{N} + \bar{N}^2 \rangle = \langle N^2 \rangle - \bar{N}^2$$

Carrying through a calculation similar to that above, one finds

$$\langle (N - \bar{N})^2 \rangle = 4 \sum_k u_k^2 v_k^2$$

Note that this is nonzero unless the occupancy cuts off discontinuously with v_k going from 1 to 0 and u_k from 0 to 1. Also note that both $\bar{N}$ and $\langle (N - \bar{N})^2 \rangle$ scale as the volume if one compares systems of various sizes but the same particle density. (This follows because the number of k values in a given energy range is proportional to the volume.) Accordingly,

$$\delta N_{\text{rms}} = \langle (N - \bar{N})^2 \rangle^{1/2} \approx \bar{N}^{1/2} \approx 10^{10} \qquad (2\text{-}17a)$$

while the fractional uncertainty is

$$\frac{\delta N_{\text{rms}}}{\bar{N}} \approx \frac{1}{\bar{N}^{1/2}} \approx 10^{-10} \qquad (2\text{-}17b)$$

Thus, as is typical of many-particle statistical situations, as $N \to \infty$, the absolute fluctuations become large, but the fractional fluctuations approach zero.

Although for practical purposes we can usually ignore exact particle-number conservation, it is of interest to note that we can project out the N-particle part of $| \psi_G \rangle$, if necessary, by a rather simple method used by P. W. Anderson, namely,

$$| \psi_N \rangle = \int_0^{2\pi} d\varphi e^{-iN\varphi/2} \prod_k (|u_k| + |v_k| e^{i\varphi} c_{k\uparrow}^* c_{k\downarrow}^*) | \phi_0 \rangle \qquad (2\text{-}18)$$

This technique takes advantage of the fact that in an isolated superconductor the relative phase φ between states with N differing by a pair of particles is arbitrary. By integrating over all values of φ, i.e., by making φ completely uncertain, we can enforce a precise specification of the number N. [The integration over φ gives zero except for those terms in the expansion of the product in (2-18) in which there are precisely $N/2$ factors of $e^{i\varphi}$, each of which is associated with the creation of a pair.]

94788

On the other hand, with φ fixed, at zero for example, we have seen that $\delta N_{rms} \approx 10^{10}$. These results illustrate the uncertainty relation

$$\Delta N \, \Delta \varphi \gtrsim 1 \qquad (2\text{-}19)$$

There is an instructive analogy with the case of the electromagnetic field. In order to have a semiclassical electric field $\mathbf{E}$ with well-defined phase and amplitude, one must have enough photons (as in a laser) so that one can tolerate a superposition of states with various numbers present.

2-4 VARIATIONAL METHOD

We have studied the structure of ψ_G with some care to bring out some of its interesting features. Now we must actually make it explicit by finding appropriate values for the $u_\mathbf{k}$ and $v_\mathbf{k}$. Our first approach will be a variational calculation, as was used in the original BCS paper. Later we shall discuss another technique which leads to the same conclusions but in somewhat more modern form.

2-4.1 Determination of the Coefficients

We make the calculation using the so-called "pairing hamiltonian" or "reduced hamiltonian"

$$\mathscr{H} = \sum_{\mathbf{k}\sigma} \epsilon_\mathbf{k} n_{\mathbf{k}\sigma} + \sum_{\mathbf{kl}} V_{\mathbf{kl}} c^*_{\mathbf{k}\uparrow} c^*_{-\mathbf{k}\downarrow} c_{-\mathbf{l}\downarrow} c_{\mathbf{l}\uparrow} \qquad (2\text{-}20)$$

presuming that it includes the terms decisive for superconductivity, although it omits many other terms which involve electrons not paired as $(\mathbf{k}\uparrow, -\mathbf{k}\downarrow)$. Such terms have zero expectation value in the BCS ground-state wavefunction, but may be important in other applications. To regulate the mean number of particles $\bar{N}$, we include a term $-\mu N_{op}$, where μ is the chemical potential (or Fermi energy) and N_{op} is the particle-number operator. We then minimize the expectation value of the sum by setting

$$\delta \langle \psi_G | \mathscr{H} - \mu N_{op} | \psi_G \rangle = 0$$

The inclusion of $-\mu N_{op}$ is mathematically equivalent to taking the zero of kinetic energy to be μ (or E_F). So, more explicitly, we set

$$\delta \langle \psi_G | \sum_{\mathbf{k}\sigma} \xi_\mathbf{k} n_{\mathbf{k}\sigma} + \sum_{\mathbf{kl}} V_{\mathbf{kl}} c^*_{\mathbf{k}\uparrow} c^*_{\mathbf{k}\downarrow} c_{-\mathbf{l}\downarrow} c_{\mathbf{l}\uparrow} | \psi_G \rangle = 0$$

where, as before, $\xi_\mathbf{k} = \epsilon_\mathbf{k} - \mu$ is the single-particle energy relative to the Fermi energy. By the method of calculation used above to find $\bar{N}$, we see at once that the first term yields

$$\langle KE - \mu N \rangle = 2 \sum_\mathbf{k} \xi_\mathbf{k} |v_\mathbf{k}|^2 \qquad (2\text{-}21)$$

Similarly, the interaction term gives

$$\langle V \rangle = \sum_{\mathbf{kl}} V_{\mathbf{kl}} u_{\mathbf{k}} v_{\mathbf{k}}^* u_{\mathbf{l}}^* v_{\mathbf{l}} \qquad (2\text{-}22)$$

as can be seen by direct calculation. Alternatively it can be seen by inspection by noting that the term $V_{\mathbf{kl}}$ scatters from a state with $(\mathbf{l}\uparrow, -\mathbf{l}\downarrow)$ to one with $(\mathbf{k}\uparrow, -\mathbf{k}\downarrow)$. This requires the initial state to have the $\mathbf{l}$ pair occupied and the $\mathbf{k}$ pair empty and vice versa for the final state. The probability *amplitude* for such an initial state is $u_{\mathbf{k}} v_{\mathbf{l}}$ and for the final state it is $v_{\mathbf{k}}^* u_{\mathbf{l}}^*$, thus leading to the above result. We should perhaps note that $V_{\mathbf{kl}}$ contributes nothing to the energy in the normal state. This is obvious at $T = 0$, since states are either 100 percent occupied or empty, so that the product of the probabilities of being full and empty is zero. At $T > 0$, the Fermi distribution does not cut off sharply, and so one might think there would be a nonzero contribution. However, in the normal state the various Slater determinants representing specific electron occupation numbers are superimposed with random relative phase so that the appropriate products of probability amplitudes (corresponding to $u_{\mathbf{k}} v_{\mathbf{k}}^* u_{\mathbf{l}}^* v_{\mathbf{l}}$ in the ordered BCS state) average to zero. Hence these scattering terms make no contribution to the average energy in the normal state.

Combining (2-21) and (2-22), and for simplicity taking $u_{\mathbf{k}}$ and $v_{\mathbf{k}}$ to be real, we have

$$\langle \psi_G | \mathscr{H} - \mu N_{\mathrm{op}} | \psi_G \rangle = 2 \sum_{\mathbf{k}} \xi_{\mathbf{k}} v_{\mathbf{k}}^2 + \sum_{\mathbf{kl}} V_{\mathbf{kl}} u_{\mathbf{k}} v_{\mathbf{k}} u_{\mathbf{l}} v_{\mathbf{l}} \qquad (2\text{-}23)$$

which is to be minimized subject to the constraint that $u_{\mathbf{k}}^2 + v_{\mathbf{k}}^2 = 1$. This constraint is conveniently imposed by letting

$$u_{\mathbf{k}} = \sin \theta_{\mathbf{k}} \qquad \text{and} \qquad v_{\mathbf{k}} = \cos \theta_{\mathbf{k}} \qquad (2\text{-}24)$$

Then, after using elementary trigonometric identities, the right member of (2-23) can be written

$$\sum_{\mathbf{k}} \xi_{\mathbf{k}} (1 + \cos 2\theta_{\mathbf{k}}) + \tfrac{1}{4} \sum_{\mathbf{kl}} V_{\mathbf{kl}} \sin 2\theta_{\mathbf{k}} \sin 2\theta_{\mathbf{l}}$$

whence $\quad \dfrac{\partial}{\partial \theta_{\mathbf{k}}} \langle \psi_G | \mathscr{H} - \mu N_{\mathrm{op}} | \psi_G \rangle = 0 = -2\xi_{\mathbf{k}} \sin 2\theta_{\mathbf{k}} + \sum_{\mathbf{l}} V_{\mathbf{kl}} \cos 2\theta_{\mathbf{k}} \sin 2\theta_{\mathbf{l}}$

$$(2\text{-}25)$$

(The extra factor of 2 enters in the second sum because both $\mathbf{k}$ and $\mathbf{l}$ indices run over any given value $\mathbf{k}'$.) Thus

$$\tan 2\theta_{\mathbf{k}} = \frac{\sum_{\mathbf{l}} V_{\mathbf{kl}} \sin 2\theta_{\mathbf{l}}}{2\xi_{\mathbf{k}}} \qquad (2\text{-}26)$$

Now we *define* the quantities

$$\Delta_{\mathbf{k}} = -\sum_{\mathbf{l}} V_{\mathbf{kl}} u_{\mathbf{l}} v_{\mathbf{l}} = -\tfrac{1}{2} \sum_{\mathbf{l}} V_{\mathbf{kl}} \sin 2\theta_{\mathbf{l}} \qquad (2\text{-}27)$$

and $$E_{\mathbf{k}} = (\Delta_{\mathbf{k}}^2 + \xi_{\mathbf{k}}^2)^{1/2} \qquad (2\text{-}28)$$

(These quantities will soon acquire a physical significance as energy-gap parameter and quasi-particle excitation energy, respectively.) Then (2-26) becomes

$$\tan 2\theta_k = -\frac{\Delta_k}{\xi_k} \qquad (2\text{-}29a)$$

so that

$$2u_k v_k = \sin 2\theta_k = \frac{\Delta_k}{E_k} \qquad (2\text{-}29b)$$

and

$$v_k^2 - u_k^2 = \cos 2\theta_k = -\frac{\xi_k}{E_k} \qquad (2\text{-}29c)$$

This choice of signs for the sine and cosine [only their relative sign is fixed by (2-29a)] gives the occupation number $v_k^2 \to 0$ as $\xi_k \to \infty$, as is required for a reasonable solution.

We can now substitute (2-29b) back into (2-27) to evaluate Δ_k, leading to the condition for self-consistency

$$\Delta_k = -\frac{1}{2}\sum_l \frac{\Delta_l}{E_l} V_{kl} = -\frac{1}{2}\sum_l \frac{\Delta_l}{(\Delta_l^2 + \xi_l^2)^{1/2}} V_{kl} \qquad (2\text{-}30)$$

We note first the trivial solution in which $\Delta_k = 0$, so that $v_k = 1$ for $\xi_k < 0$, and $v_k = 0$ for $\xi_k > 0$. The associated $|\psi\rangle$ is just the single Slater determinant with all states up to k_F occupied, the normal Fermi sea at $T = 0$. But we expect a nontrivial solution with lower energy if V_{kl} is negative. We retain the model of V_{kl} used by Cooper and by BCS, namely

$$V_{kl} = \begin{cases} -V & \text{if } |\xi_k| \text{ and } |\xi_l| \le \hbar\omega_c \\ 0 & \text{otherwise} \end{cases} \qquad (2\text{-}31)$$

with V being a positive constant. (Our theoretical discussion of V_{kl} actually suggests that the relevant energy is $|\xi_k - \xi_l|$, the energy change of the electron in the scattering process, but to get a simple solution it is necessary to make the stronger restriction that $|\xi_k|$ and $|\xi_l|$ separately are smaller than $\hbar\omega_c$.) Inserting this V_{kl} in (2-30), we find that it is satisfied by

$$\Delta_k = \begin{cases} \Delta & \text{for } |\xi_k| < \hbar\omega_c \\ 0 & \text{for } |\xi_k| > \hbar\omega_c \end{cases} \qquad (2\text{-}32)$$

Since in this model $\Delta_k = \Delta$ is actually independent of k, we may cancel it from both sides of (2-30), and our condition for self-consistency then reads

$$1 = \frac{V}{2}\sum_k \frac{1}{E_k} \qquad (2\text{-}33)$$

Upon replacing the summation by an integration from $-\hbar\omega_c$ to $\hbar\omega_c$, and using the symmetry of $\pm\xi$ values, this becomes

$$\frac{1}{N(0)V} = \int_0^{\hbar\omega_c} \frac{d\xi}{(\Delta^2 + \xi^2)^{1/2}} = \sinh^{-1}\frac{\hbar\omega_c}{\Delta} \qquad (2\text{-}33a)$$

Thus

$$\Delta = \frac{\hbar\omega_c}{\sinh\left[1/N(0)V\right]} \approx 2\hbar\omega_c e^{-1/N(0)V} \qquad (2\text{-}34)$$

where the last step is justified in the weak coupling limit $N(0)V \ll 1$. Since it will turn out that $N(0)V$ is typically $\lesssim 0.3$ and less than 0.5 for all known superconductors, the approximate equality in (2-34) is typically good to 1 percent.

Having found Δ, we may simply compute the coefficients $u_{\mathbf{k}}$ and $v_{\mathbf{k}}$ which specify the optimum BCS wavefunction. A convenient approach is to start with (2-29c) and the normalization condition $u_{\mathbf{k}}^2 + v_{\mathbf{k}}^2 = 1$. In this way one finds that the fractional occupation number $v_{\mathbf{k}}^2$ is given by

$$v_{\mathbf{k}}^2 = \frac{1}{2}\left(1 - \frac{\xi_{\mathbf{k}}}{E_{\mathbf{k}}}\right) = \frac{1}{2}\left[1 - \frac{\xi_{\mathbf{k}}}{(\Delta^2 + \xi_{\mathbf{k}}^2)^{1/2}}\right] \qquad (2\text{-}35)$$

while

$$u_{\mathbf{k}}^2 = \frac{1}{2}\left(1 + \frac{\xi_{\mathbf{k}}}{E_{\mathbf{k}}}\right) = 1 - v_{\mathbf{k}}^2$$

A plot of $v_{\mathbf{k}}^2$ is shown in Fig. 2-1. Note that $v_{\mathbf{k}}^2$ approaches unity well below the Fermi energy and zero well above, rather like the Fermi function appropriate to normal metals at finite temperatures. In fact, there is a startling resemblance between $v_{\mathbf{k}}^2$ for the BCS ground state at $T = 0$ and the normal-metal Fermi function at $T = T_c$, also plotted in Fig. 2-1 for comparison purposes. From this comparison we see that, contrary to the early ideas of Fröhlich, Bardeen, and others, the change in the metal on cooling from T_c to $T = 0$ cannot be usefully described in terms of changes in the occupation numbers of one-electron momentum eigenstates. In particular, no gap opens up in $\mathbf{k}$ space. Rather, the disorder associated with partial occupation of these states with random phases is being replaced by a *single* quantum state of the system, in which more or less the same set of many-body states with various one-electron occupancies are now superposed with a fixed phase relation.

A further remark about our result for $v_{\mathbf{k}}^2$ is that it falls off as $1/\xi_{\mathbf{k}}^2$ for $\xi_{\mathbf{k}} \gg \Delta$, the same dependence as we found earlier for $g_{\mathbf{k}}^2$ in our simple treatment of a single Cooper pair. In fact, apart from the asymmetry introduced by the artificial restriction of keeping the Fermi sea undisturbed, that simple calculation gives quite an accurate idea of how the energy-lowering correlated-pair state is formed. Finally, we note from (2-35) that Δ is the characteristic energy determining the range of $\mathbf{k}$ values involved in forming the Cooper pairs. Since Δ will turn out to be 1.76 kT_c at $T = 0$, this reinforces our expectation of a characteristic size $\sim \xi_0 \sim \hbar v_F/\Delta \sim \hbar v_F/kT_c$.

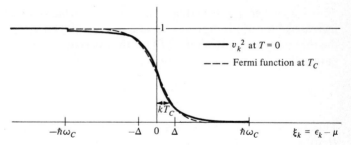

FIGURE 2-1

Plot of BCS occupation fraction v_k^2 vs. electron energy measured from the chemical potential (Fermi energy). To make the cutoffs at $\pm\hbar\omega_c$ visible, the plot has been made for a strong-coupling superconductor with $N(0)V = 0.43$. For comparison, the Fermi function for the normal state at T_c is also shown on the same scale, using the BCS relation $\Delta(0) = 1.76kT_c$.

2-4.2   Evaluation of Ground-State Energy

With $|\psi_G\rangle$ determined, we now calculate its energy, to show that it is indeed lower than the Fermi-sea state. From (2-23), using (2-27) and (2-35), we have

$$\langle\psi_G|\,\mathcal{H} - \mu N_{\mathrm{op}}\,|\psi_G\rangle = \sum_{\mathbf{k}}\left(\xi_{\mathbf{k}} - \frac{\xi_{\mathbf{k}}^2}{E_{\mathbf{k}}}\right) - \frac{\Delta^2}{V}$$

As noted above, the normal state at $T = 0$ corresponds to the BCS state with $\Delta = 0$, in which case $E_{\mathbf{k}} = |\xi|_{\mathbf{k}}$. Thus

$$\langle\psi_n|\,\mathcal{H} - \mu N_{\mathrm{op}}\,|\psi_n\rangle = \sum_{|\mathbf{k}| < k_F} 2\xi_{\mathbf{k}}$$

the terms for $|\mathbf{k}| > k_F$ giving zero since $E_{\mathbf{k}} = \xi_{\mathbf{k}}$. Thus, the difference in these energies is

$$\langle E\rangle_s - \langle E\rangle_n = \sum_{|\mathbf{k}| > k_F}\left(\xi_{\mathbf{k}} - \frac{\xi_{\mathbf{k}}^2}{E_{\mathbf{k}}}\right) + \sum_{|\mathbf{k}| < k_F}\left(-\xi_{\mathbf{k}} - \frac{\xi_{\mathbf{k}}^2}{E_{\mathbf{k}}}\right) - \frac{\Delta^2}{V}$$

$$= 2\sum_{|\mathbf{k}| > k_F}\left(\xi_{\mathbf{k}} - \frac{\xi_{\mathbf{k}}^2}{E_{\mathbf{k}}}\right) - \frac{\Delta^2}{V}$$

by symmetry about the Fermi energy. Going over to the continuum approximation and carrying out the integration, we find (in the weak coupling limit)

$$\langle E\rangle_s - \langle E\rangle_n = \left[\frac{\Delta^2}{V} - \frac{1}{2}N(0)\Delta^2\right] - \frac{\Delta^2}{V}$$

We have kept the kinetic-energy difference together inside the bracket to make explicit the cancellation of its leading term against the attractive potential-energy term. The *net* energy lowering is thus down by a factor of $N(0)V/2 \approx 0.1$ from the increase in kinetic energy or decrease in potential energy separately. Introducing the thermodynamic symbol $U(T)$ for the internal energy of the system, and anticipating that $\Delta(T)$ is temperature dependent, our final result is

$$U_s(0) - U_n(0) = -\frac{1}{2} N(0)\Delta^2(0) \qquad (2\text{-}36)$$

This is the "condensation energy" at $T = 0$, which must by definition equal $H_c^2(0)/8\pi$, where $H_c(T)$ is the thermodynamic critical field.

Isotope effect Since Δ is proportional to $\hbar\omega_c$ [from (2-34)], and since ω_c should be proportional to $M^{-1/2}$, the isotope effect ($H_c \sim M^{-1/2}$) follows from (2-36) if $N(0)$ and V are unchanged from one isotope to another. This is to be expected in the case of $N(0)$, which is a purely electronic property. On the other hand, since the parameter V is determined jointly by the electrons and the phonons, it seems unlikely to be constant. In fact, although the observed isotope-shift exponent is quite accurately $-\frac{1}{2}$ for a few classic superconductors, such as lead, recent experiments on other materials have shown that it can range all the way to zero or even change sign. It has been shown theoretically[1] that the observed variations are compatible with a more detailed theory of the interaction than we have been able to give here, but as usual it is very difficult to make a reliable a priori calculation of such a property, which depends on the details of an individual material. It is perhaps more useful to invert the procedure, and view the measured isotope effect as providing information about the detailed nature of the interaction, as has been done by McMillan.[2]

2-5 SOLUTION BY CANONICAL TRANSFORMATION

The variational method used in the original BCS treatment, which we have just sketched, is a direct approach for calculating the condensation energy of the superconducting ground state relative to the normal state. It is somewhat clumsy, however, though workable, in dealing with excited states. In this section we outline another approach, closer to the more sophisticated modern methods, which is well suited to handle excitations. This alternate method is also a self-consistent field method, but no appeal to a variational calculation is required.

We start with the observation that the characteristic BCS pair-interaction hamiltonian will lead to a ground state which is some phase-coherent superposition of many-body states with pairs of Bloch states $(\mathbf{k}\uparrow, -\mathbf{k}\downarrow)$ occupied or unoccupied as units. Because of the coherence, operators such as $c_{-\mathbf{k}\downarrow} c_{\mathbf{k}\uparrow}$ can have

[1] P. Morel and P. W. Anderson, *Phys. Rev.* **125**, 1263 (1962); J. W. Garland, Jr., *Phys. Rev. Letters* **11**, 114 (1963).
[2] W. L. McMillan, *Phys. Rev.* **167**, 331 (1968).

nonzero expectation values b_k in such a state, rather than averaging to zero as in a normal metal, where the phases are random. Moreover, because of the large numbers of particles involved, the fluctuations about these expectation values should be small. This suggests that it will be useful to express such a product of operators formally as

$$c_{-k\downarrow} c_{k\uparrow} = b_k + (c_{-k\downarrow} c_{k\uparrow} - b_k) \qquad (2\text{-}37)$$

and subsequently neglect quantities which are bilinear in the presumably small fluctuation term in parentheses. If we follow this procedure with our pairing hamiltonian (2-20), we obtain the so-called model hamiltonian

$$\mathcal{H}_M = \sum_{k\sigma} \xi_k c_{k\sigma}^* c_{k\sigma} + \sum_{kl} V_{kl}(c_{k\uparrow}^* c_{-k\downarrow}^* b_l + b_k^* c_{-l\downarrow} c_{l\uparrow} - b_k^* b_l) \qquad (2\text{-}38)$$

where the b_k are to be determined self-consistently so that

$$b_k = \langle c_{-k\downarrow} c_{k\uparrow} \rangle_{av} \qquad (2\text{-}39)$$

Note that in gaining the simplicity of eliminating quartic terms in the c_k's from the hamiltonian, we have thrown it into an approximate form which does not conserve particle number. Rather, there are now terms which create or destroy pairs of particles. This is analogous to the situation noted above in which the simple BCS product wavefunction with fixed phase contained many different numbers of particles. Only by integrating over the phase φ were we able to set an exact particle number. The corresponding situation here is that we have assigned a definite phase to b_k. In any case, as before, we can handle this situation by introducing the chemical potential μ so as to fix $\bar{N}$ at any desired value.

Now to proceed with the solution, we define

$$\Delta_k = -\sum_l V_{kl} b_l = -\sum_l V_{kl} \langle c_{-l\downarrow} c_{l\uparrow} \rangle \qquad (2\text{-}40)$$

This definition is evidently very analogous to the one given in (2-27), and it will turn out to give the gap in the energy spectrum. In terms of Δ_k, the model hamiltonian becomes (after relabeling some subscripts)

$$\mathcal{H}_M = \sum_{k\sigma} \xi_k c_{k\sigma}^* c_{k\sigma} - \sum_k (\Delta_k c_{k\uparrow}^* c_{-k\downarrow}^* + \Delta_k^* c_{-k\downarrow} c_{k\uparrow} - \Delta_k b_k^*) \qquad (2\text{-}41)$$

which is a sum of terms, each bilinear in the pair of operators corresponding to the partners in a Cooper pair. Such a hamiltonian can be diagonalized by a suitable linear transformation. As shown independently by Bogoliubov[1] and by Valatin,[2] the appropriate transformation is specified by

$$c_{k\uparrow} = u_k^* \gamma_{k0} + v_k \gamma_{k1}^*$$
$$c_{-k\downarrow}^* = -v_k^* \gamma_{k0} + u_k \gamma_{k1}^* \qquad (2\text{-}42)$$

[1] N. N. Bogoliubov, *Nuovo Cimento* **7**, 794 (1958); *Zh. Eksperim. i Teor. Fiz.* **34**, 58 (1958) [*Soviet Phys.—JETP* **7**, 41 (1958)].
[2] J. G. Valatin, *Nuovo Cimento* **7**, 843 (1958).

where the numerical coefficients u_k and v_k satisfy $|u_k|^2 + |v_k|^2 = 1$ and the γ_k are new Fermi operators. Note that γ_{k0} participates in destroying an electron with $\mathbf{k}\uparrow$ or creating one with $-\mathbf{k}\downarrow$; in both cases, the net effect is to decrease the system momentum by $\mathbf{k}$ and to reduce S_z by $\hbar$. The operator γ_{k1}^* has similar properties, so γ_{k1} itself decreases the system momentum by $-\mathbf{k}$ (i.e., increases it by $\mathbf{k}$) and has the net effect of converting a down spin to an up spin.

Substituting these new operators (2-42) into the model hamiltonian (2-38), and carrying out the indicated products taking account of the noncommutivity of the operators, we obtain

$$
\begin{aligned}
\mathcal{H}_M = \sum_k \xi_k [(\,|u_k^2| - |v_k|^2)(\gamma_{k0}^* \gamma_{k0} + \gamma_{k1}^* \gamma_{k1}) + 2\,|v_k|^2 + 2u_k^* v_k^* \gamma_{k1}\gamma_{k0} \\
+ 2u_k v_k \gamma_{k0}^* \gamma_{k1}^*] + \sum_k [(\Delta_k u_k v_k^* + \Delta_k^* u_k^* v_k)(\gamma_{k0}^* \gamma_{k0} + \gamma_{k1}^* \gamma_{k1} - 1) \\
+ (\Delta_k v_k^{*2} - \Delta_k^* u_k^{*2})\gamma_{k1}\gamma_{k0} + (\Delta_k^* v_k^2 - \Delta_k u_k^2)\gamma_{k0}^* \gamma_{k1}^* + \Delta_k b_k^*]
\end{aligned}
$$
(2-43)

Now, if we choose u_k and v_k so that the coefficients of $\gamma_{k1}\gamma_{k0}$ and $\gamma_{k0}^* \gamma_{k1}^*$ vanish, the hamiltonian is diagonalized; i.e., it is carried into a form containing only constants plus terms proportional to the occupation numbers $\gamma_k^* \gamma_k$. The coefficients of both undesired terms are zero if

$$
2\xi_k u_k v_k + \Delta_k^* v_k^2 - \Delta_k u_k^2 = 0
$$

Multiplying through by Δ_k^*/u_k^2 and solving by the quadratic formula, this condition becomes

$$
\frac{\Delta_k^* v_k}{u_k} = (\xi_k^2 + |\Delta_k|^2)^{1/2} - \xi_k \equiv E_k - \xi_k \qquad (2\text{-}44)
$$

using the definition of E_k introduced earlier. (We have chosen the positive sign of the square root so as to correspond to the stable solution of minimum rather than maximum energy.) Given the normalization requirement that $|u_k|^2 + |v_k|^2 = 1$, and knowing that $|v_k/u_k| = (E_k - \xi_k)/|\Delta_k|$ from (2-44), we can solve for the coefficients and find

$$
|v_k|^2 = 1 - |u_k|^2 = \frac{1}{2}\left(1 - \frac{\xi_k}{E_k}\right) \qquad (2\text{-}35')
$$

in exact agreement with (2-35), our variationally obtained result.

Although the phases of u_k, v_k, and Δ_k are individually arbitrary, they are related by (2-44), since $\Delta_k^* v_k/u_k$ is real. That is, the phase of v_k relative to u_k must be the phase of Δ_k. There is no loss in generality in choosing all the u_k to be real and positive. If we do so, v_k and Δ_k must have the same phase.

2-5.1   Excitation Energies and the Energy Gap

With u_k and v_k chosen so as to diagonalize the model hamiltonian (2-43), the remaining terms reduce to

$$\mathscr{H}_M = \sum_k (\xi_k - E_k + \Delta_k b_k^*) + \sum_k E_k(\gamma_{k0}^* \gamma_{k0} + \gamma_{k1}^* \gamma_{k1}) \qquad (2\text{-}45)$$

The first sum is a constant, which differs from the corresponding sum for the normal state at $T = 0$ ($E_k = |\xi_k|$, $\Delta_k = 0$) by exactly the condensation energy (2-36) found earlier. The second sum gives the increase in energy above the ground state in terms of the number operators $\gamma_k^* \gamma_k$ for the γ_k fermions. Thus, these γ_k describe the elementary quasi-particle excitations of the system, which are often called *Bogoliubons*. Evidently, the energies of these excitations are just

$$E_k = (\xi_k^2 + |\Delta_k|^2)^{1/2} \qquad (2\text{-}46)$$

Thus, as we had anticipated, Δ_k plays the role of an *energy gap* or minimum excitation energy, since even at the Fermi surface, where $\xi_k = 0$, $E_k = |\Delta_k| > 0$. Moreover, the notation E_k has now received its justification as the energy of an elementary excitation of momentum $\hbar k$.

As in our earlier variational calculation, we require self-consistency when $\langle c_{-l\downarrow} c_{l\uparrow} \rangle$ computed from our solution is inserted back into (2-40). Rewriting the c_k operators in terms of the γ_k, and dropping off-diagonal terms in quasi-particle operators $\gamma_{k0}^* \gamma_{k1}^*$ and $\gamma_{k1} \gamma_{k0}$ (since they do not contribute to averages), we have

$$\Delta_k = -\sum_l V_{kl} \langle c_{-l\downarrow} c_{l\uparrow} \rangle = -\sum_l V_{kl} u_l^* v_l \langle 1 - \gamma_{l0}^* \gamma_{l0} - \gamma_{l1}^* \gamma_{l1} \rangle \qquad (2\text{-}47)$$

At $T = 0$, when no quasi-particles are excited, this reduces to (2-27), so that exactly the same result (2-34) for $\Delta(0)$ in terms of $\hbar\omega_c$ and $N(0)V$ follows from the canonical transformation method as from the variational one. However, the present method is much more convenient for handling the extension of the calculation to $T > 0$.

2-6   FINITE TEMPERATURES

Since we have identified E_k as the excitation energy of a fermion quasi-particle, the probability of its excitation in thermal equilibrium is the usual Fermi function

$$f(E_k) = (e^{\beta E_k} + 1)^{-1} \qquad (2\text{-}48)$$

where $\beta = 1/kT$. Thus,

$$\langle 1 - \gamma_{k0}^* \gamma_{k0} - \gamma_{k1}^* \gamma_{k1} \rangle = 1 - 2f(E_k)$$

so that in general, (2-47) becomes

$$\Delta_{\mathbf{k}} = - \sum_{\mathbf{l}} V_{\mathbf{kl}} u_{\mathbf{l}}^* v_{\mathbf{l}} [1 - 2f(E_{\mathbf{l}})]$$

$$= - \sum_{\mathbf{l}} V_{\mathbf{kl}} \frac{\Delta_{\mathbf{l}}}{2E_{\mathbf{l}}} \tanh \frac{\beta E_{\mathbf{l}}}{2} \qquad (2\text{-}49)$$

Making the BCS approximation that $V_{\mathbf{kl}} = -V$, we have $\Delta_{\mathbf{k}} = \Delta_{\mathbf{l}} = \Delta$, and the self-consistency condition becomes

$$\frac{1}{V} = \frac{1}{2} \sum_{\mathbf{k}} \frac{\tanh(\beta E_{\mathbf{k}}/2)}{E_{\mathbf{k}}} \qquad (2\text{-}50)$$

where, as usual, $E_{\mathbf{k}} = (\xi_{\mathbf{k}}^2 + \Delta^2)^{1/2}$. Equation (2-50) determines the temperature dependence of the energy gap $\Delta(T)$.

2-6.1 Determination of T_c

The critical temperature T_c is the temperature at which $\Delta(T) \to 0$. In this case, $E_{\mathbf{k}} \to |\xi_{\mathbf{k}}|$, and the excitation spectrum becomes the same as in the normal state. Thus, T_c is found by replacing $E_{\mathbf{k}}$ with $|\xi_{\mathbf{k}}|$ in (2-50) and solving. After changing the sum to an integral, taking advantage of the symmetry of $|\xi_{\mathbf{k}}|$ about the Fermi level, and changing to a dimensionless variable of integration, this condition becomes

$$\frac{1}{N(0)V} = \int_0^{\beta_c \hbar \omega_c /2} \frac{\tanh x}{x} dx$$

This integral can be evaluated and yields $\ln(A\beta_c \hbar\omega_c)$, where $A = 2\gamma/\pi \approx 1.13$, γ here being Euler's constant. Consequently,

$$kT_c = \beta_c^{-1} = 1.13\hbar\omega_c e^{-1/N(0)V}. \qquad (2\text{-}51)$$

Comparing this with (2-34), we see that

$$\frac{\Delta(0)}{kT_c} = \frac{2}{1.13} = 1.764 \qquad (2\text{-}52)$$

so that the gap at $T = 0$ is indeed comparable in energy to kT_c. The numerical factor 1.76 has been tested in many experiments and found to be reasonable. That is, experimental values of 2Δ for different materials and different directions in k space generally fall in the range 3.0 to $4.5kT_c$, with most clustered near the BCS value of $3.5kT_c$.

2-6.2 Temperature Dependence of the Gap

Given (2-50), or its integral equivalent

$$\frac{1}{N(0)V} = \int_0^{\hbar\omega_c} \frac{\tanh \frac{1}{2}\beta(\xi^2 + \Delta^2)^{1/2}}{(\xi^2 + \Delta^2)^{1/2}} d\xi \qquad (2\text{-}53)$$

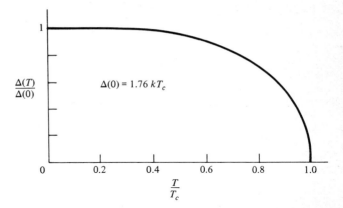

FIGURE 2-2

Temperature dependence of the energy gap in the BCS theory. Strictly speaking, this universal curve holds only in the weak-coupling limit, but it is a good approximation in most cases.

$\Delta(T)$ can be computed numerically. For weak-coupling superconductors, in which $\hbar\omega_c/kT_c \gg 1$, $\Delta(T)/\Delta(0)$ is a universal function of T/T_c which decreases monotonically from one at $T = 0$ to zero at T_c, as shown in Fig. 2-2. Near $T = 0$, the temperature variation is exponentially slow, since $e^{-\Delta/kT} \approx 0$, so that the hyperbolic tangent is very nearly unity and insensitive to T. Physically speaking, Δ is nearly constant until a significant number of quasi-particles are thermally excited. On the other hand, near T_c, $\Delta(T)$ drops to zero with a vertical tangent, approximately as

$$\frac{\Delta(T)}{\Delta(0)} \approx 1.74\left(1 - \frac{T}{T_c}\right)^{1/2} \qquad T \approx T_c \qquad (2\text{-}54)$$

The variation of the order parameter Δ with the square root of $(T_c - T)$ is characteristic of all mean-field theories. For example, $M(T)$ has the same dependence in the molecular-field theory of ferromagnetism.

2-6.3 Thermodynamic Quantities

With $\Delta(T)$ determined, the temperature-dependent set of fermion excitation energies $E_\mathbf{k} = [\xi_\mathbf{k}^2 + \Delta(T)^2]^{1/2}$ is fixed. These energies determine the quasi-particle occupation numbers $f_\mathbf{k} = (1 + e^{\beta E_\mathbf{k}})^{-1}$, which in turn determine the electronic entropy in the usual way for a fermion gas, namely

$$S_{es} = -2k \sum_\mathbf{k} [(1 - f_\mathbf{k}) \ln (1 - f_\mathbf{k}) + f_\mathbf{k} \ln f_\mathbf{k}] \qquad (2\text{-}55)$$

Given $S_{es}(T)$, the specific heat is given by

$$C_{es} = T \frac{dS_{es}}{dT} = -\beta \frac{dS_{es}}{d\beta}$$

Using (2-55), this is

$$C_{es} = 2\beta k \sum_{\mathbf{k}} \frac{\partial f_{\mathbf{k}}}{\partial \beta} \ln \frac{f_{\mathbf{k}}}{1 - f_{\mathbf{k}}} = -2\beta^2 k \sum_{\mathbf{k}} E_{\mathbf{k}} \frac{\partial f_{\mathbf{k}}}{\partial \beta}$$

$$= -2\beta^2 k \sum_{\mathbf{k}} E_{\mathbf{k}} \frac{df_{\mathbf{k}}}{d(\beta E_{\mathbf{k}})} \left(E_{\mathbf{k}} + \beta \frac{dE_{\mathbf{k}}}{d\beta} \right)$$

$$= 2\beta k \sum_{\mathbf{k}} -\frac{\partial f_{\mathbf{k}}}{\partial E_{\mathbf{k}}} \left(E_{\mathbf{k}}^2 + \tfrac{1}{2}\beta \frac{d\Delta^2}{d\beta} \right) \qquad (2\text{-}56)$$

The first term is the usual one coming from the redistribution of quasi-particles among the various energy states as the temperature changes. The second term is more unusual, and describes the effect of the temperature-dependent gap in changing the energy levels themselves.

Evidently both terms in C_{es} will be exponentially small at $T \ll T_c$, where the minimum excitation energy Δ is much greater than kT. This accounts for the exponential form (1-13) noted earlier. Another interesting limit is very near T_c. Then, as $\Delta(T) \to 0$, one can replace $E_{\mathbf{k}}$ by $|\xi_{\mathbf{k}}|$ in (2-56). The first term then reduces to the usual normal-state electronic specific heat

$$C_{en} = \gamma T = \frac{2\pi^2}{3} N(0)k^2 T \qquad (2\text{-}57)$$

which is continuous at T_c. The second term is finite below T_c, where $d\Delta^2/dT$ is large, but it is zero above T_c, giving rise to a discontinuity ΔC in the electronic specific heat at T_c. The size of the discontinuity is readily evaluated by changing the sum to an integral, as follows:

$$\Delta C = (C_{es} - C_{en})\Big|_{T_c} = N(0)k\beta^2 \left(\frac{d\Delta^2}{d\beta} \right) \int_{-\infty}^{\infty} \left(\frac{-\partial f}{\partial |\xi|} \right) d\xi$$

$$= N(0) \left(\frac{-d\Delta^2}{dT} \right)\Big|_{T_c} \qquad (2\text{-}58)$$

where we have used the fact that $\partial f/\partial |\xi| = \partial f/\partial \xi$, since $\partial f/\partial \xi$ is an even function of ξ. Using the approximate form (2-54) for $\Delta(T)$, with $\Delta(0) = 1.76 kT_c$, we obtain $\Delta C = 9.4 N(0)k^2 T_c$. Comparing with (2-57), we find that the normalized magnitude of the discontinuity is

$$\frac{\Delta C}{C_{en}} = \frac{9.4}{2\pi^2/3} = 1.43 \qquad (2\text{-}59)$$

The overall behavior of the electronic specific heat is sketched in Fig. 2-3b.

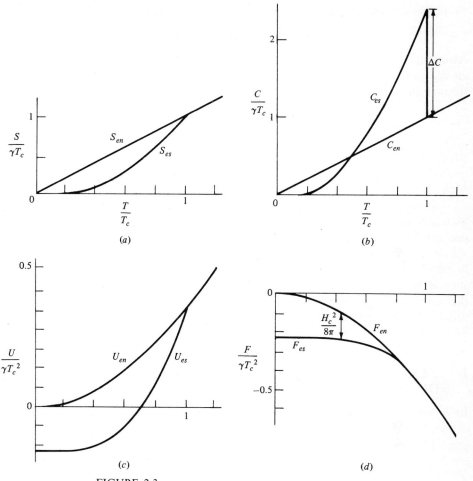

FIGURE 2-3
Comparison of thermodynamic quantities in superconducting and normal states. $U_{en}(0)$ is chosen as the zero of ordinates in (c) and (d). Because the transition is of second order, the quantities S, U, and F are continuous at T_c. Moreover, the slope of F_{es} joins continuously to that of F_{en} at T_c, since $\partial F/\partial T = -S$.

With $C_{es}(T)$ determined numerically from (2-56), one can integrate it to find the change in internal energy $U(T)$ as one decreases the temperature from T_c. At T_c it must be the same as the normal value $U_{en}(0) + \frac{1}{2}\gamma T_c^2$, since the specific heat remains finite there. Thus,

$$U_{es}(T) = U_{en}(0) + \frac{1}{2}\gamma T_c^2 - \int_T^{T_c} C_{es}\, dT \qquad (2\text{-}60)$$

From this and the entropy (2-55) we may then compute the free energy

$$F_{es}(T) = U_{es}(T) - TS_{es}(T) \qquad (2\text{-}61)$$

Assuming that the effect of the superconducting transition on the lattice free energy can be neglected, the thermodynamic critical field is then determined through the relation

$$\frac{H_c^2(T)}{8\pi} = F_{en}(T) - F_{es}(T) \qquad (2\text{-}62)$$

where $F_{en}(T) = U_{en}(0) - \frac{1}{2}\gamma T^2$. These various thermodynamic quantities are plotted in Fig. 2-3. A useful numerical tabulation has been given by Mühlschlegel.[1]

Since the critical-field curve $H_c(T)$ can be measured with greater accuracy than can typical thermodynamic quantities, it is of some importance to note that one can be derived from the other by rigorous thermodynamic computations, starting from (2-62). For example, the approximate parabolic temperature dependence of H_c quoted in (1-2) is inconsistent with an exponential variation of C_{es}, such as that quoted in (1-13). Precise measurements of the deviation of $H_c(T)$ from the parabolic approximation have been used by Mapother[2] to test the BCS predictions of thermodynamic properties.

2-7 STATE FUNCTIONS AND THE DENSITY OF STATES

By inverting (2-42), we find the expressions

$$\gamma_{k0}^* = u_k^* c_{k\uparrow}^* - v^* c_{-k\downarrow}$$

$$\gamma_{k1}^* = u_k^* c_{-k\downarrow}^* + v_k^* c_{k\uparrow} \qquad (2\text{-}63)$$

for the γ_k^* operators, which create quasi-particle excitations of the two spin directions from the superconducting ground state, in terms of the electron creation operators c_k^*. The superconducting ground state $|\psi_G\rangle$ is defined as the vacuum state of the γ particles, that is, by the relations

$$\gamma_{k0}|\psi_G\rangle = \gamma_{k1}|\psi_G\rangle = 0 \qquad (2\text{-}64)$$

The structure of $|\psi_G\rangle$ in terms of the γ particles is of no interest, but it is of interest to demonstrate that in terms of the c_k it agrees with the BCS product form (2-14). This may be verified by considering, for example,

$$\gamma_{k0}|\psi_G\rangle = (u_k c_{k\uparrow} - v_k c_{-k\downarrow}^*) \prod_l (u_l + v_l c_{l\uparrow}^* c_{-l\downarrow}^*)|\phi_0\rangle$$

[1] B. Mühlschlegel, *Z. Phys.* **155**, 313 (1959).
[2] D. E. Mapother, *Phys. Rev.* **126**, 2021 (1962).

Multiplying out the factor involving the kth pair, we find

$$u_{\mathbf{k}}^2 c_{\mathbf{k}\uparrow} + u_{\mathbf{k}} v_{\mathbf{k}} c_{\mathbf{k}\uparrow} c_{\mathbf{k}\uparrow}^* c_{-\mathbf{k}\downarrow}^* - v_{\mathbf{k}} u_{\mathbf{k}} c_{-\mathbf{k}\downarrow}^* - v_{\mathbf{k}}^2 c_{-\mathbf{k}\downarrow}^* c_{\mathbf{k}\uparrow}^* c_{-\mathbf{k}\downarrow}^*$$

All of these terms give zero when operating on $|\phi_0\rangle$ for the following reasons: the first is zero because the annihilation operator $c_{\mathbf{k}\uparrow}$ operates on the vacuum; the next two terms cancel because $c_{\mathbf{k}\uparrow} c_{\mathbf{k}\uparrow}^*$ operating on the vacuum gives a factor of unity; the last term vanishes because it contains the same creation operator twice with no intervening annihilation operator. The case $\gamma_{\mathbf{k}1} |\psi_G\rangle$ works out similarly. Thus, (2-64) is verified, and the BCS form for the ground state is in exact agreement with that of the canonical-transformation method.

Now let us look at the excited states. For example, consider

$$\gamma_{\mathbf{k}0}^* |\psi_G\rangle = (|u_{\mathbf{k}}|^2 c_{\mathbf{k}\uparrow}^* + u_{\mathbf{k}}^* v_{\mathbf{k}} c_{\mathbf{k}\uparrow}^* c_{\mathbf{k}\uparrow}^* c_{-\mathbf{k}\downarrow}^* - v_{\mathbf{k}}^* u_{\mathbf{k}} c_{-\mathbf{k}\downarrow} - |v_{\mathbf{k}}|^2 c_{-\mathbf{k}\downarrow} c_{\mathbf{k}\uparrow}^* c_{-\mathbf{k}\downarrow}^*)$$
$$\times \prod_{\mathbf{l} \neq \mathbf{k}} (u_{\mathbf{l}} + v_{\mathbf{l}} c_{\mathbf{l}\uparrow}^* c_{-\mathbf{l}\downarrow}^*) |\phi_0\rangle$$

The two middle terms give zero for reasons mentioned above. The last term may be changed to $+ |v_{\mathbf{k}}|^2 c_{\mathbf{k}\uparrow}^* c_{-\mathbf{k}\downarrow} c_{-\mathbf{k}\downarrow}^*$ using the anticommutation of the fermion operators, and then the factor $c_{-\mathbf{k}\downarrow} c_{-\mathbf{k}\downarrow}^*$ may be dropped, since, operating on $|\phi_0\rangle$, it gives a factor of unity. Combining the remainder with the first term, we have

$$\gamma_{\mathbf{k}0}^* |\psi_G\rangle = c_{\mathbf{k}\uparrow}^* \prod_{\mathbf{l} \neq \mathbf{k}} (u_{\mathbf{l}} + v_{\mathbf{l}} c_{\mathbf{l}\uparrow}^* c_{-\mathbf{l}\downarrow}^*) |\phi_0\rangle \qquad (2\text{-}65a)$$

Similarly,

$$\gamma_{\mathbf{k}1}^* |\psi_G\rangle = c_{-\mathbf{k}\downarrow}^* \prod_{\mathbf{l} \neq \mathbf{k}} (u_{\mathbf{l}} + v_{\mathbf{l}} c_{\mathbf{l}\uparrow}^* c_{-\mathbf{l}\downarrow}^*) |\phi_0\rangle \qquad (2\text{-}65b)$$

These are the excited states called "singles" in the original BCS treatment, where they were written down by inspection. They correspond to putting with certainty a single electron into one of the states of the pair $(\mathbf{k}\uparrow, -\mathbf{k}\downarrow)$, while leaving with certainty the other state of the pair empty. This effectively blocks that pair state from participation in the many-body wavefunction, and increases the system energy accordingly.

These $\gamma_{\mathbf{k}}^*$ operators change the expectation value of the particle number by $(u_{\mathbf{k}}^2 - v_{\mathbf{k}}^2)$, which ranges from -1 to $+1$ as $\xi_{\mathbf{k}}$ ranges from well below zero to well above. This net change results from a probability $u_{\mathbf{k}}^2$ of a change by $+1$ and a probability $v_{\mathbf{k}}^2$ of a change by -1. Such behavior would be inconsistent with exact number conservation in an isolated system. This apparent paradox is resolved by recalling that the $\gamma_{\mathbf{k}}^*$ operators are defined only with respect to a ground state having a definite phase of $\Delta_{\mathbf{k}}$, and such a state has a large uncertainty in N. If one wishes to consider excited states of an isolated system with precisely $N/2$ pairs of particles in the ground state, one uses a prescription like (2-18) to project out the N-particle part *after* operating with the $\gamma_{\mathbf{k}}^*$ on $|\psi_G\rangle$. Note that $\gamma_{\mathbf{k}}^* |\psi_G\rangle$ has no component with an even number of particles, and hence has no projection on the N-particle subspace. More generally, electron conservation requires that excitations always must be created or destroyed in pairs, as is expected for fermions. (It is for this reason that the "spectroscopic" gap is 2Δ, not Δ.) Considering, for example, $\gamma^*_{k0}\gamma^*_{k'0}|\psi_G\rangle$, and using (2-65a), we see that it has a well-defined N-particle projection

$$c^*_{k\uparrow}c^*_{k'\uparrow}\int_0^{2\pi} d\varphi\, e^{-i(N-2)\varphi/2}\prod_{l\neq k,\,k'}(|u_l| + |v_l|e^{i\varphi}c^*_{l\uparrow}c^*_{-l\downarrow})|\phi_0\rangle \qquad (2\text{-}66)$$

Anticipating the discussion of tunneling experiments, in which an excited state is created by actual addition or subtraction of an electron, we see that the state resulting from γ^*_{k0} operating on an N-electron system is

$$c^*_{k\uparrow}\int d\varphi\, e^{-iN'\varphi/2}\prod_{l\neq k}(|u_l| + |v_l|e^{i\varphi}c^*_{l\uparrow}c^*_{-l\downarrow})|\phi_0\rangle \qquad (2\text{-}67)$$

where $N' = N$ if an electron is added, while $N' = N - 2$ if an electron is removed. Thus, we can always write out explicit expressions for these various excited states with a definite number of particles, although there is seldom any need to. The important qualitative point is simply that pairs of electrons can be added to or subtracted from the condensate at will to achieve number conservation; for energy bookkeeping purposes, such electrons are at the chemical potential μ.

To avoid the need to deal with explicit wavefunctions such as (2-66) and (2-67), let us follow Josephson[1] in introducing an operator S which annihilates a Cooper pair, while S^* creates one. [For later reference in treating the Josephson effect, we note that S has the eigenvalue $e^{i\varphi}$ in a BCS state in which the phase of Δ (or of u^*v) is φ.] An equivalent operator p was also introduced by Bardeen.[2] We can then define a set of modified quasi-particle operators which definitely create either an electron or a hole, i.e., which either increase or decrease the number of electrons by one. Thus, the two types of operators (2-63) are replaced by four:

$$\gamma^*_{ek0} = u^*_k c^*_{k\uparrow} - v^*_k S^* c_{-k\downarrow}$$

$$\gamma^*_{hk0} = u^*_k S c^*_{k\uparrow} - v^*_k c_{-k\downarrow}$$

$$\gamma^*_{ek1} = u^*_k c^*_{-k\downarrow} + v^*_k S^* c_{k\uparrow}$$

$$\gamma^*_{hk1} = u^*_k S c^*_{-k\downarrow} + v^*_k c_{k\uparrow} \qquad (2\text{-}68)$$

Note that the hole and electron operators are related by

$$\gamma^*_{hk} = S\gamma^*_{ek} \qquad (2\text{-}69)$$

which corresponds to the fact that creating a holelike excitation is equivalent to annihilating a pair and creating an electronlike excitation. Thus, these are not independent excitations, but the same excitation with different numbers of condensed pairs.

[1] B. D. Josephson, *Phys. Letters* **1**, 251 (1962).
[2] J. Bardeen, *Phys. Rev. Letters* **9**, 147 (1962).

In dealing with tunneling processes which transfer electrons from one system to another, it is useful to reintroduce the chemical potential explicitly, since it will differ between conductors maintained at different voltages. Since all our calculations of system energy have been referred to the chemical potential μ by subtracting μN_{op}, we simply add this back in, and write

$$\mathcal{H} = \mu N_{op} + E_G + \sum_{\mathbf{k}} E_{\mathbf{k}} \gamma_{\mathbf{k}}^* \gamma_{\mathbf{k}} \qquad (2\text{-}70)$$

where E_G is the ground-state energy, and the sum runs over all the excitations. In view of (2-70), we see that the energy to create an electronlike excitation is $E_{e\mathbf{k}} = (E_{\mathbf{k}} + \mu)$, whereas that to create a hole is $E_{h\mathbf{k}} = (E_{\mathbf{k}} - \mu)$. In an isolated superconductor, the simplest number-conserving excitation consists of a hole and an electron, with total excitation energy

$$(E_{\mathbf{k}} + \mu) + (E_{\mathbf{k}'} - \mu) = E_{\mathbf{k}} + E_{\mathbf{k}'} \geq 2\Delta \qquad (2\text{-}71)$$

On the other hand, in a tunneling process in which an electron is transferred from metal 1 to 2, conservation of energy requires that

$$(E_{\mathbf{k}1} - \mu_1) + (E_{\mathbf{k}'2} + \mu_2) = 0$$

so that
$$E_{\mathbf{k}1} + E_{\mathbf{k}'2} = (\mu_1 - \mu_2) = eV_{12} \qquad (2\text{-}72)$$

It is of some historical interest to note that in the original form of the BCS theory it was necessary to give special treatment to "excited pairs," since the excited state formed by (2-66) with $\mathbf{k}'\uparrow$ replaced by $-\mathbf{k}\downarrow$ is not orthogonal to the ground state. As the appropriate orthogonal state is generated automatically by $\gamma_{\mathbf{k}1}^* \gamma_{\mathbf{k}0}^* | \psi_G \rangle$, however, no such special mathematical attention is required so long as excited states are expressed in terms of the $\gamma_{\mathbf{k}}^*$ operators.

2-7.1   Density of States

Now that we have seen that the quasi-particle excitations can be simply described as fermions created by the $\gamma_{\mathbf{k}}^*$, which are in one-to-one correspondence with the $c_{\mathbf{k}}^*$ of the normal metal, we can obtain the superconducting density of states $N_s(E)$ by equating

$$N_s(E)\, dE = N_n(\xi)\, d\xi$$

(Since we are thinking here of a single superconductor, we can safely revert to taking $\mu = 0$.) Because we are largely interested in energies ξ only a few millielectronvolts from the Fermi energy, we can take $N_n(\xi) = N(0)$, a constant. This leads directly to the simple result

$$\frac{N_s(E)}{N(0)} = \frac{d\xi}{dE} = \begin{cases} \dfrac{E}{(E^2 - \Delta^2)^{1/2}} & (E > \Delta) \\[2mm] 0 & (E < \Delta) \end{cases} \qquad (2\text{-}73)$$

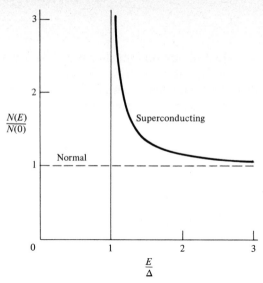

FIGURE 2-4
Density of states in superconducting compared to normal state. All **k** states whose energies fall in the gap in the normal metal are raised in energy above the gap in the superconducting state.

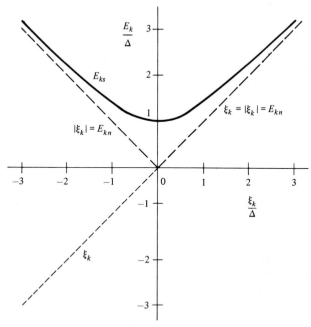

FIGURE 2-5
Energies of elementary excitations in the normal and superconducting states as functions of ξ_k, the independent-particle kinetic energy relative to the Fermi energy.

since $E_{\mathbf{k}}^2 = \Delta^2 + \xi_{\mathbf{k}}^2$. Excitations with all momenta $\mathbf{k}$, even those whose $\xi_{\mathbf{k}}$ fall in the gap, have their energies raised above Δ. Moreover, one expects a divergent state density just above $E = \Delta$, as indicated in Fig. 2-4. Of course, the total number of states is conserved because of the one-to-one correspondence between the $\gamma_{\mathbf{k}}$ and the $c_{\mathbf{k}}$. The nature of this correspondence is made more explicit by Fig. 2-5, which shows the relationship between the excitation energies in the normal and superconducting states.

It is worth noting that if the BCS model is followed literally, a narrow peak in the density of states occurs at the cutoff energy $\hbar\omega_c$ because above this energy, $\Delta = 0$ and $E_{\mathbf{k}} = \xi_{\mathbf{k}}$. As a result, $N(0)\Delta^2/2\hbar\omega_c$ extra states fall in an energy range of width $\Delta^2/2\hbar\omega_c$, causing a doubling of $N(E)$ in this range. Of course, this consequence of the model is not to be taken seriously, since it depends critically on the admittedly crude cutoff procedure. As will be described in the next section, a more rigorous treatment of the phonon-mediated interaction spreads this cutoff effect out over the entire energy range $\hbar\omega_c$, so that the actual departures from (2-73) are of order $(\Delta/\hbar\omega_c)^2$ or $(T_c/\Theta_D)^2$, to give about the same integrated effect.

2-8 ELECTRON TUNNELING

By far the most detailed experimental examination of the density of states is provided by electron tunneling. This technique was pioneered by Giaever,[1] who used it to confirm the density of states and temperature dependence of the energy gap predicted by BCS. The basic idea is that there is a nonzero probability of charge transfer by the quantum-mechanical tunneling of electrons between two conductors separated by a thin insulating barrier. This probability falls exponentially with the distance of separation and it depends on the details of the insulating material, but these aspects can be absorbed in a phenomenological tunneling-matrix element $T_{\mathbf{kq}}$. That is, we can assume a coupling term in the hamiltonian of the form

$$\mathscr{H}_T = \sum_{\sigma\mathbf{kq}} T_{\mathbf{kq}} c_{\mathbf{k}\sigma}^* c_{\mathbf{q}\sigma} + \text{herm conj} \qquad (2\text{-}74)$$

where the subscript $\mathbf{k}$ refers to one metal and $\mathbf{q}$ to the other; we assume no spin flip in the tunneling process, since there are no magnetic perturbations in the problem. The explicitly written term transfers an electron from metal $\mathbf{q}$ to $\mathbf{k}$, whereas the conjugate term does the reverse. As usual, the transition probability (and hence the current) is proportional to the square of the matrix element, so long as we exclude the coherent processes of the Josephson tunneling.

If we consider, for example, an electron transferred by (2-74) into a state $\mathbf{k}\!\uparrow$

[1] I. Giaever, *Phys. Rev. Letters* **5**, 147, 464 (1960).

in a superconductor, we must reexpress this electron state in terms of the appropriate excitations, the γ_k, using (2-68). The result is

$$c_{k\uparrow}^* = u_k \gamma_{ek0}^* + v_k^* \gamma_{hk1} \qquad (2\text{-}75)$$

If the superconductor is in its ground state, the second term gives zero, and the process contributes a current proportional to $|u_k|^2 |T_{jk}|^2$. The physical significance of the factor $|u_k|^2$ is that it is the probability that the state k is *not* occupied in the BCS function, and hence is able to receive an incoming electron. Thus, it appears on the face of it that the tunneling current will depend on the nature of the superconducting ground state as well as on the density of available excited states; but this turns out not to be true. As is evident from Fig. 2-5, there is another state k' having exactly the same energy $E_{k'} = E_k$, but with $\xi_{k'} = -\xi_k$. Using (2-75), with k replaced by k' we see that tunneling into k' contributes a current proportional to $|u_{k'}|^2 |T_{jk'}|^2 = |v_k|^2 |T_{jk'}|^2$, since $|u(-\xi)| = |v(\xi)|$. Making the reasonable assumption that the two matrix elements are nearly equal, since k and k' are both near the same point on the Fermi surface, the total current from these two channels is proportional to $(|u_k|^2 + |v_k|^2) |T_{jk}|^2 = |T_{jk}|^2$, and the characteristic coherence factors of the superconducting wavefunction, u_k and v_k, have dropped out. If we now generalize to finite temperatures, so that the quasi-particle occupation numbers f_k are nonzero, both terms of (2-75) contribute, the first as $(1 - f_k)$, the second as f_k. Again, when the degenerate channels are combined, the current is simply proportional to $|T_{jk}|^2$.

2-8.1 The Semiconductor Model

This disappearance of the coherence factors u_k and v_k makes it possible and convenient to reexpress the computation of the tunneling current in what is often called the "semiconductor model." In this method, the normal metal is represented in the familiar elementary way as a continuous distribution of independent-particle energy states with density $N(0)$, including energies below as well as above the Fermi level. The superconductor is represented by an ordinary semiconductor with a density of independent-particle states obtained from Fig. 2-4 by adding its reflection on the negative-energy side of the chemical potential, so that it will reduce properly to the normal-metal density of states as $\Delta \to 0$. At $T = 0$, all states up to μ are filled; for $T > 0$, the occupation numbers are given by the Fermi function. It is worth noting that f_k now runs from 0 to 1, whereas in our previous convention, f_k ranged only from 0 to $\frac{1}{2}$ since $E_k \geq 0$. This difference reflects the fact that in the present model f_k measures a departure from the vacuum, whereas in the previous "excitation representation" it measured a departure from the ground state of the system.

With this model, tunneling transitions are all "horizontal," that is, they occur at constant energy after adjusting the relative levels of μ in the two metals to account for the applied potential difference eV. This property facilitates summing up all contributions to the current in an elementary way, since the various parallel channels noted above do not have to be considered anew in each case. Because this scheme so greatly simplifies the computations, we shall use it here to work out the tunneling characteristics of various types of junctions, and simply refer the reader to more detailed treatments which are available in the literature.[1] It should be borne in mind, however, that this technique to some extent *over*simplifies. It is sound to treat the normal metal in this way, but it is less safe to conceal the mixing of hole and electron states which is present in the superconducting state even at $T = 0$. Although our argument above for simply adding the currents from the two degenerate channels is valid for the usual case, there can be an interference effect between them which causes an oscillatory variation of the tunnel current with voltage or sample thickness known as the Tomasch effect.[2] This comment illustrates the need for caution in using the semiconductor model. Of course, this model also is inadequate for dealing with processes in which the condensed pairs play a role, since the ground state does not appear in the energy-level diagram.

Within the independent-particle approximation, the tunneling current from metal 1 to metal 2 can be written as

$$I_{1 \to 2} = A \int_{-\infty}^{\infty} |T|^2 N_1(E) f(E) N_2(E + eV)[1 - f(E + eV)] \, dE$$

where V is the applied voltage, eV is the resulting difference in the chemical potential across the junction, and $N(E)$ is the appropriate normal or superconducting density of states. The factors $N_1 f$ and $N_2(1 - f)$ give the numbers of occupied initial states and of available (i.e., empty) final states in unit energy interval. This expression assumes a constant tunneling-matrix element T; A is a constant of proportionality. Subtracting the reverse current, the net current is

$$I = A|T|^2 \int_{-\infty}^{\infty} N_1(E) N_2(E + eV)[f(E) - f(E + eV)] \, dE \qquad (2\text{-}76)$$

We shall now use this expression to treat a number of important cases.

[1] A particularly explicit discussion of the contributions of the various channels is given by M. Tinkham, *Phys. Rev.* **B6**, 1747 (1972). Earlier treatments and reviews have been given by M. H. Cohen, L. M. Falicov, and J. C. Phillips, *Phys. Rev. Letters* **8**, 316 (1962); D. H. Douglass, Jr., and L. M. Falicov, in C. J. Gorter (ed.), "Progress in Low Temperature Physics," vol. 4, p. 97, North-Holland, Amsterdam, 1964; W. L. McMillan and J. M. Rowell, in R. D. Parks (ed.), "Superconductivity," vol. 1, chap. 11, Marcel Dekker, New York, 1969.
[2] W. J. Tomasch, *Phys. Rev. Letters* **15**, 672 (1965); **16**, 16 (1966).

2-8.2   Normal-Normal Tunneling

If both metals are normal, (2-76) becomes

$$I_{nn} = A\,|\,T\,|^2 N_1(0) N_2(0) \int_{-\infty}^{\infty} [f(E) - f(E + eV)]\, dE$$

$$= A\,|T|^2 N_1(0) N_2(0) eV \equiv G_{nn} V \tag{2-77}$$

so that the junction is "ohmic," i.e., it has a well-defined conductance G_{nn}, independent of V. Note that it is also independent of the temperature.

To help reduce any lingering confusion about the relation of this semiconductor, or independent-particle, scheme to the elementary excitation scheme, let us indicate how this simple case would have been treated in the other framework. First, at $T = 0$, all $f_k = 0$, and there are no excitations present, both metals being in their Fermi sea ground states. Thus any tunneling process must involve creating two excitations, a hole in one metal and an electron in the other, the sum of the two excitation energies being eV, as given by (2-72). The resulting current is

$$I = A\,|\,T\,|^2 \int_0^{eV} N_1(E) N_2(eV - E)\, dE$$

$$= A\,|\,T\,|^2 N_1(0) N_2(0) eV$$

exactly as found in (2-77). For $T > 0$, the current from this process is reduced by the excitations already present, which block final states, but this effect is canceled by the extra current from the tunneling of the excitations, leading to a temperature-independent result.

2-8.3   Normal-Superconductor Tunneling

A more interesting case arises if one metal is superconducting. Then (2-76) becomes

$$I_{ns} = A\,|\,T\,|^2 N_2(0) \int_{-\infty}^{\infty} N_{1s}(E)[f(E) - f(E + eV)]\, dE$$

$$= \frac{G_{nn}}{e} \int_{-\infty}^{\infty} \frac{N_{1s}(E)}{N_1(0)} [f(E) - f(E + eV)]\, dE \tag{2-78}$$

In general, numerical means are required to evaluate this expression for the BCS density of states and thus to allow quantitative comparison with experiment, although the qualitative behavior is easily sketched. As indicated in Fig. 2-6a, at $T = 0$ there is no tunneling current until $e\,|\,V\,| \geq \Delta$, since the chemical-potential difference must provide enough energy to create an excitation in the superconductor. The magnitude of the current is independent of the sign of V because hole and electron excitations have equal energies. For $T > 0$, the energy of excitations

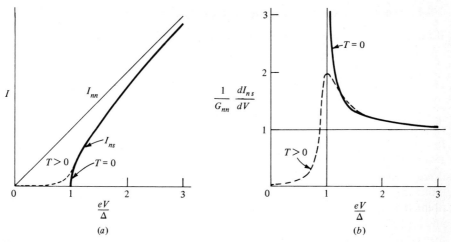

FIGURE 2-6
Characteristics of normal-superconductor tunnel junctions. (a) I-V characteristic. (b) Differential conductance. Solid curves refer to $T = 0$; dashed curves, to a finite temperature.

already present allows them to tunnel at lower voltages, giving an exponential tail of the current in the region below $eV = \Delta$.

A more direct comparison of theory and experiment can be made if one considers the differential conductance dI/dV as a function of V. From (2-78)

$$G_{ns} = \frac{dI_{ns}}{dV} = G_{nn} \int_{-\infty}^{\infty} \frac{N_{1s}(E)}{N_1(0)} \left[-\frac{\partial f(E + eV)}{\partial(eV)} \right] dE \qquad (2\text{-}79)$$

Since $-[\partial f(E + eV)/\partial(eV)]$ is a bell-shaped weighting function peaked at $E = -eV$, with width $\sim kT$ and unit area under the curve, it is clear that as $kT \to 0$, this approaches

$$G_{ns} \bigg|_{T=0} = \frac{dI_{ns}}{dV} \bigg|_{T=0} = G_{nn} \frac{N_{1s}(e|V|)}{N_1(0)} \qquad (2\text{-}80)$$

Thus, in the low-temperature limit, the differential conductance measures directly the density of states. At finite temperatures, as shown in Fig. 2-6b, the conductance measures a density of states smeared by $\sim kT$ in energy, due to the width of the weighting function. Because this function has exponential "skirts," it turns out that the differential conductance at $V = 0$ is related exponentially to the width of the gap. In the limit $kT \ll \Delta$, this relation reduces to

$$\frac{G_{ns}}{G_{nn}} \bigg|_{V=0} = \left(\frac{2\pi\Delta}{kT} \right)^{1/2} e^{-\Delta/kT} \qquad (2\text{-}81)$$

2-8.4 Superconductor-Superconductor Tunneling

If both metals are superconducting, (2-76) becomes

$$I_{ss} = \frac{G_{nn}}{e} \int_{-\infty}^{\infty} \frac{N_{s1}(E)}{N_1(0)} \frac{N_{s2}(E + eV)}{N_2(0)} [f(E) - f(E + eV)] \, dE$$

$$= \frac{G_{nn}}{e} \int_{-\infty}^{\infty} \frac{|E|}{[E^2 - \Delta_1^2]^{1/2}} \frac{|E + eV|}{[(E + eV)^2 - \Delta_2^2]^{1/2}} [f(E) - f(E + eV)] \, dE \quad (2-82)$$

In the second form it is understood that the range of integration excludes values of E such that $|E| < |\Delta_1|$ and $|E + eV| < |\Delta_2|$. Again numerical integration is required to compute complete I-V curves. However, the qualitative features are indicated in Fig. 2-7. At $T = 0$, no current can flow until $eV = \Delta_1 + \Delta_2$; at this point the potential difference supplies enough energy to create a hole on one side and a particle on the other. Since the density of states is infinite at the gap edges, it turns out that there is a discontinuous jump in I_{ss} at $eV = \Delta_1 + \Delta_2$, even at finite temperatures. At $T > 0$, current also flows at lower voltages because of the availability of thermally excited quasi-particles. This current rises sharply to a peak when $eV = |\Delta_1 - \Delta_2|$ because this voltage provides just the energy to allow thermally excited quasi-particles in the peak of the density of states at Δ_1, say, to tunnel into the peaked density of available states at Δ_2. The existence of this peak leads to a "negative-resistance region" $[(dI/dV) < 0]$ for $|\Delta_1 - \Delta_2| \leq eV \leq \Delta_1 + \Delta_2$. This region cannot be observed with the usual current-source arrangement, since there are three possible values of V for given I and the one with $dI/dV < 0$ is unstable. A voltage source must be used to maintain stable operation. The existence of sharp features at both $|\Delta_1 - \Delta_2|$ and $\Delta_1 + \Delta_2$ allows very convenient determinations of $\Delta_1(T)$ and $\Delta_2(T)$ from the tunneling curves.

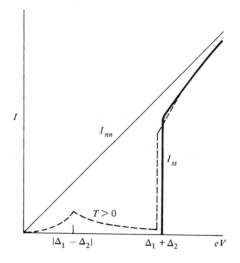

FIGURE 2-7
Superconductor-superconductor tunneling characteristic. Note that for $T > 0$ there are sharp features corresponding to both the sum and the difference of the two gap values. The peak at $|\Delta_1 - \Delta_2|$ would actually be a logarithmic singularity in the absence of gap anisotropy and level broadening due to lifetime effects.

The *s-s* tunneling method is superior to the *n-s* tunneling method in this regard because the existence of very sharply peaked densities of states at the gap edges of both materials helps counteract the effects of thermal smearing.

2-8.5 Phonon Structure

When tunneling curves are measured in materials with strong electron-phonon coupling, structure beyond that outlined above is quite readily observed. Giaever et al.[1] noticed this first in lead, and remarked that the structure occurred near energies characteristic of the phonon structure. This observation has since been greatly refined, both from experimental and theoretical viewpoints. The key point, made by Schrieffer, Scalapino, and Wilkins,[2] is that the observed density of states should be

$$N_s(E) = N(0) \, \text{Re} \frac{E}{[E^2 - \Delta^2(E)]^{1/2}} \qquad (2\text{-}83)$$

where Re indicates the real part of the expression following. This reduces to our earlier result (2-73) if Δ can be taken to be real and constant, as in the simple BCS model, but in the strong-coupling theory Δ becomes complex and energy dependent. The resulting energy-dependent phase of $\Delta(E)$ has physical content, and is entirely distinct from the single arbitrary phase choice for Δ within the BCS approximation. The imaginary part of Δ corresponds to a damping of the quasiparticle excitations by decay with creation of real phonons; hence it is large when $E \approx \hbar\omega_{ph}$. There is a corresponding resonant variation of Re Δ. Such behavior is, of course, hardly surprising given our model in which the interaction causing superconductivity is phonon mediated. The simple BCS approximation suppresses these details caused by the retarded nature of the interaction [apart from the crude manifestation of a narrow peak in $N(E)$ at $\hbar\omega_c$, mentioned at the end of Sec. 2-7], but the more exact Eliashberg[3] procedure appears to give a quantitative account of the observed phenomena. This agreement between theory and experiment in such materials as lead and mercury has eliminated any remaining doubt of the correctness of the electron-phonon mechanism for superconductivity in these materials.

McMillan and Rowell[4] have been very successful in inverting the tunneling data to find the spectrum of $\alpha^2 F$, that is, the product of the electron-phonon coupling strength and the density of phonon states, both as functions of energy.

[1] I. Giaever, H. R. Hart, and K. Megerle, *Phys. Rev.* **126**, 941 (1962).

[2] J. R. Schrieffer, D. J. Scalapino, and J. W. Wilkins, *Phys. Rev. Letters* **10**, 336 (1963); ibid., *Phys. Rev.* **148**, 263 (1966).

[3] G. M. Eliashberg, *Zh. Eksperim. i Teor. Fiz.* **38**, 966 (1960) [*Soviet Phys.—JETP* **11**, 696 (1960)].

[4] W. L. McMillan and J. M. Rowell, *Phys. Rev. Letters* **14**, 108 (1965); see also, J. M. Rowell and L. Kopf, *Phys. Rev.* **137**, A907 (1965). An excellent review is given by McMillan and Rowell in R. D. Parks (ed.), "Superconductivity," chap. 11, Marcel Dekker, New York, 1969.

Their results generally agree well with expectations based on neutron scattering from phonons where such data are available. In fact, the tunneling data give results of superior accuracy for features such as the location of van Hove singularities in the phonon density of states, which appear as peaks in the experimentally obtained trace of the second derivative of the tunneling current.

2-9   TRANSITION PROBABILITIES AND COHERENCE EFFECTS

The effect of an external perturbation on the electrons in a metal can be expressed in terms of an interaction hamiltonian

$$\mathscr{H}_1 = \sum_{k\sigma, k'\sigma'} B_{k'\sigma', k\sigma} c^*_{k'\sigma'} c_{k\sigma} \qquad (2\text{-}84)$$

where the $B_{k'\sigma', k\sigma}$ are matrix elements of the perturbing operator between the ordinary one-electron states of the normal metal. In the normal state, each term in this sum is independent, and the square of each $B_{k'\sigma', k\sigma}$ is proportional to a corresponding transition probability. This is not the case in the superconducting state, however, because it consists of a phase-coherent superposition of occupied one-electron states. As a result, there are interference terms which are not present in the normal state.

This can be seen in detail by expanding the terms in (2-84) using the γ operators. It is then seen that the terms $c^*_{k'\sigma'} c_{k\sigma}$ and $c^*_{-k-\sigma} c_{-k'-\sigma'}$ connect the same quasi-particle states. For example,

$$c^*_{k'\uparrow} c_{k\uparrow} = u_k u^*_{k'} \gamma^*_{k'0} \gamma_{k0} - v^*_{k'} v_k \gamma^*_{k1} \gamma_{k'1} + u_{k'} v_k \gamma^*_{k'0} \gamma^*_{k1} + v^*_{k'} u^*_k \gamma_{k'1} \gamma_{k0} \qquad (2\text{-}85a)$$

and

$$c^*_{-k\downarrow} c_{-k'\downarrow} = -v^*_k v_{k'} \gamma^*_{k'0} \gamma_{k0} + u_k u^*_{k'} \gamma^*_{k1} \gamma_{k'1} + u_k v_{k'} \gamma^*_{k'0} \gamma^*_{k1} + v^*_k u^*_{k'} \gamma_{k'1} \gamma_{k0} \qquad (2\text{-}85b)$$

Thus, it is clear that matrix elements of these two terms in (2-84) must be added *before* squaring, since they add coherently in determining the transition rate.

Fortunately, this addition can be done quite readily, since one expects the coefficients $B_{k'\sigma', k\sigma}$ and $B_{-k-\sigma, -k'-\sigma'}$ to differ at most in sign because both represent processes in which the momentum change of the electron is $k' - k$ and its spin change is $\sigma' - \sigma$. Thus, these terms can be combined as

$$B_{k'\sigma', k\sigma}(c^*_{k'\sigma'} c_{k\sigma} \pm c^*_{-k-\sigma} c_{-k'-\sigma'}) \qquad (2\text{-}86)$$

where the sign choice depends on the nature of $\mathscr{H}_1$.

As shown by BCS, there are two cases. Case I is typified by the electron-phonon interaction responsible for ultrasonic attenuation. Being the interaction of the electron with a simple scalar deformation potential, it depends only on the momentum change; it is independent of the sense of $\mathbf{k}$ or σ, and the two matrix elements have the same sign and add coherently. Case II is typified by the interaction of the electron with the electromagnetic field via a term $\mathbf{p} \cdot \mathbf{A}$; since this changes sign on replacing $\mathbf{k}$ by $-\mathbf{k}$ the negative sign in (2-86) is appropriate.

Neither of the above interactions has involved the spin, so that we have had $\sigma = \sigma'$. If there *is* a spin change, as with terms of the sort $I_+ S_-$ in the hyperfine interaction of the electron with a nucleus, the sign associations are formally reversed. Following BCS, we indicate this with a factor $\Theta_{\sigma\sigma'}$, which is ± 1 for $\sigma' = \pm\sigma$. Thus, upon collecting terms and taking $u_\mathbf{k}$, $v_\mathbf{k}$, and Δ real, (2-86) becomes

$$B_{\mathbf{k'}\sigma',\,\mathbf{k}\sigma}[(u_{\mathbf{k'}} u_\mathbf{k} \mp v_{\mathbf{k'}} v_\mathbf{k})(\gamma^*_{\mathbf{k'}\sigma'} \gamma_{\mathbf{k}\sigma} \pm \Theta_{\sigma'\sigma} \gamma^*_{-\mathbf{k}-\sigma} \gamma_{-\mathbf{k'}-\sigma'})$$

$$+ (v_\mathbf{k} u_{\mathbf{k'}} \pm u_\mathbf{k} v_{\mathbf{k'}})(\gamma^*_{\mathbf{k'}\sigma'} \gamma^*_{-\mathbf{k}-\sigma} \pm \Theta_{\sigma'\sigma} \gamma_{-\mathbf{k'}-\sigma'} \gamma_{\mathbf{k}\sigma})] \qquad (2\text{-}87)$$

(To facilitate writing this out for general spin directions, we have used a slightly modified notation, in which $\gamma_{\mathbf{k}\sigma} = \gamma_{\mathbf{k}0}$ for $\sigma = \uparrow$, and $\gamma_{\mathbf{k}\sigma} = \gamma_{-\mathbf{k}1}$ for $\sigma = \downarrow$.) From this expansion we see that the factor $\Theta_{\sigma\sigma'}$ really has no effect on the magnitude of transition probabilities; it affects only the relative phase of off-diagonal matrix elements connecting disparate states. The decisive point is whether the interaction is of case I or II, corresponding to the upper and lower signs in the coherence factors, i.e., the combinations of $u_\mathbf{k}$ and $v_\mathbf{k}$ in (2-87). For example, although the part $I_z S_z$ of the hyperfine coupling does not flip the spin, and hence has $\Theta_{\sigma\sigma'} = +1$ rather than -1 as above for $I_+ S_-$, both terms are governed by case II coherence factors because they are odd with respect to reversal of the spin. Generalizing from these examples, we see that cases I and II pertain to perturbations which are even and odd, respectively, under time reversal of the electronic states, which interchanges the partners in the Cooper pairing scheme.

From (2-87) we see that in the computation of transition probabilities, the squared matrix elements $|B_{\mathbf{k'}\sigma',\,\mathbf{k}\sigma}|^2$ will be multiplied by so-called *coherence factors*, namely, $(uu' \mp vv')^2$ for the scattering of quasi-particles, and $(vu' \pm uv')^2$ for the creation or annihilation of two quasi-particles. (We have made an obvious condensation of the notation here.) With u and v given by (2-35), these coherence factors can be evaluated as explicit functions of energy. For example,

$$(uu' \mp vv')^2 = \frac{1}{4}\left\{\left[\left(1 + \frac{\xi}{E}\right)\left(1 + \frac{\xi'}{E'}\right)\right]^{1/2} \mp \left[\left(1 - \frac{\xi}{E}\right)\left(1 - \frac{\xi'}{E'}\right)\right]^{1/2}\right\}^2$$

$$= \frac{1}{4}\left\{\left(1 + \frac{\xi}{E} + \frac{\xi'}{E'} + \frac{\xi\xi'}{EE'}\right) + \left(1 - \frac{\xi}{E} - \frac{\xi'}{E'} + \frac{\xi\xi'}{EE'}\right)\right.$$

$$\left. \mp 2\left[\left(1 - \frac{\xi^2}{E^2}\right)\left(1 - \frac{\xi'^2}{E'^2}\right)\right]^{1/2}\right\} = \frac{1}{2}\left(1 + \frac{\xi\xi'}{EE'} \mp \frac{\Delta^2}{EE'}\right)$$

Since E is an even function of ξ, when we sum over $\xi_\mathbf{k}$, terms appear in pairs such that the terms odd in ξ or ξ' cancel. Thus, effectively the coherence factor for *scattering* is

$$(uu' \mp vv')^2 = \frac{1}{2}\left(1 \mp \frac{\Delta^2}{EE'}\right) \qquad (2\text{-}88a)$$

Similarly, the coherence factor for *creation* or *annihilation* of a pair of quasi-particles is

$$(vu' \pm uv')^2 = \frac{1}{2}\left(1 \pm \frac{\Delta^2}{EE'}\right) \qquad (2\text{-}88b)$$

It is convenient to note that in the "semiconductor-model" sign convention, in which one of each pair of quasi-particles created or destroyed is assigned a negative energy, the coherence factors for both scattering and pair creation have the same form

$$F(\Delta, E, E') = \frac{1}{2}\left(1 \mp \frac{\Delta^2}{EE'}\right) \qquad (2\text{-}89)$$

where the upper sign corresponds to case I and the lower to case II.

Evidently, the greatest effect of these coherence factors is for energies E and E' near the gap edge Δ, in which case (2-89) is either ~ 0 or ~ 1, depending on the sign. If one considers low-energy scattering processes for $\hbar\omega \ll \Delta$, no quasi-particles are created, so E and E' have the same sign. Then for case I processes like ultrasonic attenuation, $F \ll 1$, whereas for case II processes like nuclear relaxation, $F \sim 1$. The situation is reversed for high-energy processes with $\hbar\omega \gtrsim 2\Delta$, which create pairs of quasi-particles. Then $F \sim 1$ for case I processes and $F \ll 1$ for case II processes. Of course if E and $E' \gg \Delta$, there is little difference between case I and II, and the superconducting coherence is unimportant.

These general considerations are made more clear by considering examples of the calculation of transition rates. Following the same line of argument used in reaching (2-76) for the case of tunneling, we expect a net transition rate between energy levels E and $E' = E + \hbar\omega$ to be proportional to

$$\alpha_s = \int |M|^2 F(\Delta, E, E + \hbar\omega) N_s(E) N_s(E + \hbar\omega)$$
$$[f(E) - f(E + \hbar\omega)]\, dE \qquad (2\text{-}90)$$

where M is the magnitude of a suitable one-electron matrix element. Since we will always be interested in ratios to the normal-state values, we need not know more about the actual value of M. Upon inserting the explicit expressions for N_s and F, and simplifying, (2-90) becomes

$$\alpha_s = |M|^2 N^2(0) \int_{-\infty}^{\infty} \frac{|E(E + \hbar\omega) \mp \Delta^2|\,[f(E) - f(E + \hbar\omega)]}{(E^2 - \Delta^2)^{1/2}[(E + \hbar\omega)^2 - \Delta^2]^{1/2}}\, dE$$

where it is understood that the regions with $|E|$ or $|E + \hbar\omega| < \Delta$ are excluded from the integration. In the normal state, $\Delta = 0$, and the corresponding expression reduces to $\alpha_n = |M|^2 N^2(0)\hbar\omega$. Thus, the desired ratio is

$$\frac{\alpha_s}{\alpha_n} = \frac{1}{\hbar\omega} \int_{-\infty}^{\infty} \frac{|E(E + \hbar\omega) \mp \Delta^2| [f(E) - f(E + \hbar\omega)]}{(E^2 - \Delta^2)^{1/2} [(E + \hbar\omega)^2 - \Delta^2]^{1/2}} \, dE \qquad (2\text{-}91)$$

with the upper sign referring to case I processes and the lower to case II. We now use this general expression to treat some important specific cases.

2-9.1   Ultrasonic Attenuation

As noted above, the relevant matrix elements for treating the attenuation of longitudinal sound waves have case I coherence factors, i.e., the upper sign in (2-91). (We restrict our attention to longitudinal waves to avoid the complications which arise in the transverse case because currents are generated which are screened electromagnetically, giving a mixture of effects.) We note further that in typical ultrasonic experiments the sound frequency is less than 10^9 Hz, so that $\hbar\omega \lesssim 10^{-2}\Delta(0)$; also, $\hbar\omega \ll kT$. These inequalities enable us to consider only a simple low-frequency limiting case. Inspecting (2-91) in the limit as $\hbar\omega \to 0$, we see that most of the factors cancel, leaving

$$\frac{\alpha_s}{\alpha_n} = \lim_{\hbar\omega \to 0} \frac{1}{\hbar\omega} \int [f(E) - f(E + \hbar\omega)] \, dE$$

$$= -\int \frac{\partial f}{\partial E} \, dE$$

with the integration extending from $-\infty$ to $-\Delta$ and from Δ to ∞. Thus,

$$\frac{\alpha_s}{\alpha_n} = f(-\infty) - f(-\Delta) + f(\Delta) - f(\infty)$$

$$= 2f(\Delta) = \frac{2}{1 + e^{\Delta/kT}} \qquad (2\text{-}92)$$

This very simple result, combined with our previous calculation of $\Delta(T)$, predicts the behavior shown in Fig. 2-8. In particular, the infinite slope of $\Delta(T)$ at T_c causes α_s/α_n to drop with infinite slope as T is lowered below T_c. On the other hand, for $T \ll T_c$, the excess of α_s/α_n above a residual value due to nonelectronic mechanisms becomes exponentially small as the number of thermally excited quasi-particles available to absorb energy goes to zero.

When it is possible to carry experiments to low enough values of T/T_c to establish the residual attenuation level quite exactly, it is possible to infer a value of $\Delta(T)$ from the attenuation data using (2-92). In fact, by propagating sound in

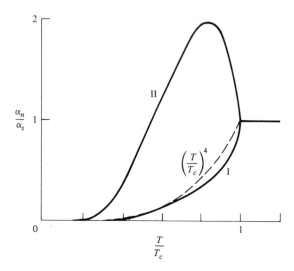

FIGURE 2-8
Temperature dependence of low-frequency absorption processes obeying case I and II coherence factors, compared with the $(T/T_c)^4$ dependence that might be expected for *all* processes from a simple two-fluid model. The curve for case I applies to ultrasonic attenuation and it is a well-defined low-frequency limit. The curve for case II, which applies to nuclear relaxation or electromagnetic absorption, has no well-defined low-frequency limit unless gap anisotropy or level broadening is taken into account. The curve drawn here corresponds to a broadening of about $0.02\Delta(0)$.

different directions in single crystals, Morse and coworkers[1] were among the first to be able to get some measure of the anisotropy of the gap $\Delta_{\mathbf{k}}$ with respect to crystalline axes. This technique suffers from the fact that a given direction of sound propagation $\hat{\mathbf{k}}_s$ measures an average of $\Delta_{\mathbf{k}}$ over a disk perpendicular to $\hat{\mathbf{k}}_s$. The reason for this is that for efficient energy transfer between the sound wave and the electrons, the component of quasi-particle velocity parallel to $\hat{\mathbf{k}}_s$ must equal the sound velocity. Since $v_{\text{sound}} \ll v_{\text{electron}}$, this means that only electrons moving almost perpendicular to $\hat{\mathbf{k}}_s$ are effective in the attenuation. Nonetheless, values of $2\Delta(0)$ in tin ranging from $3.3kT_c$ to $3.9kT_c$ have been inferred from such measurements, for example, by Claiborne and Einspruch.[2] More recently, Phillips[3] has carefully reexamined the detailed nature of elastic scattering in anisotropic superconductors and concluded that some of the earlier discussions were oversimplified.

[1] R. W. Morse, in K. Mendelssohn, (ed.), "Progress in Cryogenics," vol. I, p. 219, Heywood, London, 1959.
[2] L. T. Claiborne and N. G. Einspruch, *Phys. Rev. Letters* **15**, 862 (1965).
[3] W. A. Phillips, *Proc. Roy. Soc.* **A309**, 259 (1969).

mentioned above can be put to good use as a means of measuring the penetration depth very near T_c, where it approaches infinity. The point is that the screening becomes essentially complete as soon as $\lambda_L(T)$ is less than the ultrasonic wavelength, so that the electrodynamic attenuation becomes negligible for lower temperatures, leading to a very sharp drop in attenuation within less than 1 percent of T_c. Using this effect, Fossheim[1] has obtained what are probably the most reliable available values for λ_L of indium from measurements of α_s/α_n within a few millidegrees of T_c.

Another complication which can enter ultrasonic-attenuation measurements arises from the electronic damping of the motion of dislocations driven by the sound waves.[2] Because of dislocation pinning effects, this leads to a nonlinear response which makes definition of α_s/α_n difficult except at very low signal levels. Mason[3] showed that when these effects were taken into account, the apparently anomalous measurements on lead[4] could be reconciled with expectations based on a gap $2\Delta(0) = 4.1kT_c$, a value in reasonable agreement with results from other techniques.

2-9.2 Nuclear Relaxation

The matrix elements for nuclear spin relaxation by interaction with quasi-particles have the case II coherence factor, which corresponds to "constructive" interference in the relevant low-energy scattering processes. This causes the relaxation rate $1/T_1$ to *rise* above the normal value upon cooling through T_c before it eventually goes exponentially to zero, as it must when all quasi-particles above the gap are "frozen out." This behavior, sketched in Fig. 2-8 and confirmed experimentally by Hebel and Slichter,[5] is in sharp contrast with the *drop* with vertical tangent found in the case of ultrasonic attenuation. The ability of the BCS pairing theory, with its coherence factors, to explain this difference in a natural way was one of the key triumphs which validated the theory. By way of contrast, any simple "two-fluid" model which attributed the ultrasonic attenuation and nuclear relaxation to a certain fraction of "normal electrons" would have to give the same temperature dependence for all such properties. For example, one might expect a normal-electron density $n_n/n = (T/T_c)^4$ in a model which correlated the empirical temperature variation of the penetration depth $\lambda^{-2} \sim [1 - (T/T_c)^4]$ with a similar variation of n_s/n in the London theory. As indicated by the dashed curve in Fig. 2-8, this gives a qualitative fit to the ultrasonic attenuation result; but it cannot possibly explain the rise in $1/T_{1s}$ above $1/T_{1n}$, since that would require the

[1] K. Fossheim, *Phys. Rev. Letters* **19**, 81 (1967).
[2] B. R. Tittman and H. E. Bömmel, *Phys. Rev.* **151**, 178 (1965).
[3] W. P. Mason, *Phys. Rev.* **143**, 229 (1966).
[4] R. E. Love, R. W. Shaw, and W. A. Fate, *Phys. Rev.* **138**, A1453 (1965).
[5] L. C. Hebel and C. P. Slichter, *Phys. Rev.* **107**, 901 (1957); **113**, 1504 (1959); L. C. Hebel, *Phys. Rev.* **116**, 79 (1959).

number of normal electrons to exceed the total number of electrons, despite the existence of "superconducting electrons" as well.

Let us now examine this in more detail. We are again interested in the limit in which $\hbar\omega = \hbar\gamma H$ is much less than Δ and kT. However, we cannot proceed as simply as before, because for the case II coherence factors $\alpha_s/\alpha_n \to \infty$ as $\omega \to 0$. However, we can still simplify the Fermi function by converting to a derivative form. Thus,

$$\frac{\alpha_s}{\alpha_n} = 2 \int_\Delta^\infty \frac{E(E + \hbar\omega) + \Delta^2}{(E^2 - \Delta^2)^{1/2}[(E + \hbar\omega)^2 - \Delta^2]^{1/2}} \left(-\frac{\partial f}{\partial E}\right) dE \qquad (2\text{-}93)$$

where the factor of 2 results from the fact that the integration over negative energies $(E + \hbar\omega) \le -\Delta$ gives exactly the same contribution as the integration over positive energies $E \ge \Delta$. If we now try to set $\omega = 0$, we find

$$\frac{\alpha_s}{\alpha_n} = 2 \int_\Delta^\infty \frac{E^2 + \Delta^2}{E^2 - \Delta^2} \left(-\frac{\partial f}{\partial E}\right) dE$$

which is easily seen to diverge logarithmically from the integration at Δ. If one kept ω finite, the divergence would be replaced by a factor of order $\ln(\Delta/\hbar\omega) \sim 10$, for typical values. This is still much greater than the experimentally observed rise in $1/T_1$, typically only a factor of 2, before the exponential fall at low temperatures. Thus some other concept must be brought in to explain this quantitative discrepancy. The origin of the incipient divergence in α_s can be traced to the product of the two highly peaked, superconducting density-of-states factors in (2-93). Dropping nonsingular factors,

$$\alpha \propto \int N(E)N(E + \hbar\omega)\, dE \approx \int N^2(E)\, dE$$

for $\hbar\omega \ll \Delta$. Since $\int N(E)\, dE$ is conserved on going into the superconducting state, the sharp rise of $N_s(E)$ must cause $\int N_s^2\, dE$ to exceed $\int N_n^2\, dE$. An excessively high value of α results if we have overestimated the sharpness of the peak in state density by using the simple BCS form.

Two reasons have been advanced for the apparent extra breadth in the peak in the density of states. The first is that the anisotropy of the energy gap in real crystals leads to a range in Δ_k over the Fermi surface. Thus the peaks in $N_s(E)$ will be smeared out over a fractional energy range, typically $\frac{1}{10}$. Experiments of Masuda[1] on aluminum with varying amounts of impurity support this explanation. He observed a bigger rise of $1/T_1$ in the dirtier samples consistent with the Anderson theory of dirty superconductors, which holds that the gap should become more nearly the same for all electrons when rapid electron scattering causes electrons to sample many $\mathbf{k}$ values during the relevant time interval $\hbar/\Delta$.

[1] Y. Masuda, *Phys. Rev.* **126**, 1271 (1962).

The other explanation, due to Fibich,[1] is that the finite lifetimes of the quasi-particle states against decay into phonons limits the sharpness of the peak, as indicated by an uncertainty principle argument. Such a mechanism would be expected to be roughly independent of impurity, but more important in strong-coupling superconductors. Some data on indium seem to support this mechanism. Thus both mechanisms seem important in appropriate situations.

2-9.3 Electromagnetic Absorption

Since the interaction hamiltonian $\mathbf{p} \cdot \mathbf{A}$ also obeys the case II coherence factors, we can carry over the results of the previous section on nuclear relaxation without modification to describe the absorption of low-frequency electromagnetic radiation. The quantity α_s/α_n is now called σ_{1s}/σ_n, since for a given E field the electromagnetic energy absorption per unit volume is $\sigma_1 E^2$, where σ_1 is the real part of the complex conductivity $\sigma_1(\omega) - i\sigma_2(\omega)$. Thus, for $\hbar\omega \ll \Delta$, we expect σ_{1s}/σ_n to rise above unity just below T_c and then to fall exponentially to zero at low temperatures. As noted before, such behavior is qualitatively incompatible with a simple two-fluid picture in which $n_n \leq n$.

Unlike the case of nuclear relaxation, however, it is now possible to utilize frequencies large enough to create pairs of quasi-particles. Such processes occur, in addition to the scattering processes treated already, as soon as $\hbar\omega \geq 2\Delta$. In fact, at $T = 0$ there are no thermally excited quasi-particles present, and the *only* process allowing absorption of energy is the creation of pairs. With our "semiconductor" sign conventions, the initial state energy E must be $\leq -\Delta$, and the final state energy $E + \hbar\omega \geq \Delta$. Thus, $\sigma_1(\omega) = 0$ for $\hbar\omega < 2\Delta$, at which point there is an "absorption edge," as shown in Fig. 2-9. The absorption can be computed using (2-91), with the Fermi functions being either 0 or 1 at $T = 0$. Thus,

$$\left.\frac{\sigma_{1s}}{\sigma_n}\right|_{T=0} = \frac{1}{\hbar\omega}\int_{\Delta-\hbar\omega}^{-\Delta} \frac{|E(E + \hbar\omega) + \Delta^2|}{(E^2 - \Delta^2)^{1/2}[(E + \hbar\omega)^2 - \Delta^2]^{1/2}}\, dE \qquad (2\text{-}94)$$

As shown by Mattis and Bardeen,[2] this integral can be expressed in terms of the tabulated complete elliptic integrals E and K, namely,

$$\left.\frac{\sigma_{1s}}{\sigma_n}\right|_{T=0} = \left(1 + \frac{2\Delta}{\hbar\omega}\right)E(k) - \frac{4\Delta}{\hbar\omega}K(k) \qquad \hbar\omega \geq 2\Delta \qquad (2\text{-}95)$$

where

$$k = \frac{\hbar\omega - 2\Delta}{\hbar\omega + 2\Delta} \qquad (2\text{-}95a)$$

[1] M. Fibich, *Phys. Rev. Letters* **14**, 561 (1965).
[2] D. C. Mattis and J. Bardeen, *Phys. Rev.* **111**, 412 (1958).

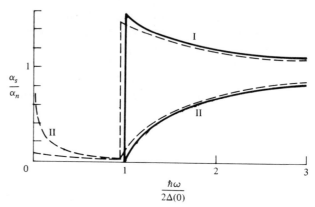

FIGURE 2-9

Frequency dependence of absorption processes obeying case I and II coherence factors at $T = 0$ (solid curves) and $T \approx \frac{1}{2}T_c$ (dashed curves).

As shown in Fig. 2-9, σ_{1s}/σ_n rises from zero with finite slope at $\hbar\omega = 2\Delta$, and slowly approaches unity for $\hbar\omega \gg 2\Delta$. At finite temperatures, $\Delta(T) < \Delta(0)$, and also the thermally excited quasi-particles contribute absorption for $\hbar\omega < 2\Delta(T)$. Numerical calculations are needed to describe the exact behavior, but the qualitative behavior for $T > 0$ is indicated by the dashed curve. The rise as $\hbar\omega \to 0$ is the logarithmic dependence discussed in connection with $1/T_1$.

Historically speaking, the first spectroscopic measurements clearly showing the existence and width of the energy gap in superconductors well below T_c were made by Glover and Tinkham[1] with far infrared radiation in the region of the absorption edge. These first measurements slightly preceded the appearance of the BCS theory, but were soon found to be in excellent accord with (2-95). With improvements in techniques over the years,[2] the quality of data has now reached the state that small deviations from the simple BCS curves observed in measurements on thin lead films can be interpreted in terms of strong-coupling effects analogous to the phonon structure observed in tunneling experiments.

It is worth remarking that with case I coherence factors, α_s/α_n rises discontinuously at $\hbar\omega = 2\Delta$ to a value greater than one, and then decreases, as shown in Fig. 2-9. One can show that the total area under the curve is conserved in this case, whereas for case II the area seems to disappear when the gap opens up. However, the oscillator-strength sum rule in the form[3]

$$\int_0^\infty \sigma_1(\omega)\, d\omega = \frac{\pi n e^2}{2m}$$

[1] R. E. Glover and M. Tinkham, *Phys. Rev.* **104**, 844 (1956).
[2] D. M. Ginsberg and M. Tinkham, *Phys. Rev.* **118**, 990 (1960); L. H. Palmer and M. Tinkham, *Phys. Rev.* **165**, 588 (1968).
[3] See, for example, R. Kubo, *J. Phys. Soc. Japan* **12**, 570 (1957).

requires that the area under the curve of $\sigma_1(\omega)$ have the same value in the super-conducting as in the normal state. Tinkham and Ferrell[1] were able to argue that the "missing area" A at finite frequencies appears as a δ function at $\omega = 0$, which physically represents the absorption of energy from a dc electric field to supply the kinetic energy of the accelerated supercurrent. The argument is based on the Kramers-Kronig relations,[2] which connect the real and imaginary parts of any causal linear-response function. Written in terms of the complex conductivity $\sigma_1 - i\sigma_2$, and with a time dependence of $e^{+i\omega t}$, they have the form

$$\sigma_1(\omega) = \frac{2}{\pi} \int_0^\infty \frac{\omega' \sigma_2(\omega')\, d\omega'}{\omega'^2 - \omega^2} + \text{const} \qquad (2\text{-}96a)$$

$$\sigma_2(\omega) = -\frac{2\omega}{\pi} \int_0^\infty \frac{\sigma_1(\omega')\, d\omega'}{\omega'^2 - \omega^2} \qquad (2\text{-}96b)$$

From (2-96b), we see that a term $\sigma_1 = A\,\delta(\omega)$ yields $\sigma_2 = 2A/\pi\omega$. For comparison, the London equation (1-3) is equivalent to $\sigma_2 = 1/\Lambda\omega = n_s e^2/m\omega = c^2/4\pi\lambda^2\omega$. Thus, we see that the penetration depth is related to the missing area by $\lambda^{-2} = 8A/c^2$, so that in principle the dc superconducting properties can be computed from the high-frequency absorption spectrum. Roughly speaking, the gap implies the superconductivity, but more carefully, it is the missing area that is important. The example of case I coherence factors, mentioned above, shows that missing area is not a necessary consequence of the energy gap; case II coherence factors are also required. Thus, the coherence factors are a more essential feature of superconductivity than is the gap in the spectrum. This point is driven home by recalling that *semi*conductors have gaps but are not superconducting (because there is no missing area), whereas it has been shown that superconductors made gapless by magnetic impurities retain superconducting properties as long as there is still missing area under the $\sigma_1(\omega)$ curve.

2-10 ELECTRODYNAMICS

The simple treatment we have just given of the absorption of electromagnetic fields gives no direct account of the dramatic supercurrent properties because we restricted ourselves to dissipative processes analogous to those of normal electrons. Rather than undertake the complications of a complete treatment of the response of a superconductor to an arbitrary electromagnetic field, we shall instead content ourselves with a treatment of the response to a static magnetic field.

[1] M. Tinkham and R. A. Ferrell, *Phys. Rev. Letters* **2**, 331 (1959).
[2] For a more detailed discussion of the application of sum-rule and Kramers-Kronig methods to superconducting electrodynamics, see the chapter on superconductivity by the author in "Low Temperature Physics," edited by C. de Witt, B. Dreyfus, and P. G. de Gennes, Gordon and Breach, New York, 1962.

This response must be nondissipative, and hence is complementary to that just treated. Since the Meissner effect and related superfluid responses are nearly independent of frequency until one reaches frequencies on the order of the gap, this treatment actually is useful in a wide range of applications.

In the cases of interest, the field can be described by a classical transverse vector potential $A(r)$ such that $B = \text{curl } A$. These are the total fields including the effects of screening by supercurrents, which will have to be introduced in a self-consistent way. In the presence of a vector potential, it is well known that the canonical momentum of both classical and quantum mechanics contains a potential-momentum term as well as the usual kinetic momentum, so that $p = mv + eA/c$, where e is the charge of the particle. The kinetic energy $\frac{1}{2}mv^2$ then becomes $(p - eA/c)^2/2m$, while the potential-energy expression is unchanged. Since we are interested only in calculating the linear response to weak fields, we can expand this kinetic-energy expression, keeping only terms linear in A. With the operator replacement $p_i \rightarrow -i\hbar\nabla_i$, the resulting perturbation term is

$$\mathcal{H}_1 = \frac{ieh}{2mc} \sum_i (\nabla_i \cdot A + A \cdot \nabla_i)$$

where the sum runs over all the particles. If the vector potential is expanded in spatial Fourier components

$$A(r) = \sum_q a(q)e^{iq \cdot r}$$

$\mathcal{H}_1$ may be written

$$\mathcal{H}_1 = -\frac{eh}{mc} \sum_{k, q} k \cdot a(q) c^*_{k+q, \sigma} c_{k, \sigma} \qquad (2\text{-}97)$$

since $q \cdot a(q) = 0$ for a transverse field, by definition. Because the dominant interaction is with the orbital motion, the spin index is unchanged. The explicit form of the $B_{k'\sigma', k\sigma}$ of (2-84) may be read out from (2-97). From the proportionality of these matrix elements to k, we see that they are of case II, as noted earlier, and following (2-87), the interaction can be written out in full as

$$\mathcal{H}_1 = -\frac{eh}{mc} \sum_{k, q} k \cdot a(q)[(u_k u_{k+q} + v_k v_{k+q})(\gamma^*_{k+q, 0}\gamma_{k0} - \gamma^*_{k1}\gamma_{k+q, 1})$$

$$+ (v_k u_{k+q} - u_k v_{k+q})(\gamma^*_{k+q, 0}\gamma^*_{k1} - \gamma_{k+q, 1}\gamma_{k0})] \qquad (2\text{-}98)$$

In the preceding section, we dealt with the energy absorption due to this operator. Now we consider the *current* induced by this perturbation. In the presence of a vector potential, the current consists of two terms, J_1 and J_2, corresponding to the two terms in $v = (p/m) - (eA/mc)$. The second term gives the simple result

$$J_2 = -\frac{ne^2}{mc} A \qquad (2\text{-}99)$$

which would correspond exactly to the London equation (1-8) if n could be interpreted as n_s, the number of "superconducting electrons." In fact, though, n is always the *total* electron density, and this term has the same value also in the normal state. Thus the first term J_1, often called the "paramagnetic" current term because it tends to cancel the diamagnetic current J_2, must play a vital role. It is not hard to see that the qth Fourier component of J_1 will be found by evaluating the operator

$$J_1(q) = \frac{e\hbar}{m} \sum_k k c^*_{k-q} c_k \qquad (2\text{-}100)$$

which may then be expanded in terms of the γ operators to give a form analogous to (2-98).

Before proceeding with detailed calculations, it is convenient to introduce a standard notation for describing the current response to the various Fourier components of a vector potential, namely,

$$J(q) = -\frac{c}{4\pi} K(q) a(q) \qquad (2\text{-}101)$$

For example, in the London theory, where

$$J(r) = -\frac{1}{c\Lambda} A(r) = -\frac{c}{4\pi\lambda_L^2} A(r) \qquad (2\text{-}102)$$

the response is independent of q, and

$$K_L(q) = K_L(0) = \frac{1}{\lambda_L^2} \qquad (2\text{-}103)$$

Using the definition $\lambda_L^2(0) = mc^2/4\pi n e^2$ and (2-99), we may write the response function, including both J_1 and J_2, as

$$K(q, T) = \lambda_L^{-2}(0)[1 + \lambda_L^2(0) K_1(q, T)] \qquad (2\text{-}104)$$

where the above discussion leads us to expect that K_1 will be negative. As we are restricting our attention to isotropic systems and transverse fields, $K(q)$ is a function only of $|q|$, and we shall normally denote it simply $K(q)$.

It is also useful to consider in general the relation between a q-dependent $K(q)$ and the corresponding nonlocal response in coordinate space described by a range function or kernel $F(R)$ in the expression

$$J(r) = C \int \frac{R[R \cdot A(r')]}{R^4} F(R) \, dr' \qquad (2\text{-}105)$$

where $R = r - r'$. By inserting $A(r') \sim e^{iq \cdot r'}$, one can show that the relation between $K(q)$ and $F(R)$ is

$$K(q) = \frac{16\pi^2 C}{3c} \int_0^\infty \left[\frac{3}{qR} j_1(qR) \right] F(R) \, dR \qquad (2\text{-}106)$$

where $j_1(x) = x^{-2} \sin x - x^{-1} \cos x$ is a spherical Bessel function, such that $[3x^{-1}j_1(x)]$ is a damped oscillatory function with the value unity for $x = 0$. Thus,

$$K(0) = \frac{16\pi^2 C}{3c} \int_0^\infty F(R)\, dR \quad (2\text{-}106a)$$

depends on the integral of $F(R)$, whereas as $q \to \infty$

$$K(q) \underset{q \to \infty}{\longrightarrow} \frac{16\pi^2 C}{3c} \frac{F(0)}{q} \int_0^\infty \frac{3}{x} j_1(x)\, dx = \frac{4\pi^3 C F(0)}{cq} \quad (2\text{-}106b)$$

which depends only on the value of $F(R)$ at the origin. Taking the ratio

$$\frac{K(q)}{K(0)} \underset{q \to \infty}{\longrightarrow} \frac{3\pi}{4qL} \quad (2\text{-}106c)$$

where $L = F^{-1}(0) \int F(R)\, dR$ is a measure of the range of the real-space kernel $F(R)$. These relations are very general; they apply to the relation between $\mathbf{J}$ and $\mathbf{E}$ in the normal state with $L = \ell$, and to the relation between $\mathbf{J}$ and $\mathbf{A}$ in the superconducting state with, as we shall soon see, $L \approx \xi_0$.

2-10.1 Calculation of $K(0, T)$ or $\lambda_L(T)$

Now let us proceed to calculate the temperature dependence of $K(q, T)$ for the simple limiting case $q = 0$, which corresponds to infinite wavelength. This will determine the temperature dependence of $\lambda_L(T)$, which is defined by

$$K(0, T) \equiv \frac{1}{\lambda_L^2(T)} \quad (2\text{-}107)$$

since the nonlocal theory reduces to the local London theory for fields which vary slowly in space. In other words, the temperature dependence of $K(0, T)$ is identified with the temperature dependence of n_s in the London theory. For $\mathbf{q} = 0$, obviously the coherence factor in the second term in (2-98) is zero, while that in the first term is unity. Moreover, for $\mathbf{q} = 0$, $\gamma_{\mathbf{k}+\mathbf{q},0}^* \gamma_{\mathbf{k}0}$ becomes the number operator $\gamma_{\mathbf{k}0}^* \gamma_{\mathbf{k}0}$. Thus, the perturbing hamiltonian (2-98) simply shifts the energies of the quasi-particle excitations, as follows:

$$E_{\mathbf{k}0} \to E_{\mathbf{k}0} - \frac{e\hbar}{mc} \mathbf{k} \cdot \mathbf{a}(0)$$

$$E_{\mathbf{k}1} \to E_{\mathbf{k}1} + \frac{e\hbar}{mc} \mathbf{k} \cdot \mathbf{a}(0) \quad (2\text{-}108)$$

Similarly, the expansion of $\mathbf{J}_1(0)$ in quasi-particle operators reduces to simply

$$\mathbf{J}_1(0) = \frac{e\hbar}{m} \sum_{\mathbf{k}} \mathbf{k}(\gamma_{\mathbf{k}0}^* \gamma_{\mathbf{k}0} - \gamma_{\mathbf{k}1}^* \gamma_{\mathbf{k}1}) \qquad (2\text{-}109)$$

$$= \frac{e\hbar}{m} \sum_{\mathbf{k}} \mathbf{k}(f_{\mathbf{k}0} - f_{\mathbf{k}1})$$

where in the second line the operators have been replaced by their expectation values $f_{\mathbf{k}0}$ and $f_{\mathbf{k}1}$, the Fermi functions corresponding to the shifted energies (2-108). In the limit of small $\mathbf{a}(0)$, $f_{\mathbf{k}0} \approx f_{\mathbf{k}1}$, and the difference may be found by taking the first term in a Taylor's series expansion, that is,

$$f_{\mathbf{k}0} - f_{\mathbf{k}1} \approx \left(-\frac{\partial f}{\partial E_{\mathbf{k}}}\right) \frac{2e\hbar}{mc} \mathbf{k} \cdot \mathbf{a}(0)$$

Inserting this in (2-109), we have

$$\mathbf{J}_1(0) = \frac{2e^2\hbar^2}{m^2 c} \sum_{\mathbf{k}} [\mathbf{a}(0) \cdot \mathbf{k}]\mathbf{k}\left(-\frac{\partial f}{\partial E_{\mathbf{k}}}\right) \qquad (2\text{-}110)$$

By symmetry, $\mathbf{J}_1(0)$ is parallel to $\mathbf{a}(0)$, and the average of the square of the component of $\mathbf{k}$ along $\mathbf{J}_1$ (or $\mathbf{a}$) will be $k_F^2/3$, since $\cos^2 \theta$ averages to $1/3$ over a sphere. Thus, the K_1 corresponding to (2-110) can be written as

$$K_1(0, T) = \frac{-4\pi \mathbf{J}_1(0)}{c\mathbf{a}(0)} = -\left(\frac{4\pi n e^2}{mc^2}\right)\left(\frac{4E_F}{3n}\right) \sum_{\mathbf{k}} \left(-\frac{\partial f}{\partial E_{\mathbf{k}}}\right)$$

Since $N(0) = 3n/4E_F$, this can be rewritten as

$$K_1(0, T) = -\lambda_L^{-2}(0) \int_{-\infty}^{\infty} \left(-\frac{\partial f}{\partial E}\right) d\xi$$

Inserting this in (2-104), the total response, including $\mathbf{J}_1$ and $\mathbf{J}_2$, is

$$K(0, T) = \lambda_L^{-2}(T) = \lambda_L^{-2}(0)\left[1 - 2\int_{\Delta}^{\infty} \left(-\frac{\partial f}{\partial E}\right) \frac{E}{(E^2 - \Delta^2)^{1/2}} dE\right] \qquad (2\text{-}111)$$

We note first that if $\Delta = 0$ (normal state, $T \geq T_c$), then the integral reduces to $f(0) = 1/2$, and $K(0, T \geq T_c) = 0$. This corresponds to the absence of any Meissner effect in the normal state because of the cancellation of the paramagnetic and diamagnetic current terms $\mathbf{J}_1$ and $\mathbf{J}_2$. As soon as $\Delta > 0$, however, the integral no longer gives exact cancellation, and there is a finite $\lambda_L(T)$ leading to a Meissner effect if the sample is large enough. As T decreases and $\Delta(T)/kT$ increases further, the integral becomes less and less, eventually becoming exponentially small when $\Delta/kT \gg 1$. Thus $\lambda_L(T)$ defined by (2-111) does reduce properly to $\lambda_L(0)$ as defined by (1-9) when $T \to 0$, justifying our notations. The general behavior is shown in

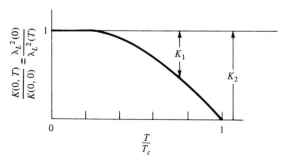

FIGURE 2-10

Temperature dependence of $K(0, T)/K(0, 0) = \lambda_L^2(0)/\lambda_L^2(T) = n_s(T)/n$ according to the BCS theory. Note that $K(0, T)$ is the result of the partial cancellation of the constant diamagnetic term K_2 by the temperature-dependent paramagnetic term K_1.

Fig. 2-10. The physical origin of K_1 is that excitations on the trailing edge of the displaced momentum distribution in **k** space have lower energy and are thus more highly populated, so that the excited quasi-particles carry a net current in the reverse direction which partially cancels the current $\mathbf{J}_2$. It is conceptually important to recognize that the *total* response, including the (negative) quasi-particle contribution, is the supercurrent. The quasi-particle contribution does not die away because it represents a distribution giving minimum free energy.

In considering Fig. 2-10, it should be borne in mind that $\lambda_L(T)$ will give the temperature dependence of the experimentally observed penetration depth only if the strength of the response to the self-consistent vector potential is adequately approximated by the response to the component with $\mathbf{q} = 0$. We now consider the q-dependence of $K(q, T)$ to clarify this point.

2-10.2   Calculation of $K(q, 0)$

For simplicity we shall examine the q-dependence only at $T = 0$, although the calculation is readily extended by use of appropriate statistical techniques. At $T = 0$, the electrons must be in the BCS ground state $|\psi_G\rangle$. Then by ordinary first-order perturbation theory, the perturbed state $|\psi\rangle$ in the presence of $\mathcal{H}_1$, is

$$|\psi\rangle = |\psi_G\rangle - \sum_n \frac{\langle\psi_n|\mathcal{H}_1|\psi_G\rangle}{E_n}|\psi_n\rangle$$

where the sum on n runs over the various excited states with excitation energy E_n. Taking the expectation value of $\mathbf{J}_1(\mathbf{q})$ over $|\psi\rangle$ we have

$$\langle\psi|\mathbf{J}_1(\mathbf{q})|\psi\rangle = \langle\psi_G|\mathbf{J}_1(\mathbf{q})|\psi_G\rangle - 2\,\mathrm{Re}\sum_n \frac{\langle\psi_G|\mathbf{J}_1(\mathbf{q})|\psi_n\rangle\langle\psi_n|\mathcal{H}_1|\psi_G\rangle}{E_n}$$

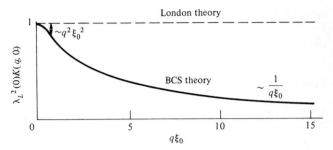

FIGURE 2-11
Comparison of q-dependent response of the nonlocal BCS theory with the q-independent response of the local London theory. In both cases, the curves are drawn for pure metals, with infinite mean free path.

The first term is zero, since the electrons all have paired momenta. In the second term, the only contribution comes from states ψ_n containing two quasi-particles generated by the term $\gamma^*_{k+q,\,0}\gamma^*_{k1}$ in (2-98). Since the same coherence factor appears in the matrix elements of $\mathbf{J}_1(\mathbf{q})$ as in the elements of $\mathcal{H}_1$, one finds

$$\langle \psi \,|\, \mathbf{J}_1(\mathbf{q}) \,|\, \psi \rangle = \frac{2e^2\hbar^2}{m^2 c} \sum_k \frac{(v_k u_{k+q} - u_k v_{k+q})^2}{E_k + E_{k+q}} [\mathbf{k} \cdot \mathbf{a}(\mathbf{q})]\mathbf{k} \qquad (2\text{-}112)$$

This term obviously gives a current parallel to $\mathbf{A}$, partially cancelling the diamagnetic current $\mathbf{J}_2$, even at $T = 0$. The amount of cancellation depends on q, going to zero at $q = 0$, as noted earlier.

Using arguments similar to those used to reach (2-111), (2-112) leads to

$$K(q, 0) = \lambda_L^{-2}(0)\left\{1 - \int_{-\infty}^{\infty} \frac{(v_k u_{k+q} - u_k v_{k+q})^2}{E_k + E_{k+q}}\, d\xi\right\} \qquad (2\text{-}113)$$

for the response function K. Without going into details, one can see that this implies that as q first increases from zero, $K(q, 0)$ decreases by a fractional amount of the order of $q^2\xi_0^2$, where, following BCS, ξ_0 is defined by

$$\xi_0 \equiv \frac{\hbar v_F}{\pi \Delta(0)} \qquad (2\text{-}114)$$

[This follows since the coherence factor in the numerator can be written for small $\mathbf{q}$ as $(\xi_{k+q} - \xi_k)\,\partial(u_k - v_k)/\partial\xi_k$; the first factor is of order $\hbar v_F q$, and the second is of order $1/\Delta(0)$ for the important part of the integration region.] On the other hand, if $q\xi_0 \gg 1$, to a first approximation the integral in (2-113) cancels the first term, leaving a residual value of

$$K(q, 0) = K(0, 0)\frac{3\pi}{4q\xi_0} \qquad q\xi_0 \gg 1 \qquad (2\text{-}115)$$

as anticipated in (2-106c). The complete dependence of $K(q, 0)$ on q is shown in Fig. 2-11. Note that the response of the normal state at $T = 0$ can be obtained as the limit $\Delta(0) \to 0$, which leads to $\xi_0 \to \infty$, so that $K(q, 0) = 0$ for all q.

2-10.3 Nonlocal Electrodynamics in Coordinate Space

Given $K(q, T)$, one can find the corresponding real-space kernel by Fourier transformation, the inverse of the transformation leading to (2-106). The result, called $J(R, T)$ by BCS, is sketched in Fig. 2-12. It is very similar to the exponential form e^{-R/ξ_0} proposed by Pippard.[1] In fact, the $J(R, T)$ of BCS is normalized so that

$$\int_0^\infty J(R, T)\, dR = \xi_0 = \int_0^\infty e^{-R/\xi_0}\, dR \qquad (2\text{-}116)$$

with ξ_0 as defined in (2-114). At $R = 0$, the value of J ranges smoothly from 1 at $T = 0$ to 1.33 at $T = T_c$, whereas in the Pippard approximation, the value of the exponential kernel at $R = 0$ remains unity at all temperatures. Apart from this minor difference, the similarity is remarkably complete.

Inserting the normalization (2-115) into (2-106a) and recalling the definition (2-107), the BCS version of the nonlocal current relation (2-105) becomes

$$\mathbf{J(r)} = -\frac{3c}{16\pi^2 \xi_0 \lambda_L^2(T)} \int \frac{\mathbf{R[R \cdot A(r')]}}{R^4} J(R, T)\, d\mathbf{r'} \qquad (2\text{-}117)$$

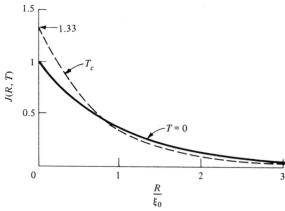

FIGURE 2-12
Schematic comparison of the BCS range function $J(R, T)$ at $T = 0$ and T_c. Note that the range of nonlocality is reduced by a factor of about 0.75 on going from $T = 0$ to T_c.

[1] A. B. Pippard, *Proc. Roy. Soc.* (*London*) **A216**, 547 (1953).

where $\mathbf{R} = \mathbf{r} - \mathbf{r}'$. This differs from the Pippard form (1-11) only by the substitution of $J(R, T)$ for the exponential e^{-R/ξ_0}. If $\mathbf{A}(\mathbf{r}')$ is constant over the range of $J(R, T)$, (2-117) reduces to the London relation

$$\mathbf{J}(\mathbf{r}) = -\frac{c}{4\pi\lambda_L^2(T)}\mathbf{A}(\mathbf{r}) \qquad (2\text{-}118)$$

However, if $\mathbf{A}(\mathbf{r})$ is varying significantly on the scale of ξ_0, this reduction can not be made, and the nonlocality of the electrodynamics must be taken into account.

2-10.4 Effect of Impurities

Our results so far have been for the case of a pure metal, in which $\mathbf{k}$ is a good quantum number, allowing the simple formulation of the BCS pairing theory which we have given. Although generalization of the microscopic theory to include the effects of impurity scattering can be done by Green function techniques, we will use instead a more phenomenological approach. By analogy with the corresponding expression of Chambers for the nonlocal response of normal electrons to an electric field, we expect a factor of $e^{-R/\ell}$ to multiply the kernel $J(R, T)$, where ℓ is the mean free path. This has the effect of making the electrodynamic response more local, and if ℓ is sufficiently small on the scale of spatial variation of $\mathbf{A}$, one can always simplify to a local response of the London type (2-118), but with a modified (reduced) coefficient characterized by a $\lambda_{\text{eff}}(> \lambda_L)$. The appropriate value of λ_{eff} is determined by

$$\frac{\lambda_L^2(T)}{\lambda_{\text{eff}}^2(\ell, T)} = \frac{K(0, T, \ell)}{K(0, T, \infty)} = \frac{\int_0^\infty J(R, T)e^{-R/\ell}\,dR}{\xi_0} \qquad (2\text{-}119)$$

where we have used (2-106a) and (2-116). [Note that we restrict use of the notation $\lambda_L(T)$ to characterize the response of pure metal at $q = 0$.]

To avoid numerical evaluation of (2-119), we first consider the "extreme dirty limit," $\ell \ll \xi_0$, in which the integral in the numerator reduces simply to $J(0, T)\ell$. Then

$$\lambda_{\text{eff}}(\ell, T) = \lambda_L(T)\left(\frac{\xi_0}{\ell}\right)^{1/2}[J(0, T)]^{-1/2} \qquad (2\text{-}120)$$

Since $J(0, T)$ ranges only from 1 at $T = 0$ to 1.33 at T_c, the final factor introduces only a small correction.

Another convenient approximation is to use the Pippard exponential e^{-R/ξ_0} for $J(R, T)$. If this is done, the integral in (2-119) becomes simply ξ, where

$$\xi^{-1} = \xi_0^{-1} + \ell^{-1} \qquad (2\text{-}121)$$

and then

$$\lambda_{\text{eff}}(\ell, T) = \lambda_L(T)\left(\frac{\xi_0}{\xi}\right)^{1/2} = \lambda_L(T)\left(1 + \frac{\xi_0}{\ell}\right)^{1/2} \qquad (2\text{-}122)$$

This approximation is useful even when ℓ is not very small, since it reduces to λ_L in the pure limit, and as a result it has been commonly used. However, it is inexact in that it is missing the factor $[J(0, T)]^{-1/2}$ of the microscopic theory in the dirty limit (2-120).

   This defect can be remedied by a simple improvement. We retain the exponential approximation for the form of $J(R, T)$, but we replace e^{-R/ξ_0} by $J(0, T) \exp -[J(0, T)R/\xi_0]$. This new form agrees with the microscopic $J(R, T)$ in its initial value at $R = 0$ as well as in its integral over R. With this improvement, (2-122) is replaced by

$$\lambda_{\text{eff}}(\ell, T) = \lambda_L(T)\left(\frac{\xi_0}{\xi'}\right)^{1/2} = \lambda_L(T)\left(1 + \frac{\xi_0}{\ell}\right)^{1/2} \qquad (2\text{-}123)$$

where the modified Pippard coherence lengths ξ' and ξ_0 are defined by the relation

$$\frac{1}{\xi'} \equiv \frac{1}{\xi_0} + \frac{1}{\ell} \equiv \frac{J(0, T)}{\xi_0} + \frac{1}{\ell} \qquad (2\text{-}123a)$$

Clearly, (2-123) reduces properly to (2-120), but also behaves well when ℓ is not small. Thus, (2-123) gives a very good approximation for the $q \approx 0$ response over the entire range of its arguments. It will also give a good approximation to the actual penetration depth if $\xi \ll \lambda_{\text{eff}}$, where ξ is given by (2-121). Note that (2-123) reduces to (2-122) at $T = 0$, and to

$$\lambda_{\text{eff}}(\ell, T) = \lambda_L(T)\left(1 + 0.75\frac{\xi_0}{\ell}\right)^{1/2} \qquad T \approx T_c \qquad (2\text{-}123b)$$

for $T \approx T_c$. The latter form is often used in connection with the Ginzburg-Landau theory, which is valid near T_c.

2-10.5   Complex Conductivity

In the section on transition probabilities and absorption, we worked out an expression [(2-91) with the upper sign] for the ratio of the real part of the complex conductivity in the superconducting state to that of the normal state, σ_{1s}/σ_n, as a function of frequency. Now that we have worked out the low-frequency lossless response to a vector potential, we can embed the conductivity result in a more general picture. Since we can write

$$\mathbf{E} = -\frac{1}{c}\frac{\partial \mathbf{A}}{\partial t} = -\frac{i\omega\mathbf{A}}{c}$$

for a periodic electromagnetic field, we can define a complex conductivity proportional to $K(q, \omega, T)$. Equating two expressions for the current

$$\mathbf{J}(q, \omega, T) = \sigma(q, \omega, T)\mathbf{E}(q, \omega) = -\frac{c}{4\pi} K(q, \omega, T)\mathbf{A}(q, \omega)$$

we have

$$\sigma(q, \omega, T) = \frac{ic^2}{4\pi\omega} K(q, \omega, T) \qquad (2\text{-}124)$$

Of course, we have worked out only some special cases of $K(q, \omega, T)$ [or $\sigma(q, \omega, T)$] in any detail, but these serve to outline much of the general behavior.

A simple analytic form can be given for the temperature variation of the low-frequency limit of σ_2/σ_n, namely,

$$\frac{\sigma_2}{\sigma_n} = \frac{\pi\Delta}{\hbar\omega} \tanh\frac{\Delta}{2kT} \qquad \hbar\omega \ll 2\Delta \qquad (2\text{-}125)$$

which has the limiting forms

$$\frac{\sigma_2}{\sigma_n} \rightarrow \begin{cases} \dfrac{\pi\Delta}{\hbar\omega} & T \ll T_c \quad (2\text{-}125a) \\[2ex] \dfrac{\pi}{2\,kT\hbar\omega}\Delta^2 & T \approx T_c \quad (2\text{-}125b) \end{cases}$$

Relating these expressions to the London theory, in which

$$\sigma_{2L} = \frac{n_s e^2}{m\omega} \qquad (2\text{-}126)$$

we see that these results correspond to $n_s \sim \Delta$ for $T \ll T_c$, but to $n_s \sim \Delta^2$ for $T \approx T_c$. Given Gor'kov's demonstration that near T_c the phenomenological ψ function of the Ginzburg-Landau theory is proportional to the gap Δ in the BCS theory, the latter result corresponds to the central identification of $n_s \sim |\psi|^2$.

In terms of our present, more complete picture of a nonlocal or q-dependent response, we must ask what is the region of validity of the expression for σ_{1s}/σ_n as given by (2-91) or (2-95). It turns out that these expressions are valid when the response is determined by the value of the nonlocal response kernel for $R = 0$. This will be the case in the dirty limit, $\ell \ll \xi_0$, where the factor $e^{-R/\ell}$ cuts off before $J(R, \omega, T)$ changes very much. It will also be the case when $q\xi_0 \gg 1$, the "extreme anomalous limit," since in this case the appropriate Fourier transform (2-106b) gives weight mainly to a region extending only to about π/q from the origin. In both these limits, σ_n is real, so that $\sigma_n = \sigma_{1n}$. In the dirty limit, both σ_s and σ_n are proportional to ℓ; in the extreme anomalous limit, both are proportional to $1/q$. Thus, in both limits the ratio σ_s/σ_n is dependent only on ω and T, permitting us to drop q and ℓ from the notation so long as we remember the limitation to $q\xi_0 \gg 1$ or $\ell \ll \xi_0$.

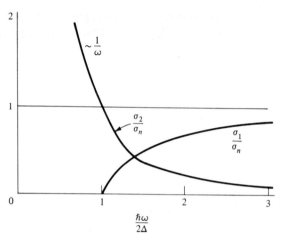

FIGURE 2-13
Complex conductivity of superconductors in extreme anomalous limit (or extreme dirty limit), at $T = 0$. The rise of σ_2 as $1/\omega$ below the gap describes the accelerative supercurrent response. Its coefficient is proportional to the "missing area" under the $\sigma_1(\omega)$ curve at finite frequencies (see text at top of page 59).

An explicit form for σ_2 at $T = 0$ analogous to (2-95) for σ_1 can be worked out, namely,

$$\frac{\sigma_{2s}}{\sigma_n} = \frac{1}{2}\left(1 + \frac{2\Delta}{\hbar\omega}\right)E(k') - \frac{1}{2}\left(1 - \frac{2\Delta}{\hbar\omega}\right)K(k') \quad (2\text{-}127)$$

where $k' = (1 - k^2)^{1/2}$ and $k = |(2\Delta - \hbar\omega)/(2\Delta + \hbar\omega)|$. The frequency dependence of this function is sketched in Fig. 2-13. Of particular importance is the proportionality of σ_2 to $1/\omega$ for $\hbar\omega \ll 2\Delta$. This corresponds to the fact that $K(q, \omega)$ is essentially independent of frequency there, as required for a frequency-independent penetration depth. More simply, this dependence is a consequence of the free-acceleration aspect of the supercurrent response as described by the London equation $\mathbf{E} = \partial(\Lambda \mathbf{J}_s)/\partial t$. For $\hbar\omega \gtrsim 2\Delta$, σ_2 falls to zero more rapidly than $1/\omega$, but at $\hbar\omega = 2\Delta$, $K(q, \omega)$ has decreased by only a factor of $2/\pi$ from its dc value. Thus, the superfluid response is essentially independent of frequency until microwave frequencies are reached.

The real and imaginary parts of σ both enter into the determination of the response of a superconducting system to time-dependent electromagnetic fields. For example, in the transmission experiments with very thin films, cited above in Sec. 2-9, the fractional transmissivity is readily shown to be

$$T = \left[\left(1 + \frac{\sigma_1 \, dZ_0}{n + 1}\right)^2 + \left(\frac{\sigma_2 \, dZ_0}{n + 1}\right)^2\right]^{-1} \quad (2\text{-}128)$$

where n is the index of refraction of the substrate, Z_0 (377 ohms per square in mks units) is the impedance of free space, and d is the film thickness. From (2-128) we see that $T_s \to 0$ as $\omega \to 0$, since $\sigma_{2s} \sim 1/\omega \to \infty$, leading to complete reflection. On the other hand, for $\hbar\omega \gg 2\Delta$, $\sigma_{2s} \to 0$ and $\sigma_{1s} \to \sigma_n$, so that $T_s \to T_n$. In between, for $\hbar\omega \approx 2\Delta$, both σ_{1s} and σ_{2s} are smaller than σ_n, as shown in Fig. 2-13, and there is a peak in transmission at which $T_s > T_n$. Observation of these characteristic features gave strong support to an energy-gap model of superconductivity and quantitative support to the BCS model in particular.

2-11 CONCLUDING SUMMARY

In this chapter we have set out the basic features of the BCS theory: the Cooper pairing due to a weak, phonon-mediated attraction between the electrons, the nature of the superconducting ground state and its condensation energy relative to the normal state, the excited quasi-particle states above the energy gap, and the temperature dependence of the gap and of the thermodynamic quantities. We then showed how electron tunneling has been able to confirm the density of excited states above the gap in detail, and even confirm the quantitative correctness of the electron-phonon mechanism for setting up the superconductive state in a number of metals. Our analysis of the transition probabilities for quasi-particle excitation and scattering gave further testimony to the correctness of the theory, since very different temperature dependences were predicted and observed for ultrasonic attentuation and nuclear relaxation processes, which would have the same temperature dependence in a simple two-fluid model without the coherence factors derived from the pairing model. Finally, we have treated the electrodynamics of superconductors, first computing the absorption due to quasi-particle processes, and then computing the lossless supercurrent response in the presence of a vector potential. The latter computation confirmed the correctness of the Pippard nonlocal generalization of the London electrodynamics, apart from some minor quantitative alterations.

Having seen that the BCS theory provides a microscopic foundation for the relatively simple semimicroscopic, or phenomenological, theory of superconductivity, we now are prepared to leave the microscopic theory behind and use the latter to explore numerous superconducting phenomena. This program will occupy most of the following chapters.

3

MAGNETIC PROPERTIES OF TYPE I SUPERCONDUCTORS

In the preceding chapter we found that the microscopic theory of BCS predicted a nonlocal electrodynamics very similar to that proposed phenomenologically by Pippard. In the present chapter we shall work out some of the operational consequences of these basic electrodynamic properties. The first step will be to find the law of penetration of the magnetic field at the surface of a superconductor which replaces the simple London exponential approximation. Next we treat the interesting case of thin films. Finally, we shall take up the so-called intermediate state, in which superconducting and normal regions coexist in a superconductor of arbitrary shape in the presence of a magnetic field.

Since in this chapter we shall need to be quite explicit in distinguishing the various magnetic field quantities, we recall that we are following the convention of de Gennes (and many other authors) in using the symbol h to refer to the local microscopic value of B, the quantity which enters the London equation. This allows us to retain the symbol B to refer to the macroscopic average of $h(r)$.

3-1   THE PENETRATION DEPTH

3-1.1   Gauge Choice for the Vector Potential

In our previous discussion, we introduced the vector potential **A** in a rather casual way to simplify the two London equations

$$\mathbf{h} = -\frac{4\pi}{c} \operatorname{curl} (\lambda_L^2 \, \mathbf{J}_s)$$

$$\mathbf{E} = \frac{4\pi}{c^2} \frac{\partial(\lambda_L^2 \, \mathbf{J}_s)}{\partial t} \tag{3-1}$$

into the single equation

$$\mathbf{J}_s(\mathbf{r}) = -\frac{c}{4\pi\lambda_L^2} \mathbf{A}(\mathbf{r}) \tag{3-2}$$

which, in turn, was generalized to a nonlocal relation of the BCS-Pippard sort. It is well known, however, that the vector potential is, in general, arbitrary with respect to a change of gauge. That is, if one goes from $\mathbf{A}(\mathbf{r})$ to $\mathbf{A}'(\mathbf{r}) = \mathbf{A}(\mathbf{r}) + \nabla\chi(\mathbf{r})$, this leaves the magnetic field $\mathbf{h} = \operatorname{curl} \mathbf{A}$ unchanged, since the curl of a gradient is zero. Yet it would change the current (3-2). Thus, we evidently must determine a unique gauge if we are to use expressions such as (3-2) which are *not* manifestly gauge-invariant.

Proceeding in the London approximation, for simplicity, the continuity-of-current requirement (div $\mathbf{J}_s = 0$) implies that

$$\operatorname{div} \mathbf{A} = 0 \tag{3-3}$$

with any satisfactory gauge choice, since $\mathbf{J}$ is proportional to $\mathbf{A}$. This restricts any gauge changes from a satisfactory $\mathbf{A}$ to those generated by a $\chi(\mathbf{r})$ for which $\nabla \cdot \nabla\chi = \nabla^2\chi = 0$, so that $\chi(\mathbf{r})$ must satisfy Laplace's equation. Further, the condition of continuity of current at the boundary determines the normal component of $\mathbf{A}$ at the surface, and hence the normal derivative of χ must vanish at the surface. The only solution to this boundary-value problem is $\chi = \text{constant}$, for which $\nabla\chi = 0$, so $\mathbf{A}$ is uniquely specified by (3-3) and the boundary condition. This is called the London gauge choice. It is useful to note that with this choice, $\mathbf{A}$ goes to zero in the interior of bulk superconductors, where there are no currents.

With the nonlocal electrodynamics, the same qualitative arguments hold, although the mathematical details are less straightforward. However, for the simple case of a plane surface in a parallel magnetic field (which we shall treat to determine the penetration law), it is clear that a gauge choice with $\mathbf{A}$ parallel to the surface but perpendicular to $\mathbf{h}$ and falling off to zero in the interior of the bulk superconductor will satisfy the boundary conditions and be uniquely defined.

If one wishes to deal with a thin superconductor, one must replace the requirement that **A** go to zero in the interior by a more specific one. For example, if the same field is applied parallel to both sides of a thin, flat slab, symmetry requires that **A** pass through zero in the midplane. In less symmetrical situations, a variational calculation may be used to adjust the gauge of **A** for minimum energy. Alternatively, we shall later show how to recast the theory in terms of gauge-invariant forms. The point is that $J_s/n_s e = \langle \mathbf{v}_s \rangle = \langle \mathbf{p} \rangle/m - e\mathbf{A}/mc$ is gauge-invariant, so the different choices of gauge for **A** must be accompanied by compensating changes in $\langle \mathbf{p} \rangle$. Our present approach is to work in a particular gauge corresponding to $\langle \mathbf{p} \rangle = 0$, that is, to a wavefunction comprised of zero-momentum pairs, even in the presence of **A**. In simple geometries, this is usually the most convenient choice, but cases will arise later in which other choices may be indicated.

3-1.2 Preliminary Estimate of λ

Before going into the rigorous solution for the penetration depth with nonlocal electrodynamics, we first give an elementary argument which yields the form of the results for "Pippard superconductors," i.e., those with $\xi_0 \gg \lambda_L$.

Even with nonlocal electrodynamics, we expect approximately exponential penetration of the magnetic field, but with a modified penetration depth λ, to be determined. If

$$h_y \approx B_0 e^{-z/\lambda}$$

then in the appropriate gauge the vector potential inside the material ($z > 0$) will be

$$A_x \approx \lambda B_0 e^{-z/\lambda}$$

which can be roughly approximated by the constant value $\bar{A} = \lambda B_0$ over a layer of thickness λ. In calculating the resulting average current density in the surface layer using the nonlocal form (2-17), we will get a value reduced from the London value for this average $\bar{A}$ by a factor $\sim \lambda/\xi_0$, which is the fraction of the effective integration volume ($\sim \xi_0^3$) in which $\bar{A}$ exists ($\sim \xi_0^2 \lambda$). Thus, we expect

$$\bar{J} \approx -\frac{c}{4\pi\lambda_L^2}\frac{\lambda}{\xi_0}\bar{A} = -\frac{c\lambda^2 B_0}{4\pi\lambda_L^2\xi_0}$$

Applying Maxwell's equations,

$$\frac{B_0}{\lambda} \approx |\overline{\text{curl } \mathbf{h}}| = 4\pi\frac{\bar{J}}{c}\frac{\lambda^2 B_0}{\lambda_L^2\xi_0}$$

Solving, we find

$$\lambda \approx (\lambda_L^2\xi_0)^{1/3} \tag{3-4}$$

Thus, when nonlocality is important, i.e., when $\xi_0 > \lambda_L$, the actual penetration depth will exceed λ_L by a factor of order $(\xi_0/\lambda_L)^{1/3}$. If $\xi_0 < \lambda_L$, of course, the above argument does not apply because the response is local, and $\lambda \approx \lambda_L$.

Note that (3-4) has a qualitative implication about the temperature dependence of λ. Very near T_c, all superconductors become local, since $\lambda(T) > \xi_0$, so $\lambda(T) \approx \lambda_L(T)$. At lower temperatures, when $\lambda_L(T) < \xi_0$, we have $\lambda(T) \approx [\lambda_L^2(T)\xi_0]^{1/3}$. Thus there will be a change in the detailed temperature dependence of λ at the temperature at which $\lambda_L(T) \approx \xi_0$. Since this occurs at different values of T/T_c for different superconductors because of their different values of $\lambda_L(0)/\xi_0$, there cannot be a universal temperature dependence for $\lambda(T)$. It follows that the famous empirical approximation $\lambda(T) = \lambda(0)[1 - (T/T_c)^4]^{-1/2}$ cannot be expected to apply to all materials equally well.

3-1.3 Exact Solution by Fourier Analysis

A convenient technique for obtaining an exact solution is to apply Fourier analysis to $\mathbf{J}$ and $\mathbf{A}$, and use (2-101) to obtain a self-consistent solution. Only a one-dimensional Fourier analysis is required, since J_x and A_x are functions only of z for the penetration of a magnetic field B_y parallel to a planar surface. Some care is needed in handling the surface, however, since our expressions for $K(q)$ are valid only in an infinite medium. This problem is handled by the mathematical artifice of introducing externally supplied source currents in the interior of the infinite medium to simulate the field applied at a surface.

Consider, for example, the case in which electrons are assumed to be *specularly reflected* at the surface. If one introduces a current sheet

$$J_{x,\,\text{ext}} = -\frac{c}{2\pi} B_0\, \delta(z) \qquad (3\text{-}5)$$

this introduces a discontinuity $2B_0$ in h_y. This can be taken symmetric about zero, so that h_y switches from $-B_0$ to $+B_0$. Now when the superconductive medium is introduced, its diamagnetic currents will screen out these fields in a length λ (to be determined). Note that electrons passing through this plane at $z = 0$ without scattering have had a past exposure along their trajectory to a vector potential exactly the same as that seen by electrons specularly reflected at the surface in the actual case, since $\mathbf{A}(-z) = \mathbf{A}(z)$. (See Fig. 3-1.) Thus, the net supercurrent induced in them should also be the same, and the simulation should be effective.

Having replaced the surface by a current sheet in an infinite medium, we now may proceed to use the response function $K(q)$ worked out for that case. We first note that

$$\nabla^2 \mathbf{A} = -\text{curl curl } \mathbf{A} = -\text{curl } \mathbf{h} = -\frac{4\pi}{c} \mathbf{J}_{\text{total}} = -\frac{4\pi}{c} (\mathbf{J}_{\text{ext}} + \mathbf{J}_{\text{med}})$$

For the qth Fourier component, this becomes

$$q^2 \mathbf{a}(q) = \frac{4\pi}{c} \mathbf{J}_{\text{ext}}(q) - K(q)\mathbf{a}(q)$$

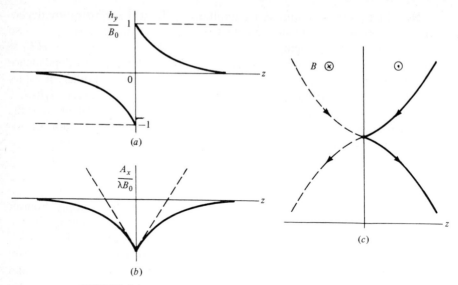

FIGURE 3-1
Simulation of surface with specular reflection by source-current sheet. (a) Magnetic field in normal (dashed) and superconducting (solid) states. (b) Vector potential in normal (dashed) and superconducting (solid) states. London gauge is used in superconducting state. (c) Electron trajectories. Solid curve shows trajectory with specular reflection; dashed parts show extensions into the other half space, with current-sheet simulation.

Solving for $\mathbf{a}(q)$, we have the general result

$$\mathbf{a}(q) = \frac{(4\pi/c)\mathbf{J}_{\text{ext}}(q)}{K(q) + q^2} \qquad (3\text{-}6)$$

For the current sheet [Eq. (3-5)], $J_{\text{ext}}(q) = -cB_0/4\pi^2$. Thus,

$$a(q) = \frac{-B_0/\pi}{K(q) + q^2}$$

We are more interested in $\mathbf{h} = \text{curl } \mathbf{A}$, so that $h_y(q) = iqa(q)$. Integrating over all the Fourier components,

$$h(z) = \frac{B_0}{i\pi} \int_{-\infty}^{\infty} \frac{qe^{iqz}\,dq}{K(q) + q^2} = \frac{2B_0}{\pi} \int_0^{\infty} \frac{q \sin qz\,dq}{K(q) + q^2} \qquad (3\text{-}7)$$

For any $K(q)$, Eq. (3-7) gives the true dependence of h on z, which will not be exactly exponential unless $K(q) = $ constant, as in the London theory. For example, the $h(z)$ computed with the $K(q)$ for either the Pippard or BCS theory actually *changes sign* deep in the interior, where $|h(z)| \ll B_0$.

To get the penetration depth, as usually defined, we integrate Eq. (3-7)

$$\lambda = B_0^{-1} \int_0^\infty h(z)\, dz = \frac{2}{\pi} \int_0^\infty \frac{q \sin qz\, dq\, dz}{K(q) + q^2}$$

or

$$\lambda_{\text{spec}} = \frac{2}{\pi} \int_0^\infty \frac{dq}{K(q) + q^2} \qquad (3\text{-}8)$$

(In carrying out the integration on z, one can replace $\int_0^Z q \sin qz\, dz = 1 - \cos qZ$ by its average value, unity, since as $Z \to \infty$, the oscillatory part effectively averages to zero in the subsequent integration over q.)

Given Eq. (3-8), we can compute λ_{spec} for any model of superconductivity which determines a $K(q)$. For example, in the London theory, $K(q) = 1/\lambda_L^2$. Then

$$\lambda_{\text{London, spec}} = \frac{2}{\pi} \int_0^\infty \frac{dq}{\lambda_L^{-2} + q^2} = \lambda_L \qquad (3\text{-}9)$$

In the Pippard theory, one has

$$K_p(q) = \frac{1}{\lambda_L^2} \frac{\xi}{\xi_0} \left\{ \frac{3}{2(q\xi)^3} [(1 + q^2\xi^2)\tan^{-1} q\xi - q\xi] \right\} \qquad (3\text{-}10)$$

This is found from Eq. (2-117) with $J_p(R, T) = e^{-R/\xi}$ by using the general relation given by Eq. (2-106). If instead one approximates the BCS kernel even more closely by $J(R, T) \approx J(0, T) \exp - [J(0, T)R/\xi_0]$, as discussed in the argument leading to Eq. (2-123), the effect is simply to replace ξ_0 by $\xi_0' = \xi_0/J(0, T)$ everywhere in Eq. (3-10), including in the definition [Eq. (2-121)] of ξ. As remarked in the previous chapter, these rather convenient, generalized Pippard forms provide quite a serviceable approximation to the exact numerical results of BCS and we shall frequently use them. However, even with the analytic expression (3-10) for $K(q)$, numerical integration is required to compute the penetration depth using (3-8). A detailed numerical investigation of the penetration depth in pure and impure superconductors using the exact BCS results was made by Miller.[1]

In order to avoid numerical calculations, considerable attention has been given to two limiting cases in which analytic results can be obtained, even though the true situation usually lies in between.

The *local approximation* replaces $K(q)$ for all q by $K(0)$, a constant, thus reducing the problem to the London form, but in general with a modified penetration depth. Using the generalized Pippard approximation

$$K(0, T) = \lambda_L^{-2} \left[1 + \frac{\xi_0}{J(0, T)\ell} \right]^{-1} \qquad (3\text{-}11)$$

one finds

$$\lambda(T) = \lambda_L(T) \left[1 + \frac{\xi_0}{J(0, T)\ell} \right]^{1/2} \qquad (3\text{-}12)$$

[1] P. B. Miller, *Phys. Rev.* **113**, 1209 (1959).

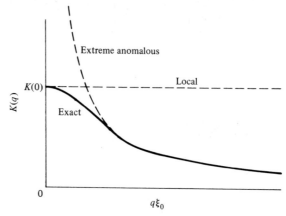

FIGURE 3-2
Schematic comparison of local and extreme anomalous approximations to the exact nonlocal response function $K(q)$.

as anticipated in (2-123). This approximation is reasonably well justified in dirty superconductors [if $\ell < \lambda(T)$] and even in pure superconductors very near T_c where $\xi_0 < \lambda(T)$.

The other approximation is the *extreme anomalous limit*, in which $K(q)$ is replaced for all q values by its asymptotic form for $q \to \infty$, where $K(q) \sim 1/q$. This approximation is reasonably well justified if $\lambda_L \gg \xi_0$, because then the dominant contribution to Eq. (3-8) will come from the q values in which this asymptotic form is valid. Figure 3-2 illustrates the two different approximations to $K(q)$. Since both approximations exceed the true $K(q)$ for some q and never err in the other direction, both will lead to lower bounds to the true value for λ.

Let us now carry out the calculation in the extreme anomalous limit. For complete generality we write

$$K(q) = \frac{a}{q}$$

where in the Pippard theory $a = 3\pi/4\lambda_L^2 \xi_0$, while in either the BCS theory or the generalized Pippard theory, ξ_0 is replaced by ξ_0' so that a is increased by a factor of $J(0, T)$. Introducing the standard notation λ_∞ for the value of λ in this limit, (3-8) becomes

$$\lambda_{\infty,\,\text{spec}} = \frac{2}{\pi} \int_0^\infty \frac{dq}{(a/q) + q^2}$$

Making a change of variable to $x = q^3/a$, this becomes

$$\lambda_{\infty,\,\text{spec}} = \frac{2}{3\pi a^{1/3}} \int_0^\infty \frac{x^{-1/3}\,dx}{1 + x} = \frac{4}{3\sqrt{3}\,a^{1/3}}$$

Inserting the value for a, we have

$$\lambda_{\infty,\,\text{spec}} = \frac{8}{9} \frac{3^{1/6}}{(2\pi)^{1/3}} (\lambda_L^2 \xi_0')^{1/3} = 0.58(\lambda_L^2 \xi_0')^{1/3} \qquad (3\text{-}13)$$

which has exactly the form anticipated in Eq. (3-4) by an elementary argument. Since the BCS correction factor $(\xi_0'/\xi_0)^{1/3} = [J(0, T)]^{-1/3}$ to the simple Pippard form varies smoothly from 1 at $T = 0$ to 0.91 at T_c, it has little effect on the behavior of the result.

If the surface scattering is taken as *diffuse* instead of specular, formulas are obtained that differ only in detail from those given above. In this case the prescription for handling the surface is simply to cut off the integration over $\mathbf{r}'$ at the surface in the coordinate-space form [Eq. (2-117)] of the response function. The physical reasoning is that electrons coming to $\mathbf{r}$ from the surface do so with no memory of any previous exposure to the field. When this prescription is transcribed into Fourier transform language, it turns out that Eq. (3-8) is replaced by

$$\lambda_{\text{diff}} = \frac{\pi}{\int_0^\infty \ln\left[1 + K(q)/q^2\right] dq} \qquad (3\text{-}14)$$

Although this looks quite different from Eq. (3-8), it actually gives exactly the same result [Eq. (3-12)] in the local approximation, and for λ_∞ it differs from Eq. (3-13) only in that the factor of 8/9 is missing. Thus, there is little difference in the results for these two different limiting assumptions about the surface scattering. This is fortunate, since the actual nature of the scattering is uncertain and probably neither limit is really justified.

Some perspective on the applicability of these results to real metals can be gained by consideration of numerical parameters for several pure materials well below T_c. Aluminum is well approximated by λ_∞, since $\lambda_L \approx 160$ Å, while $\xi_0 \approx 15,000$ Å, so that $\xi_0/\lambda_L \approx 100$. On the other hand, for tin $\lambda_L \approx 350$ Å and $\xi_0 \approx 3,000$ Å, so that λ_∞ is only a moderately good approximation. For lead, $\lambda_L \approx \xi_0/2$, and the London local approximation is actually better than λ_∞. Thus, for pure metals it is almost always necessary to resort to numerical integration of Eq. (3-8) or Eq. (3-14) to get really quantitative results. On the other hand, for alloys the local limit is often quite a good approximation.

3-1.4 Temperature Dependence of λ

As noted above, there can be no universal temperature dependence of $\lambda(T/T_c)/\lambda(0)$ in the BCS theory because of the variation of the ratio $\xi_0/\lambda_L(0)$ for different metals. In general, numerical calculations are required for each case. On the other hand, the differences are not so great as to be inconsistent with the approximately universal temperature dependence observed experimentally, especially when one takes notice of the fact that the measurements are not extremely accurate, and cannot be extended to $T = 0$. For pure metals, the two obvious limiting cases are for $\xi_0/\lambda_L(0) = 0$ and $\xi_0/\lambda_L(0) \to \infty$, corresponding to the local and extreme anomalous limits treated above. In the local limit, the temperature dependence is simply that of $\lambda_L(T)$, as given by (2-111). In the extreme anomalous limit, (3-13) implies

$$\frac{\lambda_\infty(T)}{\lambda_\infty(0)} = \left[\frac{\lambda_L^2(T)}{\lambda_L^2(0)J(0, T)}\right]^{1/3} = \left[\frac{\Delta(T)}{\Delta(0)}\tanh \tfrac{1}{2}\beta\,\Delta(T)\right]^{-1/3} \tag{3-15}$$

where the second form involves the relation (2-125) and provides a definition of $J(0, T)$ in terms of $\lambda_L(T)$ and $\Delta(T)$, quantities already defined by (2-111) and (2-53) and included in the tabulation of Mühlschlegel.[1] Finally, in the dirty $(\ell \ll \xi_0)$ local limit (3-12) reduces to

$$\lambda_{\text{eff}}(T) = \lambda_L(T)\left[\frac{\xi_0}{J(0, T)\ell}\right]^{1/2} \tag{3-16}$$

as anticipated in (2-120), so that the expected temperature dependence is

$$\frac{\lambda_{\text{eff}}(T)}{\lambda_{\text{eff}}(0)} = \frac{\lambda_L(T)}{\lambda_L(0)J^{1/2}(0, T)} = \left[\frac{\lambda_\infty(T)}{\lambda_\infty(0)}\right]^{3/2} \tag{3-17}$$

In Fig. 3-3 these various dependences are compared with the empirical "two-fluid" approximation

$$\frac{\lambda(T)}{\lambda(0)} = \left[1 - \left(\frac{T}{T_c}\right)^4\right]^{-1/2} \equiv y\left(\frac{T}{T_c}\right) \tag{3-18}$$

Evidently the temperature dependence of λ_∞ comes closest of the various theoretical forms to agreeing with (3-18). This is satisfying, since we expect λ_∞ to be the most appropriate approximation for the typical pure superconductors such as tin on which the most careful measurements have been made. Recalling that very near T_c, $\lambda(T)$ must eventually exceed ξ_0, it is clear that the approximation λ_∞ must break down there. In particular, the infinite slope of λ_∞^{-2} at T_c would be replaced by the finite slope of λ_L^{-2} in the exact calculation. Note that the different normalization of λ_L and λ_∞ in Fig. 3-3 implies that the slope of λ_L^{-2} near T_c should be increased by a factor $\sim 0.4[\xi_0/\lambda_L(0)]^{2/3}$ before being compared with λ_∞^{-2}. For the typical value $\xi_0/\lambda_L(0) = 10$, this almost doubles the slope, bringing it very close to the empirical form. Thus the full microscopic theory would give a temperature dependence considerably closer to the empirical law than λ_∞ by itself.

[1] B. Mühlschlegel, Z. Phys. **155**, 313 (1959).

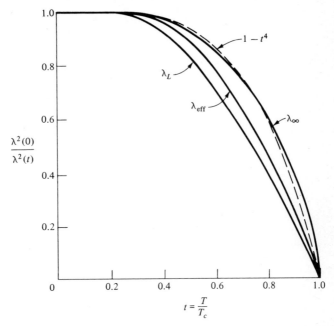

FIGURE 3-3

Comparison of the predicted temperature dependences for $1/\lambda^2$ in various limiting cases of the BCS theory. The dashed curve depicts the empirical approximation (3-18).

3-2   PENETRATION DEPTH IN THIN FILMS

The theory of the penetration depth in thin films of nonlocal superconductors is of interest on several accounts. For one, many of the earlier measurements of $\lambda(T)$ were made intentionally on thin films of thickness $d \sim \lambda$ so that there would be sizable changes in the magnetic moment with changes of T. One must ask how representative of bulk samples such measured values of λ would be. Another reason for interest is that many experiments on superconductors in magnetic fields are performed on thin films, and one wishes to understand the interaction between field and film as accurately as possible. Finally, analysis of the behavior of thin films illustrates further consequences of the nonlocal electrodynamics which are not so apparent in the classic analysis of the field penetration at the surface of a bulk superconductor.

For simplicity, we shall consider only cases in which the film is sufficiently thin that one may approximate the self-consistent vector potential by that of the applied field alone. We consider two cases: that of a field parallel to the surface

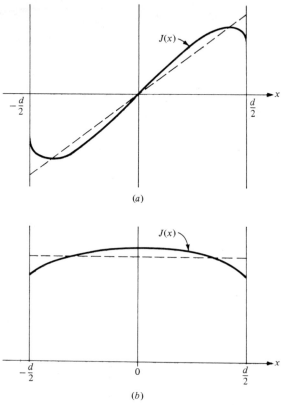

FIGURE 3-4
Distribution of current density through the thickness of a film subjected to a magnetic field. Solid curve is for nonlocal theory with $\xi \gg d$; dashed curve is for a local superconductor (so that $J \propto A$) with same λ_{eff}. (a) Refers to the case with field parallel to the film; (b) refers to the perpendicular-field case.

and that of a field perpendicular to the surface. In both cases we take x as the direction normal to the surface of the film, which is bounded by planes at $x = \pm d/2$. The field direction will be called z in the parallel case and, of course, x in the perpendicular case. In the parallel case, the vector-potential gauge choice is the symmetric one mentioned above, namely,

$$A_y = Hx \qquad \mathbf{H} \parallel \text{film} \qquad (3\text{-}19)$$

Since this changes sign on going through the film, it is clear that care will be required in evaluating the nonlocal current response, which tends to average $\mathbf{A}(\mathbf{r}')$ over a volume. In the perpendicular case, the gauge choice is less obvious, pending further study of the overall response of a film to a perpendicular field. However, since any variation of **A** with x would not affect the field (in the x direction), and since we take the thickness d to be very small, it is reasonable to take **A** to be independent of x. We also assume that it varies sufficiently slowly in the yz plane, on the scale of ξ, that we may simply take[1]

$$A_y = |\mathbf{A}| = \text{const} \qquad \mathbf{H} \perp \text{film} \qquad (3\text{-}20)$$

The resulting current distributions for the two orientations are shown in Fig. 3-4. In the London, or local, approximation, $J_y(x)$ is proportional to $A_y(x)$, as indicated by the dashed lines. It is not so simple, however, in the nonlocal theory, and we now work out how the curves plotted in Fig. 3-4 were found.

3-2.1 Diffuse Surface Scattering

Starting from the coordinate-space form of the nonlocal response (2-117), we can reduce the computation of $J_y(x)$ to a one-dimensional quadrature by first carrying out the integration over the yz plane once and for all. With the x integration cut off at $\pm d/2$, as is appropriate for diffuse surface scattering [see discussion leading to (3-14)], we have

$$J_y(x) = -\int_{-d/2}^{d/2} K(|x - x'|)A_y(x')\,dx' \qquad (3\text{-}21)$$

where
$$K(X) = \frac{3c}{16\pi^2 \xi_0 \lambda_L^2} \iint \frac{Y^2 J(R, T)}{R^4}\,dY\,dZ$$

After carrying out the trivial angular integration in cylindrical coordinates with $\rho^2 = Y^2 + Z^2 = R^2 - X^2$, and then reexpressing R as βX, $K(X)$ can be written as

$$K(X) = \frac{3c}{16\pi\xi_0 \lambda_L^2} \int_1^\infty (\beta^{-1} - \beta^{-3})J(\beta X, T)\,d\beta \qquad (3\text{-}22)$$

If we now make the generalized Pippard approximation [see discussion leading to Eq. (2-123)] that

$$J(R, T) = \left(\frac{\xi_0}{\xi'_0}\right)e^{-R/\xi'} \qquad (3\text{-}23)$$

Eq. (3-22) can be expressed in terms of the tabulated exponential-integral functions E_i as

$$K(X) = \frac{3c}{16\pi\xi'_0 \lambda_L^2}\left[E_1\left(\frac{X}{\xi'}\right) - E_3\left(\frac{X}{\xi'}\right)\right] \qquad (3\text{-}24)$$

---

[1] In the vortex state, **A** may vary on the scale of the Ginzburg-Landau length $\xi(T)$, but near T_c this is still much greater than the Pippard length ξ.

Taking asymptotic forms for these integrals, we find that

$$K(X) \rightarrow \frac{3c}{16\pi\xi_0'\lambda_L^2}\left[\ln\left(\frac{\xi'}{X}\right) - 1.077 + \frac{2X}{\xi'} + \cdots\right] \qquad \frac{X}{\xi'} \rightarrow 0 \qquad (3\text{-}24a)$$

$$K(X) \rightarrow \frac{3c}{8\pi\xi_0'\lambda_L^2}\frac{\xi'^2}{X^2}e^{-X/\xi'} \qquad \frac{X}{\xi'} \rightarrow \infty \qquad (3\text{-}24b)$$

Thus, $K(X)$ is a peaked function, which diverges logarithmically at $X = 0$ and vanishes exponentially for large X.

Having reduced the problem to the form given by Eq. (3-21) and having evaluated $K(X)$, we can now calculate $J(x)$ for any given $A(x')$. Since $J(x)$ will modify $A(x')$, in general one must search for a self-consistent solution for J and A. However, since we content ourselves with the case in which the film is very thin, we may take $A(x)$ to be simply the vector potential due to the applied field, as discussed above. Using these, the current distributions sketched in Fig. 3-4 were calculated for the limiting case $d \ll \xi'$. Note that in this case, $X \le d \ll \xi'$, so that the limiting form given in Eq. (3-24a) is always a good approximation.

Now that we have found these current distributions, we are prepared to attack the original question: how well can we characterize this response in terms of a single parameter playing the role of a penetration depth? We choose to do this in terms of an effective London (or local) penetration depth λ_{eff} which would give a current distribution with the same energy, since that is the fundamental quantity in such contexts as the Ginzburg-Landau theory. In general, the work done in setting up a current distribution is

$$W = \iint \mathbf{J}(\mathbf{r}, t) \cdot \mathbf{E}(\mathbf{r}, t) \, dt \, d\mathbf{r} = -\frac{1}{c}\iint \mathbf{J}(\mathbf{r}, t) \cdot \frac{\partial \mathbf{A}(\mathbf{r}, t)}{\partial t} \, dt \, d\mathbf{r}$$

$$= -\frac{1}{2c}\int \mathbf{J}(\mathbf{r}) \cdot \mathbf{A}(\mathbf{r}) \, d\mathbf{r} \qquad (3\text{-}25)$$

if we have a linear problem so that $\mathbf{J} \propto \mathbf{A}$. For our nonlocal case, then, the energy per unit area of film is

$$W = \frac{1}{2c}\iint A(x)A(x')K(|x - x'|) \, dx \, dx' \qquad (3\text{-}26)$$

On the other hand, our energetically equivalent London superconductor has

$$J(x) = -\frac{c}{4\pi\lambda_{\text{eff}}^2}A(x) \qquad (3\text{-}27)$$

so that (3-25) leads to an energy per unit area of

$$W = (8\pi\lambda_{\text{eff}}^2)^{-1}\int A^2(x) \, dx \qquad (3\text{-}28)$$

Equating this to (3-26), we find

$$\frac{1}{\lambda_{\text{eff}}^2} = \frac{4\pi}{c} \frac{\iint A(x)A(x')K(|x-x'|) \, dx' \, dx}{\int A^2(x) \, dx} \qquad (3\text{-}29)$$

Approximating K by the leading term of (3-24a), this becomes

$$\frac{\lambda_L^2}{\lambda_{\text{eff}}^2} = \frac{3}{4\xi_0'} \frac{\iint A(x)A(x') \ln (\xi'/|x-x'|) \, dx' \, dx}{\int A^2(x) \, dx} \qquad (3\text{-}30)$$

Since the integrands in the double and single integrals are of the same order of magnitude, the ratio of the integrals will be of the order of the range of integration, d. Thus, we expect

$$\lambda_{\text{eff}} \sim \lambda_L \left(\frac{\xi_0'}{d}\right)^{1/2} \qquad (3\text{-}31)$$

a result that is reminiscent of that (3-16) for a dirty superconductor with mean free path $\ell \sim d$.

Carrying through the evaluation of (3-30) more carefully for the perpendicular-field case, $A(x)$ is a constant and simply cancels out. Upon performing the integrations, one finds that the leading term gives $d \ln (\xi'/d)$, so that

$$\lambda_{\text{eff}, \perp} = \lambda_L \left[\frac{4\xi_0'}{3d \ln (\xi'/d)}\right]^{1/2} \qquad d \ll \xi' \qquad (3\text{-}32a)$$

On the other hand, if one carries through the integration of (3-30) for the parallel-field case, where $A = Hx$, one finds

$$\lambda_{\text{eff}, \parallel} = \left(\frac{4\lambda_L}{3}\right)\left(\frac{\xi_0'}{d}\right)^{1/2} \qquad d \ll \xi' \qquad (3\text{-}32b)$$

This differs from the previous expression by a factor of

$$\frac{\lambda_\parallel}{\lambda_\perp} = \left[\frac{4}{3} \ln \left(\frac{\xi'}{d}\right)\right]^{1/2} \qquad (3\text{-}33)$$

which always exceeds unity in its range of validity, $d \ll \xi'$.

Little detailed practical significance should be attached to this result, since it was derived using a leading term that is valid only in the limiting case of very thin films; moreover, it assumes ideal geometry with strictly plane-parallel diffuse-reflecting surfaces, which is very unrealistic for such thin films. Rather, our objective in working this example through in such detail was to illustrate the impossibility of rigorously characterizing the nonlocal electrodynamic response by a single parameter λ_{eff}. Not only have we had to make an arbitrary definition of λ_{eff} in terms of the energy, but the result depends on the configuration of the field, not just on the nature of the superconducting sample. Despite this limitation on any *rigorous* characterization by a λ_{eff}, we can in *practice* use a relation of the sort (3-31) to describe quite well the reduced electrodynamic response of a thin film of a nonlocal superconductor, because (3-32a) and (3-32b) actually do not differ very much from each other. Moreover, we expect the difference to be much less in a real film, in which volume scattering by impurities and grain boundaries actually keeps ξ' from exceeding d by any large factor and in which surface irregularities preclude the very long free paths in the plane of the film which give the logarithmic factor in the ideal problem.

3-2.2 Specular Surface Scattering

For completeness, we now sketch the alternate treatment which would be used if one assumed specular rather than diffuse scattering at the surface. As in our treatment of the penetration depth at the surface of a bulk sample, we must convert the problem to an equivalent one defined by suitable current sheets inserted in an infinite medium. For the parallel-field geometry, this requires an infinite succession of current sheets, of alternating sense and spaced d apart, which produce a square-wave $h(x)$ and saw-tooth $A(x)$ in the absence of the superconductive screening. Since this field is periodic, the Fourier integral reduces to a Fourier series, and it is a good approximation to retain only the lowest term in the series, for which $q = \pi/d$. The response is then determined by $K(\pi/d) = \lambda_L^{-2}(3d/4\xi_0')$ in the extreme anomalous limit of the BCS or generalized Pippard forms. Identifying this value of K with $1/\lambda_{\text{eff}}^2$, we have

$$\lambda_{\text{eff, spec, }\|} = \lambda_L \left(\frac{4\xi_0'}{3d}\right)^{1/2} \qquad (3\text{-}34a)$$

which differs from the corresponding result for the diffuse case [Eq. (3-32b)] by a factor of $(3/4)^{1/2}$, rather similar to the factor of 8/9 for the bulk-sample case.

For the perpendicular-field case, however, the situation is radically different. In this case, $\mathbf{A}$ is parallel to the film and constant over its thickness, so the continuation to an equivalent problem without boundaries leads to a situation with $\mathbf{A}$ constant over all space. Thus, $\mathbf{q} = 0$, and

$$\lambda_{\text{eff, spec, }\perp} = \lambda_L \qquad (3\text{-}34b)$$

if the material is pure; more generally, λ will be given by Eq. (3-12). This result for a superconducting film corresponds to the fact that specular boundary scattering in a normal film does not contribute to its ordinary electrical resistance either. Experimental evidence indicates that specular reflection seldom plays an important role in films, either in the normal or the superconducting states. Thus, our earlier results [Eq. (3-32)] for diffuse scattering are more representative of the true state of affairs.

3-3 MEASUREMENT OF λ

Before leaving the subject of the penetration depth, let us very briefly review some of the experimental techniques which have been used to measure this quantity.

The earliest experiments[1] generally involved use of large numbers of colloidal particles or thin films with a small dimension d comparable with λ. Varying the temperature [and hence $\lambda(T)$] then caused substantial fractional changes in the magnetic susceptibility, which could be measured. To the extent that the particle-size distribution of the sample was known, this permitted inferences to be made about $\lambda(d, T)$, but evidently there were quantitative uncertainties.

It was pointed out by Casimir[2] as early as 1940 that an ac susceptibility technique should be sufficiently sensitive to allow the change in field penetration $\lambda(T) - \lambda(T \approx 0)$ at the surface of a single bulk sample to be measured. This experiment was first carried out successfully by Laurmann and Shoenberg[3] using a mutual-inductance bridge operating at 70 Hz (hertz). Their sensitivity was not great enough to follow the changes in $\lambda(T)$ very far below T_c, but their results generally confirmed the empirical form (3-18) which had been noted in the earlier experiments on small particles. A tenfold increase in sensitivity (changes of λ as small as 2 Å could be detected) was obtained by Pippard[4] by raising the frequency to 10^9 to 10^{10} Hz and using microwave techniques to measure the temperature-dependent shift in resonant frequency of a cavity. This allowed him to follow changes of λ down to quite low temperatures, but did require corrections for normal-electron effects which became increasingly serious at the higher frequencies. By using the normal-state skin depth (which was quite well determined from other data) as a reference, he was even able to obtain a measurement of the absolute value of λ, not just of the changes of $\lambda(T)$, but this was of lesser accuracy.

After the BCS theory appeared, Schawlow and Devlin[5] remeasured the temperature dependence of λ in tin, this time working at about 10^5 Hz, where the finite-frequency corrections are negligible but where the sensitivity (using a frequency counter to measure small changes in frequency of the resonator in an oscillator circuit) was still as high as in the microwave measurements of Pippard. While their results generally followed the empirical relation $\lambda \propto y$, where $y = (1 - t^4)^{-1/2}$ with $t = T/T_c$, a plot of $d\lambda/dy$ versus y showed a rather sharp rise of this quantity below $y \approx 1.5$. Exactly this sort of behavior was predicted by BCS, because of the exponential cutoff of excitations due to the gap. Although there is not complete quantitative agreement between theory and experiment, this probably can be attributed to the fact that the theory is for an idealized isotropic

[1] For example, D. Shoenberg, *Proc. Roy. Soc.* (*London*) **A175**, 49 (1940); J. M. Lock, *Proc. Roy. Soc.* (*London*) **A208**, 391 (1951).

[2] H. B. G. Casimir, *Physica* **7**, 887 (1940).

[3] E. Laurmann and D. Shoenberg, *Nature* **160**, 747 (1947); *Proc. Roy. Soc.* (*London*) **A198**, 560 (1949).

[4] A. B. Pippard, *Proc. Roy. Soc.* (*London*) **A191**, 399 (1947); **203**, 98 (1950).

[5] A. L. Schawlow and G. E. Devlin, *Phys. Rev.* **113**, 120 (1959).

metal, whereas the real metal has a complex Fermi surface, with anisotropy of gap and penetration depth. Waldram[1] has given a careful review of the results of the various high-frequency measurements of the penetration depth and surface resistance of superconductors.

It is important to bear in mind that the most precise methods measure only *changes* in $\lambda(T)$ with temperature. It has proved very difficult to devise any method which can determine the absolute value of $\lambda(T)$ with comparable accuracy. Thus, when comparing theory with experiment, one is usually comparing values of $[\lambda(T) - \lambda(0)]$. As a result, most quoted "experimental" values of $\lambda(0)$ are actually values inferred from fitting data to a theory of $\lambda(T)$. Most commonly, it has been the coefficient in the old empirical relation $\lambda(t) = \lambda(0)y(t)$, with y as defined above. More recently, efforts have been made to fit to curves computed with the BCS theory. This is much less convenient, however, since it involves a numerically computed function with *two* adjustable parameters $\lambda(0)$ and ξ_0 even in the simple isotropic approximation.

A final remark: the advent of superconducting quantum-interference devices, to be discussed later, has increased the sensitivity of dc flux measurements to such an extent that dc measurements of $\Delta\lambda(T)$ can be made with sensitivity equal to that of the microwave techniques. This technique, as well as the high-frequency ones, is now being used to extend the limited amount of data presently available on the anisotropy of the penetration depth with crystallographic direction. Such data should allow a further test of the theory of superconductivity as generalized to deal with real, anisotropic metals.

3-4 SUPERCONDUCTORS IN STRONG MAGNETIC FIELDS: THE INTERMEDIATE STATE

We now consider the effect of fields which are strong enough to destroy superconductivity, rather than simply induce screening currents to keep the field out of the interior of the sample. The effect of such fields depends on the *shape* of the sample. The simplest case is that of a long, thin cylinder or sheet parallel to the field, because in this case the field everywhere along the surface is just equal to the applied field H_a. For other geometries, in which the demagnetizing factor of the sample is not zero, the field over part of the surface will exceed the applied field, as is illustrated in Fig. 3-5, causing some normal regions to appear while H_a is still less than H_c.

Let us now consider the relevant free energies in some detail, restricting attention at first to the simple case of zero demagnetizing factor. When the sample (of volume V) is normal, the total Helmholtz free energy is given by

$$F_n = Vf_{n0} + V\frac{H_a^2}{8\pi} + V_{\text{ext}}\frac{H_a^2}{8\pi} \qquad (3\text{-}35)$$

[1]J. R. Waldram, *Advan. Phys.* **13**, 1 (1964).

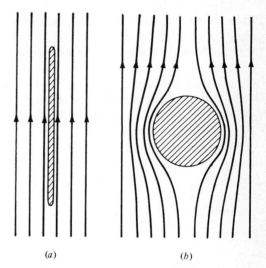

FIGURE 3-5
Contrast of exterior-field pattern (a) when demagnetizing coefficient is nearly zero and (b) when it is $\frac{1}{3}$, for a sphere. In (b) the equatorial field is $\frac{3}{2}$ the applied field for the case of full Meissner effect, which is shown.

(a)

(b)

where f_{n0} is the free-energy density in the normal state in the absence of the field, and the terms in H_a^2 denote the energy of the field inside and outside the sample, respectively. When the sample is superconducting, the Meissner effect excludes the field from the interior, so

$$F_s = Vf_{s0} + V_{ext}\frac{H_a^2}{8\pi} \qquad (3\text{-}36)$$

where f_{s0} is the free-energy density in the superconducting state. (We assume macroscopic sample dimensions, so that it is permissible to ignore the effects of field penetration and currents in a layer λ deep on the surface.) Taking the difference, we have

$$F_n - F_s = V(f_{n0} - f_{s0}) + \frac{VH_a^2}{8\pi}$$

$$= V\left(\frac{H_c^2}{8\pi}\right) + V\left(\frac{H_a^2}{8\pi}\right) \qquad (3\text{-}37)$$

In the second form we have used the defining relation for the thermodynamic critical field H_c

$$f_{n0} - f_{s0} = \frac{H_c^2}{8\pi} \qquad (3\text{-}38)$$

In particular, when $H_a = H_c$, (3-37) becomes

$$F_n - F_s \big|_{H_c} = V\left(\frac{H_c^2}{4\pi}\right) \qquad (3\text{-}39)$$

Thus, at the transition from the superconducting to the normal state, the free energy F increases by $H_c^2/4\pi$ per unit volume. Where does the energy come from? It comes from the energy source maintaining the constant field doing work against the back emf induced as the flux enters the sample, as can be shown by an elementary computation.

The reason for the awkward necessity of considering the energy of the source of the field is that we carried out the above discussion in terms of the Helmholtz free energy. This free energy is appropriate for situations in which $\mathbf{B}$ is held constant rather than $\mathbf{H}$, because if $\mathbf{B}$ is constant, there is no induced emf and no energy input from the current generator. The appropriate thermodynamic potential for the case of constant $\mathbf{H}$ is the Gibbs free energy G. This differs from F by the term $-V(BH/4\pi)$, which essentially accounts automatically for the work done by the generator. Thus, we consider the free-energy density

$$g = f - \frac{hH}{4\pi} \qquad (3\text{-}40)$$

This leads to

$$G_n = Vf_{n0} - \frac{VH_a^2}{8\pi} - \frac{V_{\text{ext}}H_a^2}{8\pi} \qquad (3\text{-}41)$$

since $\mathbf{h} = \mathbf{B} = \mathbf{H}$ in the normal state and outside the sample, while

$$G_s = Vf_{s0} - \frac{V_{\text{ext}}H_a^2}{8\pi} \qquad (3\text{-}42)$$

since $\mathbf{h} = \mathbf{B} = 0$ in the superconducting state. Taking the difference

$$G_n - G_s = V(f_{n0} - f_{s0}) - \frac{VH_a^2}{8\pi} \qquad (3\text{-}43)$$

Since the requirement for phase equilibrium is the equality of the normal and superconducting values of the appropriate thermodynamic potential, which is G for the case of fixed $\mathbf{H}$, Eq. (3-43) together with the definition of H_c given in Eq. (3-38) implies that $H_a = H_c$ is the condition for coexistence of superconducting and normal phases in equilibrium.

3-4.1 Nonzero Demagnetizing Factor

Our above discussion was, of course, an idealization for the sake of simplicity. Any real sample will have a nonzero demagnetizing factor, which will cause the field at the surface to be different from H_a, the uniform applied field at large distances from the sample. As a concrete example, let us treat the case of a spherical superconducting sample of radius R. On a macroscopic scale, we still have $\mathbf{B} = 0$ inside the superconductor, at least for $H_a \ll H_c$. Outside, the field satisfies

$$\nabla \cdot \mathbf{B} = \nabla \times \mathbf{B} = \nabla^2\mathbf{B} = 0 \qquad (3\text{-}44a)$$

with boundary conditions

$$\mathbf{B} \to \mathbf{H}_a \quad \text{as} \quad r \to \infty \quad (3\text{-}44b)$$

$$B_n = 0 \quad \text{at} \quad r = R \quad (3\text{-}44c)$$

where B_n is the normal component of **B**. This is a standard boundary-value problem, with exterior solution

$$\mathbf{B} = \mathbf{H}_a + \frac{H_a R^3}{2} \nabla \left(\frac{\cos \theta}{r^2} \right) \quad (3\text{-}45)$$

where θ is the polar angle measured from the direction of $\mathbf{H}_a$. It can readily be verified that Eq. (3-45) satisfies all the conditions of Eq. (3-44), including $B_n = 0$. Similarly, a direct calculation shows that the surface tangential component of **B** is

$$(B_\theta)_R = \tfrac{3}{2} H_a \sin \theta \quad (3\text{-}46)$$

Note that this exceeds H_a over an equatorial band of angles from $\theta = 42°$ to $138°$. At the equator, $B_\theta = 3H_a/2$, so that the equatorial field reaches H_c as soon as H_a reaches $2H_c/3$. Therefore, for even slightly higher H_a, certain regions of the sphere must go normal. Still, the whole sphere cannot go normal, since if it did, the diamagnetism would disappear completely, leaving $H = H_a \approx 2H_c/3$ everywhere, a value insufficient to keep the superconductivity from reappearing. Thus for fields in the range

$$\frac{2H_c}{3} < H_a < H_c$$

there must be a coexistence of superconducting and normal regions, which, following historical usage, is called the *intermediate state*.

Generalizing to other ellipsoidal shapes (for which alone a demagnetizing factor is well defined), one expects an intermediate state whenever the applied field lies in a range

$$1 - \eta < \frac{H_a}{H_c} < 1 \quad (3\text{-}47)$$

The demagnetizing factor η as defined here ranges from zero for the limit of a long, thin cylinder or thin plate in a parallel field, to $\frac{1}{3}$ for a sphere, $\frac{1}{2}$ for a cylinder in a transverse field, and finally unity for an infinite flat slab in a perpendicular field. Because the slab in a perpendicular field always shows the intermediate state, we treat that limiting case in the greatest detail. It is also the configuration in which most experimental studies of the intermediate state have been carried out.

3-4.2   Intermediate State in a Flat Slab

Let us consider an infinite flat slab of thickness $d \gg \lambda$ in a perpendicular field, a problem first treated in a classic paper of Landau.[1] In this case, it is appropriate to

---

[1] L. D. Landau, *Phys. Z. Sowjet.* **11**, 129 (1937).

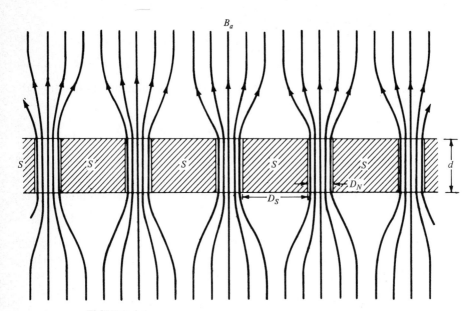

FIGURE 3-6
Schematic diagram showing magnetic flux channeling through the normal laminae in the intermediate state. Flux density is B_a at large distances, and zero or h_n ($\approx H_c$) in the cross section of the slab.

consider the average *flux* per unit area in the slab, to be fixed by the source at a value B_a, which will equal the external field far enough from the slab that any inhomogeneities induced by the slab will average out. (See Fig. 3-6.) Since $h = 0$ in the superconducting regions, the fraction ρ_n of normal material must then be related to the flux density h_n in it by

$$\rho_n = \frac{B_a}{h_n} \qquad (3\text{-}48)$$

Neglecting surface-energy effects, $h_n = H_c$, as expected from (3-43), but there will be corrections to this value.

To find the scale and shape of the superconducting and normal regions, it is necessary to consider surface energies in the energy balance. We group these into two terms: F_1, arising from the interfaces within the slab between superconducting (S) and normal (N) domains, and F_2, dominantly dependent on the behavior near the interface between the sample and the space outside.

Since the BCS theory, as discussed so far, is limited to situations in which the density of superconducting pairs is constant in space, we must anticipate our later development of the Ginzburg-Landau theory (which can deal with spatial gradients) by introducing a phenomenological surface-energy term associated with the NS interface. This is usually expressed in terms of a length, which we shall denote[1] $\delta(T)$, such that the additional energy per unit area of interface is

$$\gamma = \frac{H_c^2}{8\pi} \delta \qquad (3\text{-}49)$$

Since it will turn out that $\delta \approx \xi - \lambda$, δ is positive and of order 10^{-5} to 10^{-4} cm for typical pure superconductors. Physically speaking, δ is the thickness of the interface region which is neither fully superconductive nor fully normal. Of course, if the surface energy were negative, one could not have a stable equilibrium of macroscopic volumes of the two phases. Instead, the interfaces would proliferate to gain negative surface energy until limited by the domain-wall thickness. This is essentially what we will find happens in type II superconductors. For the present, then, we take γ to be positive.

Given a positive interface energy γ, the domain walls will assume a configuration of minimum area, all other considerations being equal. In particular, given a pattern of normal regions at the surface (which determines the energy F_2 mentioned above), the domain walls will to a good approximation run straight through the slab, perpendicular to the surface, since any other choice would increase F_1. It is less clear what the two-dimensional domain pattern should be, since it involves optimizing the trade-off between interface energy within the sample (F_1) and field energy just outside the sample (F_2). No rigorous solutions have been found in any generality; rather, various models are compared. From such studies[2] it has become clear that the free-energy differences are small between even radically different geometries (such as laminae and tubes of normal material), so long as the scale of the structure is optimized in each case. This suggests that many different configurations will be found, depending on the exact experimental conditions and the sample quality, and this expectation is borne out in practice.

Because of its analytical simplicity, and because it is representative of actual observed structures, we shall concentrate on an analysis of the laminar model of the intermediate state. In this model, there is a one-dimensional array of alternating N and S domains, of thickness D_n and D_s, with period $D = D_n + D_s$, as illustrated in Fig. 3-6. The interface energy F_1 per unit area of the slab is then readily seen to be

$$F_1 = \frac{2d\gamma}{D} = \frac{2d\delta}{D} \frac{H_c^2}{8\pi} \qquad (3\text{-}50)$$

By itself, this term favors a very coarse structure, but the exterior term F_2 works in the opposite direction, as we now show.

Although numerical calculations of F_2 were carried out by Landau and his coworkers, we shall content ourselves with a simple physical argument which

---

[1] Another common notation is $\Delta(T)$; we avoid this to prevent confusion with the energy gap.
[2] See, for example, E. R. Andrew, *Proc. Roy. Soc.* (*London*) **A194**, 98 (1948); R. N. Goren and M. Tinkham, *J. Low Temp. Phys.* **5**, 465 (1971).

gives quite similar results. The dominant contribution to F_2 is the energy of the nonuniform external magnetic field outside the domain structure of the intermediate state relative to that of the uniform field which is there if the sample is in the normal state or has an infinitely finely divided domain pattern. At the surface, the average energy density of the field is

$$\frac{\rho_n h_n^2}{8\pi}$$

since a fraction ρ_n of the volume has a field h_n, whereas, using (3-48), the energy density of the uniform field is

$$\frac{B_a^2}{8\pi} = \frac{\rho_n^2 h_n^2}{8\pi}$$

Thus the average *excess* energy density at the surface due to the domain structure is

$$\frac{(\rho_n - \rho_n^2)h_n^2}{8\pi} = \frac{\rho_n \rho_s h_n^2}{8\pi} \qquad (3\text{-}51)$$

where $\rho_s = 1 - \rho_n$ is the superconducting fraction. Above the surface, the field inhomogeneity leading to this excess energy will be substantially reduced in a "healing length" L, which will be of the order of the lesser of the lengths D_n and D_s. A convenient mathematical form embodying this observation is

$$L = (D_n^{-1} + D_s^{-1})^{-1} = \frac{D}{\rho_n^{-1} + \rho_s^{-1}} = D\rho_s \rho_n$$

Approximating F_2 by the excess energy density (3-51) out to a distance L on either side of the slab, we have

$$F_2 = \frac{2\rho_n^2 \rho_s^2 D h_n^2}{8\pi} \qquad (3\text{-}52)$$

If we now minimize the sum of F_1 and F_2 with respect to D, we find

$$D = \frac{(d\delta)^{1/2} H_c}{\rho_n \rho_s h_n} \approx \frac{(d\delta)^{1/2}}{\rho_n \rho_s} \qquad (3\text{-}53)$$

as the period of the domain structure. Note that its order of magnitude is set by the geometric mean of a macroscopic dimension, the sample thickness d, and a microscopic dimension, the domain-wall thickness δ. For typical values, $D \approx 10^{-2}$ cm. Another characteristic feature is that the number of domains becomes small (that is, D becomes large) when either ρ_n or ρ_s is small, that is, near $B_a = 0$ or H_c.

Such domain patterns have been observed experimentally[1] by such varied

---

[1] Access to this extensive literature is provided by the review of J. D. Livingston and W. DeSorbo, in R. D. Parks (ed.), "Superconductivity," chap. 21, Marcel Dekker, New York, 1969.

techniques as: (1) moving a tiny magnetoresistive or Hall effect probe over the surface; (2) making powder patterns with either ferromagnetic or superconducting (diamagnetic) powders which outline the flux bearing regions; and (3) using the Faraday magneto-optic effect in magnetic glasses in contact with the surface. Orderly laminar patterns are favored if the magnetic field is applied at an angle to the normal, causing laminae aligned with the field direction to have less domain-wall area and hence lower energy. From measurements on such structures, values of the surface-energy parameter δ have been obtained which are in satisfactory agreement with theoretical expectations based on the Ginzburg-Landau theory. In fact, these measurements played an important role in establishing that theory in the first place.

In addition to determining the scale of the domain structure, the surface energy also depresses the critical field in the intermediate state to a value H_{cI} which is somewhat below H_c, the critical field for the case of zero demagnetizing factor. We may estimate the size of this effect by computing the surface energy $F_1 + F_2$ with the optimized domain size D given by (3-53), and adding it to the volume energy terms, appropriately weighted with ρ_n or ρ_s. The resulting average free energy per unit volume of sample is

$$f_I = \rho_s f_{s0} + \rho_n\left(f_{s0} + \frac{H_c^2}{8\pi} + \frac{h_n^2}{8\pi}\right) + \frac{F_1 + F_2}{d}$$

$$= f_{s0} + \rho_n\frac{H_c^2}{8\pi} + \frac{B_a^2}{\rho_n 8\pi} + 4(1 - \rho_n)\left(\frac{\delta}{d}\right)^{1/2}\frac{H_c B_a}{8\pi} \qquad (3\text{-}54)$$

We note first that if we neglect the surface terms, f_I has its minimum when $\rho_n = B_a/H_c$, or $h_n = H_c$, as expected in that case. When the surface energy is included, f_I has its minimum when

$$\rho_n = \left(\frac{B_a}{H_c}\right)\left[1 - 4\left(\frac{\delta}{d}\right)^{1/2}\left(\frac{B_a}{H_c}\right)\right]^{-1/2} \qquad (3\text{-}55)$$

At low fields this starts out as (B_a/H_c), but the correction becomes more important at higher fields. Since we define H_{cI} as the value of B_a for which $\rho_s \to 0$ or $\rho_n \to 1$, we have, upon solving by the quadratic formula,

$$H_{cI} = H_c\left[\left(1 + \frac{4\delta}{d}\right)^{1/2} - 2\left(\frac{\delta}{d}\right)^{1/2}\right]$$

$$\approx H_c\left[1 - 2\left(\frac{\delta}{d}\right)^{1/2}\right] \qquad d \gg \delta \qquad (3\text{-}56)$$

The numerical coefficient of the correction term depends on this particular detailed model, with all its approximations, but the general form of the result seems to hold for quite a variety of models.

In concluding our discussion of this very simplified model, we note that the flux density h_n in the normal regions is given by B_a/ρ_n. Using (3-55), we see that h_n decreases from H_c to H_{cI} as the applied field is increased from zero to its critical value. Thus, it is generally true that the field in the normal regions is somewhat less than H_c. Although this result may appear paradoxical, it simply reflects the role of the surface energies neglected in zeroth-order energy arguments, which consider only terms proportional to the volume.

Refinements The above discussion of a simplified model outlines the major features of the intermediate state in a way which is certainly semiquantitatively correct in its predictions of domain size and of H_{cI}. However, it does not deal with one important qualitative feature, namely, the spreading out of the flux before it leaves the sample. This can occur most simply by having the normal domains fan out near the surface, or by more complex branching or corrugation of the domains. All of these refinements increase the interface energy somewhat, in order to decrease the field energy by a greater amount.

   In his original treatment of the problem, Landau took account of the fanning out of the normal domains within his laminar model. The result of his numerical calculations was to replace the factor $\rho_n \rho_s h_n/H_c = \rho_s B_a/H_c$ in the denominator of (3-53) by a computed function $\phi(B_a/H_c)$ which has a qualitatively similar dependence on the applied field. Thus, this refinement has little effect on the domain size. Because of the fanning out at the surface, the flux density in the normal regions at the surface is less than the interior value, as is partially anticipated by the result of our simple model that $h_n < H_c$. It is also possible to estimate the depression of H_{cI} below H_c by considering the stability of a single isolated superconducting domain (the last one, say), taking into account the fact that the surface tension of the curved interface with the surrounding normal material effectively helps the magnetic field to destroy the last bit of superconductivity.

   Curiously, Landau seems to have been somewhat unclear on the stability of this interface and he proposed a second model,[1] in which the normal domains were assumed to branch into two, repeating as necessary, in order to spread out the flux at the surface without having the flux density in the normal regions fall below H_c. Subsequent work has shown that for samples of reasonable thickness, even a single branching would raise the free energy above that of the unbranched model because the reduction in field energy is less than the increase in interface energy. However, features resembling a Landau branching structure have been reported by Solomon and Harris[2] in lead, which has a particularly low surface-energy parameter.

   Extensive experimental observations by Faber[3] showed that a complex maze structure of corrugated normal domains was often seen. Presumably, thin

[1] L. D. Landau, *Nature* **141**, 688 (1938); *J. Phys. U.S.S.R.* **7**, 99 (1943).
[2] P. R. Solomon and R. E. Harris, *Phys. Rev.* **B3**, 2969 (1971).
[3] T. E. Faber, *Proc. Roy. Soc. (London)* **A248**, 460 (1958).

normal laminae, flat in the interior, develop a corrugation of increasing amplitude as they approach the surface. This accomplishes the effective dispersal of the emerging flux over a band whose width is equal to the amplitude of the corrugation in a way which appears to be more economical of interface energy than is the branching model of Landau. Obviously such corrugations affect the interpretation of observed domain sizes in terms of a surface-energy parameter.

It should also be mentioned that flux spots or tubes, rather than laminae, may be observed under suitable circumstances. For example, Landau pointed out that normal tubes should be more favorable at low flux density, whereas superconducting tubes should be more favorable for the last superconducting material near H_c. In recent experiments,[1] Träuble and Essmann have observed a regular array of flux spots in lead foils in a perpendicular field, whereas Kirchner[2] has observed both flux spots and lamina-like "meanders" in rather similar samples. The evolution of the flux pattern with increasing field from flux tubes to corrugations, then branches, and finally into superconducting tubes at high fields is particularly clearly demonstrated by motion pictures taken by various groups[3] using the magneto-optic technique. Evidently, the richness of the phenomena observed in the intermediate state poses a severe challenge to any complete theoretical understanding. Yet another dimension of complexity is added in the time-dependent phenomena of the dynamic intermediate state, but we shall not go into that aspect here.

3-4.3   Intermediate State of a Sphere

To illustrate the application of our results in a more general geometry, we now return to the case of the sphere. As found above, the intermediate state will exist when $2/3 < H_a/H_c < 1$. In this range, the volume of the sphere is subdivided into S and N laminae, which fan out near the surface and may branch or become corrugated, but we shall ignore these refinements in the present discussion. Moreover, we shall assume that the radius of the sphere is large enough compared to the domain-wall thickness δ that we can ignore the difference between H_{cI} and H_c. Then, the flux density in the N laminae is always exactly H_c, and the normal fraction ρ_n is B/H_c, where **B** is the average of $\mathbf{h(r)}$ over the laminar structure. In the macroscopic Maxwell equations, this average serves for **B** everywhere inside the sphere. On the other hand, the magnitude of the Maxwell **H** in the sphere throughout the intermediate state is just H_c. This follows, since $H = h = H_c$ in the normal laminas, and the tangential component of **H** is continuous across the interface between laminae since the only currents there are internal ones associated with the medium in thermodynamic equilibrium. Thus, as is

[1] H. Träuble and U. Essmann, *Phys. Stat. Sol.* **25**, 395 (1968).
[2] H. Kirchner, *Phys. Letters* **26A**, 651 (1968).
[3] See, for example, P. R. Solomon and R. E. Harris, *Proc. 12th Intl. Conf. on Low Temp. Phys.*, Kyoto, Japan, 1970, p. 475. The films described here are available on loan from these authors.

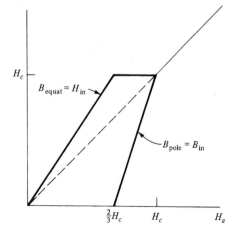

FIGURE 3-7
Internal values of B and H in superconducting sphere in an applied field H_a. As indicated, these can be measured externally by measuring the surface field at the pole and the equator, respectively. The sphere is in the intermediate state for $2H_c/3 < H_a < H_c$.

the case in more familiar examples, the macroscopic fields inside the sphere are uniform, while those outside are the sum of the applied field plus a dipole field, namely,

$$\mathbf{B} = \mathbf{H} = \mathbf{H}_a + \frac{H_1 R^3}{2} \nabla \left(\frac{\cos \theta}{r^2} \right) \qquad (3\text{-}57)$$

This has the same form as the expression [Eq. (3-45)] which we found to hold in the linear regime below $2H_c/3$; in that case the parameter H_1 was chosen to equal H_a, so as to match $B_n = 0$ at $r = R$. In the intermediate state, $B_n \neq 0$. Rather, we determine H_1 by equating the internal and external values of B_n and of H_{tang}:

$$B_n = B \cos \theta = H_a \cos \theta - H_1 \cos \theta \qquad (3\text{-}58)$$

$$H_{\text{tang}} = H_c \sin \theta = H_a \sin \theta + \frac{1}{2} H_1 \sin \theta \qquad (3\text{-}59)$$

Solving, we find $H_1 = \dfrac{2(H_c - B)}{3}$, so that

$$B = 3H_a - 2H_c \qquad \frac{2}{3} \leq \frac{H_a}{H_c} \leq 1 \qquad (3\text{-}60)$$

Thus, the magnetic induction of the sphere increases linearly from zero to H_c as the applied field H_a increases from $2H_c/3$ to H_c, as depicted in Fig. 3-7.

Because B_n is continuous, B can be measured external to the sphere by measuring B at the pole, $\theta = 0$. Similarly, the continuity of H_{tang} implies that the internal value of H can be measured externally by measuring the equatorial surface field, $B_{\text{equat}} = H_{\text{equat}}$. The predicted dependence of this quantity is also shown in Fig. 3-7. Experimental data on clean samples actually follow these predictions quite well.

3-5   CRITICAL CURRENT OF A SUPERCONDUCTING WIRE

As our final example of the electrodynamics of type I superconductors, we now discuss the appearance of resistance in a superconducting wire above its critical current. Consider a wire of radius a carrying a current I. By Maxwell's equation, the magnetic field at the surface of the wire is $2I/ca$. When this equals H_c, the wire can no longer be entirely superconducting. This defines a critical current

$$I_c = \frac{H_c ca}{2} \qquad (3\text{-}61)$$

based on Silsbee's rule that the critical current cannot exceed that which produces a critical magnetic field at the superconductor. (The critical current may be much *less* than is given by this criterion, especially if the thickness of the superconductor is much less than λ.) If $I > I_c$, then the surface field exceeds H_c, and the surface (at least) must become normal.

But if a surface layer were to go normal and leave a fully superconducting core, the current would all go through the core, leading to a still greater field at *its* surface, which would *a fortiori* be greater than H_c. Thus no stable configuration exists with a solid superconducting core surrounded by normal material. What if the sample went entirely normal? In this case, the current density J would be uniform across the cross section, leading to

$$H(r) = \frac{2Ir}{ca^2}$$

Since this drops below H_c as $r \to 0$, the core could not be wholly normal either.

These observations suggest a core region (of radius $r_1 < a$) in an intermediate state, surrounded by a normal layer which also carries current. The latter requires a longitudinal electric field, which is compatible with an intermediate-state structure, so long as its layers are oriented transverse to the axis.

The nature of the intermediate-state structure is dictated by the requirement that, neglecting surface energies, $H(r) = H_c$ for $r \le r_1$. Since $H(r) = 2I(r)/cr$, where $I(r)$ is the total current inside radius r, we need $I(r) = crH_c/2$. This requires a current density

$$J(r) = \frac{1}{2\pi r}\frac{dI}{dr} = \frac{cH_c}{4\pi r} \qquad (3\text{-}62)$$

Yet the longitudinal electric field E is independent of r, as can be seen since curl $\mathbf{E} = -(1/c)(\partial \mathbf{B}/\partial t) = 0$, if we assume the structure is stable in time. These requirements are approximately reconciled by the configuration shown in Fig. 3-8, first proposed by F. London,[1] in which the fractional path length (parallel to the axis

[1] F. London, "Une Conception nouvelle de la supraconductibilite," Hermann & Cie., Paris, 1937. A more accessible discussion may be found in London's book "Superfluids," vol. I, Wiley, New York, 1950.

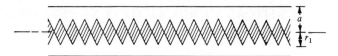

FIGURE 3-8
Intermediate-state structure in a wire carrying a current in excess of I_c. The shaded region is superconducting. The core radius r_1 is a at I_c, and ideally approaches zero only asymptotically as $I \to \infty$.

of the wire) of resistive material is r/r_1. If the normal resistivity is ρ, this leads to

$$J(r) = \frac{Er_1}{\rho r}$$

for $r < r_1$. Combining this with (3-62), we see that

$$r_1 = \frac{\rho c H_c}{4\pi E} \qquad (3\text{-}63)$$

Since the current inside the core I_1 must generate a field H_c at the surface of the core, we have

$$I_1 = \frac{cr_1 H_c}{2} = \frac{c^2 H_c^2 \rho}{8\pi E}$$

The current in the outer normal layer is

$$I_2 = \frac{E}{\rho}\pi(a^2 - r_1^2) = \frac{\pi a^2 E}{\rho} - \frac{c^2 H_c^2 \rho}{16\pi E}$$

Adding this to I_1, the total current in the wire is

$$I = \frac{\pi a^2 E}{\rho} + \frac{c^2 H_c^2 \rho}{16\pi E} \qquad (3\text{-}64)$$

Solving the quadratic equation for $E(I)$, and using (3-61), we find

$$E = \frac{\rho I}{2\pi a^2}\left\{1 \pm \left[1 - \left(\frac{I_c}{I}\right)^2\right]^{1/2}\right\} \qquad (3\text{-}65)$$

The plus sign must be chosen if E is to increase with increase of I, as required for stability. Note also that in the normal state $E = \rho I/\pi a^2$, by the usual Ohm's law relation. Thus we can write our results in terms of an apparent fractional resistance

$$\frac{R}{R_n} = \begin{cases} 0 & I < I_c \\ \frac{1}{2}\left\{1 + \left[1 - \left(\frac{I_c}{I}\right)^2\right]^{1/2}\right\} & I > I_c \end{cases} \qquad (3\text{-}66)$$

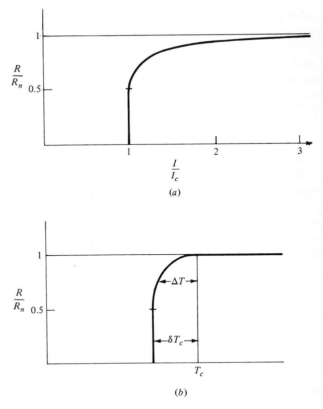

FIGURE 3-9

Resistance of wire in the intermediate state. (*a*) Current dependence at constant temperature. (*b*) Temperature dependence at constant current, showing the broadening and depression of the apparent transition temperature. The parameter $\delta T_c = I(dI_c/dT)^{-1}$.

From this we see that half the resistance appears discontinuously at I_c, at which point the intermediate-state pattern suddenly fills the entire wire. With further increase of current, the resistance increases continuously as the intermediate-state region shrinks to a smaller and smaller central core, the asymptotic behavior being $r_1/a = I_c/2I$. In principle, some superconducting material will continue to exist in the core for all finite currents. In practice, however, the Joule heating above I_c makes it hard to carry out an isothermal experiment to confirm this property in detail.

Experimental data are in good qualitative agreement with the theoretical result (3-66), which is plotted in Fig. 3-9, but there are quantitative discrepancies. In particular, the discontinuous jump in resistance typically goes to 0.7 to 0.8R_n, rather than to $\frac{1}{2}R_n$ as predicted by the simple theory.[1] This has led to a number of reexaminations of the London model. For example, Gorter[2] considered a dynamic model, with continually moving phase boundaries. On the other hand, Baird and Mukherjee[1] have made more detailed numerical studies of static models similar to London's. They were able to find the optimum ratio of domain period to wire diameter (~ 0.7), and to find curved domain-wall profiles on which $H = H_c$ and which come closer than the London model to truly satisfying the condition that $H = H_c$ throughout (rather than doing so only on the average over the domain structure). In this improved static model, they did find a larger jump in R/R_n (to 0.69) at I_c, and generally improved agreement at higher currents as well. A third approach, by Andreyev and Sharvin,[3] combines these ideas to consider a broad class of moving structures. This approach also leads to a larger jump in R/R_n at I_c of about the correct size, though apparently depending on different parameters than the static theory mentioned above. Thus, although the theoretical position is still somewhat unclear, there is no doubt that the London theory of the intermediate state in a current-carrying wire is oversimplified. Nonetheless, it provides a useful semiquantitative treatment of the problem.

This theory can also be applied to predict the temperature dependence of the resistance of a superconducting wire as it goes through its transition near T_c. In this temperature region, $I_c \propto H_c \propto (T_c - T)$, so we can write

$$I_c = \frac{dI_c}{dT}\bigg|_{T_c} \Delta T \approx ca H_c(0) \frac{\Delta T}{T_c}$$

With the London approximation (3-66), this leads to

$$\frac{R}{R_n} = \frac{1}{2}\left\{ 1 + \left[1 - \left(\frac{\Delta T}{\delta T_c}\right)^2 \right]^{1/2} \right\} \qquad (3\text{-}67)$$

where $\Delta T = T_c - T$, and $\delta T_c = I(dI_c/dT)^{-1}$. For example, if a measuring current of 1 A (ampere) were used in a 1-mm-diameter tin wire, the first onset of resistance would occur about $0.03°\mathrm{K}$ below T_c. Thus, in a critical-temperature measurement, I must be kept small enough so that the δT_c from this source is negligible compared to the intrinsic breadth of the transition as limited by sample inhomogeneity. The shape of the resistive transition due to finite current is illustrated in Fig. 3-9b.

There is also a current-induced intermediate state in thin-film superconductors. Although the geometry is much less simple to handle theoretically because of

[1] D. C. Baird and B. K. Mukherjee, *Phys. Letters* **25A**, 137 (1967), and references cited therein. See also B. K. Mukherjee, J. F. Allen, and D. C. Baird, *Proc. 11th Intl. Conf. on Low Temp. Phys.*, St. Andrews (1968) p. 827.
[2] C. J. Gorter, *Physica* **23**, 45 (1957).
[3] A. F. Andreyev and Yu. V. Sharvin, *Zh. Eksperim. i Teor. Fiz.* **53**, 1499 (1967); see also, A. F. Andreyev, *Proc. 11th Intl. Conf. on Low Temp. Phys.* St. Andrews (1968), p. 831.

edge effects, this configuration has the advantage that the intermediate state structure can be viewed by a magneto-optic technique. Recent experiments by Huebener and collaborators[1] have shown that the resistance increases in discrete increments, each associated with the appearance of an additional channel for the motion of magnetic-flux tubes across the strip. The fact that the flux pattern is moving can be demonstrated by the observation of an induced voltage in another adjacent superconducting film in a thin-film sandwich. In these experiments, a time of flight of the order of 10^{-3} sec could be inferred from noise-spectrum measurements. In other experiments, such as those of L. Rinderer, and those of Solomon and Harris cited on page 97, motion pictures have been taken of the moving domain patterns under situations in which the motion is much slower. These recent experimental results make it seem likely that some sort of time-dependent structure is characteristic of resistive regimes in superconductors. We shall return to this point in connection with dissipative effects in type II superconductors in Chap. 5.

[1] R. P. Huebener and R. T. Kampwirth, *Solid State Comm.* **10**, 1289 (1972); R. P. Huebener and D. E. Gallus, *Phys. Rev.* **B7**, 4089 (1973).

4

GINZBURG-LANDAU THEORY

The BCS microscopic theory described in Chap. 2 gives an excellent account of the data in those cases to which it is applicable, namely, those in which the energy gap Δ is constant in space. However, there are many situations in which the entire interest derives from the existence of spatial inhomogeneity. For example, in treating the intermediate state of type I superconductors, we had to consider the interface where the superconducting state joined onto the normal state. This sort of spatial inhomogeneity becomes all-pervasive in the mixed state of type II superconductors. In such situations, the fully microscopic theory becomes very difficult, and much reliance is placed on the more macroscopic Ginzburg-Landau[1] (GL) theory.

As originally proposed, this theory was a triumph of physical intuition, in which a pseudowavefunction $\psi(\mathbf{r})$ was introduced as a complex-order parameter. $|\psi(\mathbf{r})|^2$ was to represent the local density of superconducting electrons, $n_s(\mathbf{r})$. The theory was developed by applying a variational method to an assumed expansion of the free-energy density in powers of $|\psi|^2$ and $|\nabla\psi|^2$, leading to a pair of coupled differential equations for $\psi(\mathbf{r})$ and the vector potential $\mathbf{A}(\mathbf{r})$. The result

[1]V. L. Ginzburg and L. D. Landau, *Zh. Eksperim. i Teor. Fiz.* **20**, 1064 (1950).

was a generalization of the London theory to deal with situations in which n_s varied in space, and also to deal with the nonlinear response to fields strong enough to change n_s. The local approximation of the London electrodynamics was retained, however. Although quite successful in explaining intermediate-state phenomena, where the need for a theory capable of dealing with spatially inhomogeneous superconductivity was evident, this theory was generally given limited attentic 1 in the Western literature because of its phenomenological foundation.

This situation changed in 1959 when Gor'kov[1] showed that the GL theory was in fact derivable as a rigorous limiting case of the microscopic theory, suitably reformulated in terms of Green functions to allow treating a spatially inhomogeneous regime. The conditions for validity of the GL theory were shown to be a restriction to temperatures sufficiently near T_c and to spatial variations of ψ and $\mathbf{A}$ which were not too rapid. In this reevaluation of the GL theory, $\psi(\mathbf{r})$ turned out to be proportional to the gap parameter $\Delta(\mathbf{r})$, both being in general complex quantities. At first it was thought that $|\Delta(\mathbf{r})|$, found from solving the newly interpreted GL equations, was simply a BCS gap which might vary in space or with applied magnetic fields, or both. This led to a period in which experiments were (incorrectly) interpreted in this overly simple way. It has now become clear, however, that a solution to the GL equations for a given problem is only a useful first step toward understanding the spectral density of excitations. The key point is that fields, currents, and gradients act as "pairbreakers" which tend to blur out the sharp edge of the BCS gap as well as reducing the value of Δ. Detailed discussion of these effects will be deferred until Chap. 8.

The greatest value of the theory remains in treating the macroscopic behavior of superconductors, in which the overall free energy is important instead of the detailed spectrum of excitations. Thus, it will be quite reliable in predicting critical fields and the spatial structure of $\psi(\mathbf{r})$ in nonuniform situations. It also provides the qualitative framework for understanding the dramatic supercurrent behavior as a consequence of quantum properties on a macroscopic scale.

Although one could in principle now give a *derivation* of the GL theory following Gor'kov, this would require techniques beyond the level of our presentation. Instead, we shall follow Ginzburg and Landau in phenomenologically postulating the form of the theory on grounds of plausibility, and then simply appealing to the results of microscopic theory (or experiment) to evaluate the few parameters of the theory by considering simple special cases.

4-1 THE GINZBURG-LANDAU FREE ENERGY

The basic postulate of GL is that if ψ is small and varies slowly in space, the free-energy density f can be expanded in a series of the form

$$f = f_{n0} + \alpha|\psi|^2 + \frac{\beta}{2}|\psi|^4 + \frac{1}{2m^*}\left|\left(\frac{\hbar}{i}\nabla - \frac{e^*}{c}\mathbf{A}\right)\psi\right|^2 + \frac{h^2}{8\pi} \qquad (4\text{-}1)$$

[1] L. P. Gor'kov, *Zh. Eksperim. i Teor. Fiz.* **36**, 1918 (1959) [*Soviet Phys.—JETP* **9**, 1364 (1959)].

Evidently, if $\psi = 0$, this reduces properly to the free energy of the normal state $f_{n0} + h^2/8\pi$, where $f_{n0}(T) = f_{n0}(0) - \frac{1}{2}\gamma T^2$. We now consider the remaining three terms describing the superconducting effects.

In the absence of fields and gradients, we have

$$f_s - f_n = \alpha |\psi|^2 + \tfrac{1}{2}\beta |\psi|^4 \qquad (4\text{-}2)$$

which can be viewed as a series expansion in powers of $|\psi|^2$ or n_s, in which only the first two terms are retained.[1] These two terms should be adequate so long as one stays near the second-order phase transition at T_c, where the order parameter $|\psi|^2 \to 0$. Inspection of (4-2) shows that β must be positive if the theory is to be useful; otherwise, the lowest free energy would occur for arbitrarily large values of $|\psi|^2$, where the expansion is surely inadequate.

As is illustrated in Fig. 4-1, two cases arise, depending on whether α is positive or negative. If α is positive, the minimum free energy occurs at $|\psi|^2 = 0$, corresponding to the normal state. On the other hand, if $\alpha < 0$, the minimum occurs when

$$|\psi|^2 = |\psi_\infty|^2 \equiv -\frac{\alpha}{\beta} \qquad (4\text{-}3)$$

where the notation ψ_∞ is conventionally used because ψ approaches this value infinitely deep in the interior of the superconductor, where it is screened from any surface fields or currents. When this value of ψ is substituted back into (4-2), one finds

$$f_s - f_n = \frac{-H_c^2}{8\pi} = \frac{-\alpha^2}{2\beta} \qquad (4\text{-}4)$$

using the definition of the thermodynamic critical field H_c.

Evidently $\alpha(T)$ must change from positive to negative at T_c, since by definition T_c is the highest temperature at which $|\psi|^2 \neq 0$ gives a lower free energy than $|\psi|^2 = 0$. Making a Taylor's series expansion of $\alpha(T)$ about T_c, and keeping only the leading term, one has

$$\alpha(t) = \alpha'(t - 1) \qquad \alpha' > 0 \qquad (4\text{-}5)$$

where $t = T/T_c$. Note that in view of (4-4), this assumption is consistent with the linear variation of H_c with $(1 - t)$, if β is regular at T_c. Putting these temperature variations of α and β into (4-3), we see that

$$|\psi|^2 \propto (1 - t) \qquad (4\text{-}6)$$

near T_c. This is consistent with correlating $|\psi|^2$ with n_s, the density of superconducting electrons in the London theory, since $n_s \propto \lambda^{-2} \propto (1 - t)$ near T_c.

---

[1] Some considerations which restrict the choice to this form of expansion are the following: An expansion in powers of ψ itself is excluded, since f must be real. This difficulty cannot be avoided by taking the real part of ψ, since f should not depend on the absolute phase of ψ. Odd powers of $|\psi|$ are excluded because they are not analytic at $\psi = 0$.

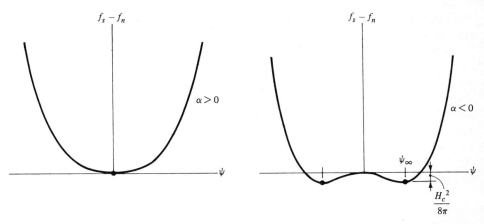

FIGURE 4-1
Ginzburg-Landau free-energy functions for $T > T_c (\alpha > 0)$ and for $T < T_c (\alpha < 0)$. Heavy dots indicate equilibrium positions. For simplicity, ψ has been taken to be real.

To make these considerations quantitative, we now consider the remaining term in the expansion (4-1), the term dealing with fields and gradients. If we write $\psi = |\psi| e^{i\varphi}$, it takes on the more transparent form

$$\frac{1}{2m^*}\left[\hbar^2 (\nabla |\psi|)^2 + \left(\hbar \nabla \varphi - \frac{e^* \mathbf{A}}{c} \right)^2 |\psi|^2 \right] \qquad (4\text{-}7)$$

The first term gives the extra energy associated with gradients in the *magnitude* of the order parameter, as in a domain wall. The second term gives the kinetic energy associated with supercurrents in a gauge-invariant form. In the London gauge, φ is constant, and this term is simply $e^{*2} A^2 |\psi|^2 / 2m^* c^2$. Equating this to the energy density for a London superconductor as given in (3-28), namely $A^2 / 8\pi \lambda_{\text{eff}}^2$, we find

$$\lambda_{\text{eff}}^2 = \frac{m^* c^2}{4\pi |\psi|^2 e^{*2}} \qquad (4\text{-}8)$$

With the identification $n_s^* = |\psi|^2$, this agrees with the usual definition of the London penetration depth, except for the presence of the starred effective number, mass, and charge values. The kinetic-energy density term can then be written as $n_s^* (\frac{1}{2} m^* v_s^2)$, where the supercurrent velocity is given by

$$m^* \mathbf{v}_s = \mathbf{p}_s - \frac{e^* \mathbf{A}}{c} = \hbar \nabla \varphi - \frac{e^* \mathbf{A}}{c} \qquad (4\text{-}9)$$

It should be noted that by writing the energy associated with the vector potential in the simple form (4-7), we have restricted the theory to the approximation of *local* electrodynamics. An expression of the sort found in (3-26) would be required to describe properly a *nonlocal* superconductor.

Now let us deal with the starred effective parameters. In the original formulation of the theory, it was thought that e^* and m^* would be the normal electronic values. However, experimental data turned out to be fitted better if $e^* \approx 2e$. The microscopic pairing theory of superconductivity makes it unambiguous that $e^* = 2e$ exactly, the charge of a pair of electrons. In the free-electron approximation, it would then be natural to take $m^* = 2m$ and $n_s^* = \frac{1}{2}n_s$, where n_s is the number of single electrons in the condensate. With these conventions, $n_s^* e^{*2}/m^* = n_s e^2/m$, so the London penetration depth is unchanged by the pairing.

The situation is more complicated in real metals. Band structure and phonon "dressing" effects may lead to an effective mass for a single electron in the normal state which typically differs from the free-electron mass by 50 percent. Moreover, the most important class of applications of GL theory is to dirty superconductors, in which $\lambda_{\mathrm{eff}}^2 \approx \lambda_L^2(\xi_0/\ell) \gg \lambda_L^2$. These increased penetration depths can be attributed formally to either an increase in m^* or a decrease in n_s. It might appear that an independent mass determination could be made by an experiment[1] in which a superconductor is rotated at angular velocity ω, and the resulting magnetic moment measured. However, as shown by Alben,[2] using a general argument based on Larmor's theorem, the induced moment must be the same as that due to a magnetic field $H_L = (2mc/e)\omega$, with the free-electron value of e/m, as indeed Brickman[1] had found. In other words, the rotation experiment adds nothing to the information available from purely magnetic measurements.

In view of the experimental inaccessibility of m^*, we can assign it an arbitrary value, and it is probably most convenient to choose twice the mass of the free electron. (This arbitrariness was emphasized by de Gennes, who suggested that one could equally well take the mass of the sun!) With $m^* = 2m$ fixed, all variations of λ, whether due to temperature, band structure, phonons, impurities, or even nonlocal electrodynamics, are taken up by an appropriate value of $|\psi_\infty|^2 = n_s^* = \frac{1}{2}n_s$. Even at $T = 0$ this number will no longer correspond to any obvious integral number of electrons per atom. Rather, our point of view is that n_s simply measures that part of the oscillator strength in the sum rule

$$\int_0^\infty \sigma_1(\omega)\, d\omega = \frac{\pi n e^2}{2m} \qquad (4\text{-}10)$$

which is located in the superfluid response at $\omega = 0$ in the form of a term $(\pi n_s e^2/2m)\, \delta(\omega)$.

An upper bound on n_s even at low temperatures is set by the oscillator strength of the conduction electrons in the normal state, which is spread over

---

[1] N. F. Brickman, *Phys. Rev.* **184**, 460 (1969).
[2] R. Alben, *Phys. Letters* **29A**, 477 (1969).

frequencies up to the collision rate $1/\tau$ for $q = 0$, and up to qv_F for $q \neq 0$ even if $\tau \to \infty$. More specifically, for spatially uniform fields, we expect the usual Drude frequency dependence of σ_1

$$\sigma_1(\omega, 0) = \frac{ne^2\tau/m}{1 + \omega^2\tau^2} \quad (4\text{-}10a)$$

while if $\tau = \infty$, we have the Lindhard[1] result that

$$\sigma_1(\omega, q) = \frac{3\pi}{4} \frac{ne^2}{mv_F q}\left(1 - \frac{\omega^2}{v_F^2 q^2}\right) \quad (4\text{-}10b)$$

for $\omega < qv_F$, and $\sigma_1 = 0$ for $\omega > qv_F$. Naturally, both of these expressions satisfy the sum rule (4-10). Speaking qualitatively, the normal-state oscillator strength lying at frequencies below the gap frequency $\omega_g = 2\Delta/\hbar$ will be converted to n_s in the transition to the superconducting state, while that above the gap will be relatively unaffected. If we consider a superconductor with local electrodynamics, so that the approximation $q = 0$ can be used, we see from (4-10a) that almost all of the oscillator strength will appear as n_s in a pure metal, where $\omega_g\tau > 1$; in this case, $n_s \approx n$, and $\lambda \approx \lambda_L$. On the other hand, n_s/n will be reduced to something of the order of $\omega_g\tau \approx \ell/\xi_0$ if the metal is dirty enough to have $\omega_g\tau < 1$. As a result, in dirty superconductors we have $\lambda/\lambda_L = (n/n_s)^{1/2} \approx (\xi_0/\ell)^{1/2}$, a result obtained more rigorously in (2-123). If we consider instead a pure, nonlocal superconductor, (4-10b) implies that $n_s(q)/n \approx \omega_g/qv_F \approx 1/q\xi_0$. This q-dependent superfluid fraction is a reflection of the fall of $K(q)$ as $1/q$ in Fig. 3-2. Taking a typical value $q \approx 1/\lambda$ for the currents in the penetration layer, this implies that $\lambda_L^2/\lambda^2 = n_s/n \approx \lambda/\xi_0$, or $\lambda \approx (\lambda_L^2\xi_0)^{1/3}$, as found more rigorously in (3-13). For use in the GL theory, which is local, one must take n_s to be such an average value appropriate to the actual penetration depth, not to λ_L. The survey in this paragraph reminds us of the power of the sum rule–energy gap argument in making simple physical estimates of the effective superfluid density in diverse situations.

Having noted that $e^* = 2e$, and taking the convention that $m^* = 2m$, we can now evaluate the parameters of the theory by solving (4-3), (4-4), and (4-8). The results are

$$|\psi_\infty|^2 \equiv n_s^* \equiv \frac{n_s}{2} = \frac{m^*c^2}{4\pi e^{*2}\lambda_{\text{eff}}^2} = \frac{mc^2}{8\pi e^2\lambda_{\text{eff}}^2} \quad (4\text{-}11a)$$

$$\alpha(T) = -\frac{e^{*2}}{m^*c^2}H_c^2(T)\lambda_{\text{eff}}^2(T) = -\frac{2e^2}{mc^2}H_c^2(T)\lambda_{\text{eff}}^2(T) \quad (4\text{-}11b)$$

$$\beta(T) = \frac{4\pi e^{*4}}{m^{*2}c^4}H_c^2(T)\lambda_{\text{eff}}^4(T) = \frac{16\pi e^4}{m^2c^4}H_c^2(T)\lambda_{\text{eff}}^4(T) \quad (4\text{-}11c)$$

where e and m are now the usual free-electron values, and λ_{eff} and H_c are measured values, or those computed from the microscopic theory.

[1] J. Lindhard, *Kgl. Danske Videnskab. Selskab., Mat.-fys. Medd.* **28**, no. 8, 1954.

Since the true electrodynamics of superconductors is nonlocal, it is evident that this prescription in terms of an effective London λ is straightforward only sufficiently near T_c that $\lambda_L(T) > \xi_0$, or in samples dirty enough that $\xi \approx \ell < \lambda(T)$, that is, where the nonlocality is unimportant. It is only under these conditions that the GL theory is really exact. Fortunately, the qualitative conclusions of the theory seem to have much wider validity; semiquantitative results can usually be obtained even when nonlocality is important by using a suitable λ_{eff}, such as we computed for films, (3-29) or (3-31). For pure bulk samples, as noted above, it is probably most appropriate to take $\lambda_{\text{eff}} = \lambda_{\text{exp}}$, the experimental value, if one attempts to apply the theory far enough below T_c that the nonlocality of the electrodynamics makes $\lambda_{\text{exp}} > \lambda_L$.

It is worth noting that if we insert the empirical approximations $H_c \propto (1 - t^2)$ and $\lambda^{-2} \propto (1 - t^4)$ into (4-11), we find

$$|\psi_\infty|^2 \propto 1 - t^4 \approx 4(1 - t)$$

$$\alpha \propto \frac{1 - t^2}{1 + t^2} \approx 1 - t$$

$$\beta \propto \frac{1}{(1 + t^2)^2} \approx \text{const} \qquad (4\text{-}12)$$

Since the theory is usually exactly valid only very near T_c, it is customary to carry only the leading dependence on temperature; that is, $|\psi_\infty|^2$ and α are usually taken to vary as $(1 - t)$ and β is taken to be constant, as anticipated in our preliminary discussion. Still, the more complete forms in (4-12) give some idea of how the theory can be extended over a wider range of temperature, and they have a certain amount of experimental support.

Finally, we recall that, although our discussion of (4-7) has centered on the kinetic energy of the supercurrent, this term also describes the energy associated with gradients in the magnitude of ψ. Moreover, no additional parameters are introduced, since gauge invariance requires a particular combination of $\mathbf{V}$ and $\mathbf{A}$ in (4-1). Thus the coefficients in the theory are completely determined by the values of $\lambda_{\text{eff}}(T)$ and $H_c(T)$. Since we showed earlier how the microscopic theory determines these parameters, we have effectively shown how the GL theory is set up to serve as an extension of BCS to the case of gradients and strong fields, but with a restriction to $T \approx T_c$.

4-2   THE GINZBURG-LANDAU DIFFERENTIAL EQUATIONS

In the absence of boundary conditions which impose fields, currents, or gradients, the free energy is minimized by having $\psi = \psi_\infty$ everywhere. On the other hand, when fields, currents, or gradients are imposed, $\psi(\mathbf{r}) = |\psi(\mathbf{r})| e^{i\varphi(\mathbf{r})}$ adjusts itself to minimize the overall free energy, given by the volume integral of (4-1). This variational problem leads, by standard methods, to the celebrated GL differential equations

$$\alpha\psi + \beta\,|\,\psi\,|^2\psi + \frac{1}{2m^*}\left(\frac{\hbar}{i}\nabla - \frac{e^*}{c}\mathbf{A}\right)^2\psi = 0 \qquad (4\text{-}13)$$

and

$$\mathbf{J} = \frac{c}{4\pi}\,\text{curl}\,\mathbf{h} = \frac{e^*\hbar}{2m^*i}(\psi^*\nabla\psi - \psi\nabla\psi^*) - \frac{e^{*2}}{m^*c}\psi^*\psi\mathbf{A} \qquad (4\text{-}14)$$

or

$$\mathbf{J} = \frac{e^*}{m^*}\,|\,\psi\,|^2\left(\hbar\nabla\varphi - \frac{e^*}{c}\mathbf{A}\right) = e^*\,|\,\psi\,|^2\mathbf{v}_s \qquad (4\text{-}14a)$$

where in the last step we have repeated the identification (4-9). Note that the current expression (4-14) has exactly the form of the usual quantum-mechanical expression for particles of mass m^*, charge e^*, and wavefunction $\psi(\mathbf{r})$. Similarly, apart from the nonlinear term, the first equation has the form of Schrödinger's equation for such particles, with energy eigenvalue $-\alpha$. The nonlinear term acts like a repulsive potential of ψ on itself, tending to favor wavefunctions $\psi(\mathbf{r})$ which are spread out as uniformly as possible in space.

In carrying through the variational procedure, boundary conditions must be provided. A possible choice, which assures that no current passes through the surface is

$$\left(\frac{\hbar}{i}\nabla - \frac{e^*}{c}\mathbf{A}\right)\psi\,\bigg|_n = 0 \qquad (4\text{-}15)$$

This is the boundary condition used by GL, and it is appropriate at an insulating surface. Using the microscopic theory, de Gennes[1] has shown that for a metal-superconductor interface with no current, (4-15) must be generalized to

$$\left(\frac{\hbar}{i}\nabla - \frac{e^*}{c}\mathbf{A}\right)\psi\,\bigg|_n = \frac{i}{b}\psi \qquad (4\text{-}15a)$$

where b is a real constant. It is easily seen that if $A_n = 0$, $\hbar b$ is the extrapolation length to the point outside the boundary at which ψ would go to zero if it maintained the slope it had at the surface. The value of b will depend on the nature of the material to which contact is made, approaching zero for a magnetic material and infinity for an insulator, with normal metals lying in between.

4-2.1 The Ginzburg-Landau Coherence Length

To help get a feeling for the differential equation (4-13), we first consider a simplified case in which no fields are present. Then $\mathbf{A} = 0$, and we can take ψ to be real, since the differential equation has only real coefficients. If we introduce a

---

[1] P. G. de Gennes, "Superconductivity of Metals and Alloys," p. 227, W. A. Benjamin, New York, 1966.

normalized wavefunction $f = \psi/\psi_\infty$, where $\psi_\infty^2 = -\alpha/\beta > 0$, the equation becomes (in one dimension)

$$\frac{\hbar^2}{2m^*|\alpha|}\frac{d^2f}{dx^2} + f - f^3 = 0 \qquad (4\text{-}16)$$

This makes it natural to define the characteristic length $\xi(T)$ for variation of ψ by

$$\xi^2(T) = \frac{\hbar^2}{2m^*|\alpha(T)|} \propto \frac{1}{1-t} \qquad (4\text{-}17)$$

Note that this $\xi(T)$ is certainly not the same length as Pippard's ξ, which we used in our discussion of the nonlocal electrodynamics, since this $\xi(T)$ diverges at T_c whereas the electrodynamic ξ is essentially constant. In fact, on the face of it, it is not clear why they should even be related. We retain this traditional notation, despite its considerable power to confuse, because it is almost invariably used in the literature, and because it does turn out that $\xi(T) \approx \xi_0$ for pure materials well away from T_c. In terms of $\xi(T)$, (4-16) becomes

$$\xi^2(T)\frac{d^2f}{dx^2} + f - f^3 = 0 \qquad (4\text{-}18)$$

The significance of $\xi(T)$ as a characteristic length for variation of ψ (or f) can be made even more evident by considering a linearized form of (4-18), in which we set $f(x) = 1 + g(x)$, where $g(x) \ll 1$. Then we have, to first order in g,

$$\xi^2 g''(x) + (1 + g) - (1 + 3g + \cdots) = 0$$

or

$$g'' = \left(\frac{2}{\xi^2}\right)g$$

so that

$$g(x) \sim e^{\pm\sqrt{2}\,x/\xi(T)} \qquad (4\text{-}19)$$

which shows that a small disturbance of ψ from ψ_∞ will decay in a characteristic length of order $\xi(T)$.

Now that we have an idea of the significance of the length $\xi(T)$, let us see what its value is. Substituting the value of α from (4-11b) into the definition (4-17), we find

$$\xi(T) = \frac{\Phi_0}{2\sqrt{2}\,\pi H_c(T)\lambda_{\mathrm{eff}}(T)} \qquad (4\text{-}20)$$

where

$$\Phi_0 = \frac{hc}{e^*} = \frac{hc}{2e} \qquad (4\text{-}21)$$

is the fluxoid quantum which will play an important role in our future discussions.

The fact that this $\xi(T)$ is at least related to the ξ_0 of Pippard and BCS is shown by the existence of the relation

$$\Phi_0 = \left(\frac{2}{3}\right)^{1/2} \pi^2 \xi_0 \lambda_L(0) H_c(0) \qquad (4\text{-}22)$$

which follows readily from our earlier BCS results $\xi_0 = \hbar v_F/\pi\Delta(0)$ and $H_c^2(0)/8\pi = \frac{1}{2}N(0)\Delta^2(0)$, if one assumes the free-electron relation between $N(0)$ and n. Combining (4-20) and (4-22), we can write

$$\frac{\xi(T)}{\xi_0} = \frac{\pi}{2\sqrt{3}} \frac{H_c(0)}{H_c(T)} \frac{\lambda_L(0)}{\lambda_{\text{eff}}(T)} \qquad (4\text{-}23)$$

From this we can see that near T_c

$$\xi(T) = 0.74 \frac{\xi_0}{(1-t)^{1/2}} \qquad \text{pure} \qquad (4\text{-}24a)$$

$$\xi(T) = 0.855 \frac{(\xi_0\ell)^{1/2}}{(1-t)^{1/2}} \qquad \text{dirty} \qquad (4\text{-}24b)$$

The precise coefficients here were determined using the exact results of BCS in the limit of $T \approx T_c$, namely,

$$H_c(t) = 1.73 H_c(0)(1-t) \qquad (4\text{-}25)$$

$$\lambda_L(t) = \frac{\lambda_L(0)}{[2(1-t)]^{1/2}} \qquad (4\text{-}26a)$$

$$\lambda_{\text{eff}}(t)\bigg|_{\substack{\text{dirty} \\ \text{limit}}} = \lambda_L(t)\left(\frac{\xi_0}{1.33\ell}\right)^{1/2} \qquad (4\text{-}26b)$$

The relation (4-24a), giving $\xi(T)$ for pure superconductors, has clear validity only in the extremely narrow temperature range near T_c in which the local electrodynamics are valid; outside this range, the appropriate effective value of ξ will be dependent on the sample configuration. Equation (4-24b) has a much broader range of validity for dirty superconductors, because there the local approximation remains good.

It is also useful to introduce the famous dimensionless Ginzburg-Landau parameter κ, which is defined as the ratio of the two characteristic lengths

$$\kappa = \frac{\lambda_{\text{eff}}(T)}{\xi(T)} = \frac{2\sqrt{2}\pi H_c(T)\lambda_{\text{eff}}^2(T)}{\Phi_0} \qquad (4\text{-}27)$$

With the empirical approximations $H_c \propto (1-t^2)$ and $\lambda^{-2} \propto (1-t^4)$, we see that κ should vary as $(1+t^2)^{-1}$. Of course this is only a rough approximation, but we can safely conclude that κ is regular at T_c, and varies only slowly with temperature. Using the numerical results above, we find the following results in the pure and dirty limits at T_c:

$$\kappa = 0.96\frac{\lambda_L(0)}{\xi_0} \qquad \text{pure} \qquad (4\text{-}27a)$$

$$\kappa = 0.715\frac{\lambda_L(0)}{\ell} \qquad \text{dirty} \qquad (4\text{-}27b)$$

In typical pure superconductors, $\kappa \ll 1$, but in dirty superconductors κ may be much greater than 1. As will be discussed later in more detail, the value $\kappa = 1/\sqrt{2}$ separates superconductors of types I and II.

4-3 CALCULATION OF THE DOMAIN-WALL-ENERGY PARAMETER

We have now developed the methods required to compute the surface-energy parameter $\gamma = H_c^2\delta/8\pi$ that we used in our discussion of the intermediate state in Chap. 3. The one-dimensional variations of $\psi(x)$ and of $h(x)$ in the domain wall are sketched in Fig. 4-2, contrasting the cases with $\kappa \ll 1$ and $\kappa \gg 1$. Qualitatively we see that the surface energy is positive for $\kappa \ll 1$, since there is a region of thickness $\sim (\xi - \lambda)$ from which the magnetic field is held out (contributing to the positive diamagnetic energy) while not enjoying the full condensation energy associated with ψ_∞. The argument is reversed for $\kappa \gg 1$, leading to a negative surface energy. Let us now see how this argument is made quantitative using the GL theory.

We seek solutions of the differential equations (4-13) and (4-14) subject to the boundary conditions

$$\psi = 0 \qquad \text{and} \qquad h = H_c \qquad \text{as } x \to -\infty$$

$$\psi = \psi_\infty \qquad \text{and} \qquad h = 0 \qquad \text{as } x \to +\infty$$

Because the problem is one-dimensional, it is possible to take ψ real. Doing so simplifies the equations by eliminating cross terms of $\mathbf{V}$ and $\mathbf{A}$ and also by reducing the second equation to the simple form $\mathbf{J} \propto |\psi(x)|^2\mathbf{A}$. Nonetheless, complete solutions can only be obtained numerically. We start by formally simplifying the expression for the surface energy.

First, we recognize that the appropriate quantity to calculate is the Gibbs free energy, since H is fixed at H_c, while B depends on the location of the domain wall. As indicated in the discussion leading to (3-43), the density of Gibbs free energy at H_c has the same value in superconducting or normal material, namely, f_{s0}, the Helmholtz free energy of the superconductor in the absence of fields or

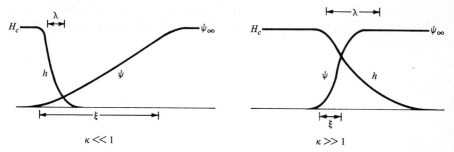

FIGURE 4-2
Schematic diagram of variation of h and ψ in a domain wall. The case $\kappa \ll 1$ refers to a type I superconductor (positive wall energy); the case $\kappa \gg 1$ refers to a type II superconductor (negative wall energy).

currents. Thus, the surface energy we seek is the excess of G over what it would be if its density were f_{s0} everywhere, that is,

$$\gamma = \int_{-\infty}^{\infty} (g_{sH} - f_{s0})\, dx = \int_{-\infty}^{\infty} \left(f_{sH} - \frac{hH_c}{4\pi} - f_{s0} \right) dx$$

$$= \int_{-\infty}^{\infty} \left[\alpha |\psi|^2 + \frac{\beta}{2} |\psi|^4 + \frac{1}{2m^*} \left| \left(\frac{\hbar \nabla}{i} - \frac{e^* \mathbf{A}}{c} \right) \psi \right|^2 + \frac{(h - H_c)^2}{8\pi} \right] dx \qquad (4\text{-}28)$$

using the fact that $f_{n0} - f_{s0} = H_c^2/8\pi$. This can be further simplified by noting that if one multiplies the GL differential equation (4-13) by ψ^* and integrates over all x by parts, one obtains the identity

$$0 = \int_{-\infty}^{\infty} \left[\alpha |\psi|^2 + \beta |\psi|^4 + \frac{1}{2m^*} \left| \left(\frac{\hbar \nabla}{i} - \frac{e^* \mathbf{A}}{c} \right) \psi \right|^2 \right] dx$$

Subtracting this from (4-28), we obtain the concise form

$$\gamma = \int_{-\infty}^{\infty} \left[-\frac{\beta}{2} |\psi|^4 + \frac{(h - H_c)^2}{8\pi} \right] dx \qquad (4\text{-}29)$$

which is to be equated to $(H_c^2/8\pi)\delta$. Finally, using (4-11), this can be written as

$$\delta = \int_{-\infty}^{\infty} \left[\left(1 - \frac{h}{H_c} \right)^2 - \left(\frac{\psi}{\psi_\infty} \right)^4 \right] dx \qquad (4\text{-}30)$$

This form clearly displays how δ is determined by the balance between the positive diamagnetic energy and the negative condensation energy due to the superconductivity, as argued qualitatively in connection with Fig. 4-2.

Numerical solutions for $h(x)$ and $\psi(x)$ must be made in order to evaluate (4-30), except in limiting cases. For these cases, the following exact results have been obtained:

$$\delta = \frac{4\sqrt{2}\,\xi}{3} = 1.89\xi \qquad \kappa \ll 1 \qquad (4\text{-}31a)$$

$$\delta = \frac{-8(\sqrt{2}-1)\lambda}{3} = -1.104\lambda \qquad \kappa \gg 1 \qquad (4\text{-}31b)$$

These results support our qualitative reasoning that δ should be of the order of $(\xi - \lambda)$.

Special consideration is required to show that the exact crossover from positive to negative surface energy occurs for $\kappa = 1/\sqrt{2}$. This was found by numerical integration by Ginzburg and Landau in their original paper, and they already anticipated that a conventional laminar intermediate state would only occur for lower values of κ. But until Abrikosov's path-breaking paper,[1] no one fully anticipated the radically different behavior that resulted from the negative surface energy at higher values of κ. In one stroke, his paper created the study of type II superconductivity, the name he gave to materials with $\kappa > 1/\sqrt{2}$. Since this is the subject of the next chapter, for the present we shall simply remark that the negative surface energy causes the flux-bearing (normal) regions to subdivide until a quantum limit is reached in which each quantum of flux $\Phi_0 = hc/2e$ passes through the sample as a distinct flux tube. These flux tubes form a regular array, and $\psi \to 0$ along the axis of each one. Unlike the intermediate state of type I superconductors, this so-called "mixed state" of type II superconductors occurs over a substantial field range even if the sample demagnetizing factor is zero.

4-4 CRITICAL CURRENT OF A THIN WIRE OR FILM

Having taken a quick look at the calculation of the interface energy, in which one immediately finds that numerical solutions are required, let us now step back and treat a number of important simpler examples in which exact analytic solutions are possible. In this way we will develop some familiarity with the GL theory before returning to more complex problems.

The very simplest applications are those in which the perturbing fields and currents are so weak that $|\psi| = \psi_\infty$ everywhere, and the GL theory reduces to the London theory.

A more interesting class of examples is that in which strong fields or currents change $|\psi|$ from ψ_∞, but in which $|\psi|$ has the same value everywhere. This will be the case if the sample is a thin wire or film so oriented with respect to any

[1] A. A. Abrikosov, *Zh. Eksperim. i Teor. Fiz.* **32**, 1442 (1957) [*Soviet Phys.—JETP* **5**, 1174 (1957)].

external field that any variation of $|\psi|$ would need to occur in a thickness $d \ll \xi(T)$. In that case, the term in the free energy proportional to $(\nabla|\psi|)^2$ would give an excessively large contribution if any substantial variations occurred. As a result they do not, and we can approximate $\psi(\mathbf{r})$ by $|\psi|e^{i\varphi(\mathbf{r})}$, where $|\psi|$ is constant. In this case, the expressions for the current and free-energy densities take on the simple forms

$$\mathbf{J}_s = \frac{2e}{m^*}|\psi|^2\left(\hbar\nabla\varphi - \frac{2e}{c}\mathbf{A}\right) = 2e|\psi|^2\mathbf{v}_s \qquad (4\text{-}32)$$

$$f = f_{n0} + \alpha|\psi|^2 + \frac{\beta}{2}|\psi|^4 + |\psi|^2\frac{1}{2}m^*v_s^2 + \frac{h^2}{8\pi} \qquad (4\text{-}33)$$

Although it is our standard convention to set $m^* = 2m$, we retain the more general formulas here to permit other normalizations of ψ to be used, if desired, and also as a reminder of the conventional nature of the parameter m^*.

Let us now apply these equations to treat the case of a uniform current density through a thin film or wire. Since the total energy due to the field term $h^2/8\pi$ is less than the kinetic energy of the current by a factor of the order of the ratio of the cross-sectional area of the conductor to λ^2, we can always neglect it for a sufficiently thin conductor. Then, for a given v_s, we can minimize (4-33) to find the optimum value of $|\psi|^2$. The result is

$$|\psi|^2 = \psi_\infty^2\left(1 - \frac{m^*v_s^2}{2|\alpha|}\right) = \psi_\infty^2\left[1 - \left(\frac{\xi m^* v_s}{\hbar}\right)^2\right] \qquad (4\text{-}34)$$

where the second form is stated in terms of ξ and m^*v_s, quantities invariant under changes in conventions. The corresponding current is

$$J_s = 2e\psi_\infty^2\left(1 - \frac{m^*v_s^2}{2|\alpha|}\right)v_s \qquad (4\text{-}35)$$

As indicated in Fig. 4-3, this has a maximum value when $\partial J_s/\partial v_s = 0$, namely, when $\frac{1}{2}m^*v_s^2 = |\alpha|/3$ and $|\psi|^2/\psi_\infty^2 = 2/3$. We identify this maximum current with the critical current. Thus,

$$J_c = 2e\psi_\infty^2\frac{2}{3}\left(\frac{2}{3}\frac{|\alpha|}{m^*}\right)^{1/2} = \frac{cH_c(T)}{3\sqrt{6}\pi\lambda(T)} \propto (1 - t)^{3/2} \qquad (4\text{-}36)$$

where, again, the second form is entirely in terms of operationally significant quantities and the indicated proportionality to $(1 - t)^{3/2}$ holds near T_c. The corresponding critical momentum is

$$p_c = m^*v_c = \frac{\hbar}{\sqrt{3}\xi(T)} \qquad (4\text{-}37)$$

The critical velocity itself is poorly defined, since it depends on the conventional choice of m^*.

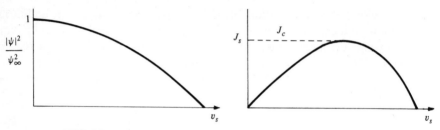

FIGURE 4-3
Variation of $|\psi|^2$ and of J_s with the superfluid velocity v_s.

It may be noted that we have taken v_s rather than $J_s \propto |\psi|^2 v_s$ as our independent variable. This was not a capricious choice; it was necessary since we are using the Helmholtz free energy, which is appropriate only if there are no induced emfs to effect energy interchanges with the source of current. This corresponds to specifying v_s, since an emf is needed to change that. If we wish to use the *current* as independent variable, then we must introduce a Legendre transformation on the free energy, as we did in (3-40) in dealing with magnetic energies. The appropriate term to subtract here to take account of work done by the generator is $m^* \mathbf{v}_s \cdot \mathbf{J}_s/2e$, so we could consider a Gibbs free-energy density

$$g = f - \frac{m^* v_s J_s}{2e} \qquad (4\text{-}38a)$$

or

$$g = f_{n0} + \alpha |\psi|^2 + \frac{\beta}{2} |\psi|^4 - \frac{m^* J_s^2}{8e^2 |\psi|^2} + \frac{h^2}{8\pi} \qquad (4\text{-}38b)$$

where we have used (4-32) to eliminate v_s. Minimizing this with respect to $|\psi|^2$ for given J_s leads to a cubic equation in $|\psi|^2$. Although algebraically more awkward, this condition is consistent with what we found above. For example, we can write it in the form

$$\frac{m^* J_s^2}{8e^2} = -\alpha |\psi|^4 - \beta |\psi|^6 \qquad (4\text{-}39)$$

whose maximum value occurs when $|\psi|^2 = -\frac{2}{3}(\alpha/\beta) = \frac{2}{3}\psi_\infty^2$, at which J_s has the critical value J_c found in (4-36).

It is of interest to compare this GL critical current with that of the London theory, where it is found by equating the density of kinetic energy to that of condensation energy

$$\tfrac{1}{2} n_s m v_s^2 = \frac{2\pi}{c^2} \lambda^2 J_s^2 = \frac{H_c^2}{8\pi}$$

so that

$$J_c = \frac{c H_c}{4\pi\lambda} \qquad (4\text{-}40)$$

This exceeds the more exact GL result (4-36) by a factor of $(3\sqrt{6}/4) = 1.84$ because it fails to take account of the decrease in $|\psi|^2$ with increasing current given by the nonlinear treatment.

It is also of interest to compare the GL result with that of the microscopic theory, where, of course, numerical computations must be made except in special cases. Bardeen has given a very useful review[1] of such calculations. Near T_c, the GL results are recovered, as expected. In the zero-temperature limit, the situation is quite different. In the presence of a uniform velocity v_s, the quasi-particle energies are shifted by $\hbar\mathbf{k} \cdot \mathbf{v}_s$. [This may be seen from (2-108) by noting that a velocity $ea(0)/mc$ is induced by a uniform vector potential $\mathbf{a}(0)$.] Thus, the gap goes to zero for some states when

$$v_s = \frac{\Delta(0)}{\hbar k_F} = \frac{\hbar}{\pi m \xi_0} \qquad (4\text{-}41)$$

Below this "depairing velocity," all electrons contribute to the supercurrent, and J_s is strictly proportional to v_s. Above this depairing velocity, some excitations occur at zero energy, the gap drops precipitously, and the maximum possible current is only 2 percent more than that at the velocity where depairing begins. The resulting $J_s(v_s)$ curve for $T = 0$ therefore shows a linear rise followed by a very steep drop to zero, in marked contrast with the GL result plotted in Fig. 4-3, appropriate near T_c.

Experimental confirmation of these results is most straightforward if both transverse dimensions of the conductor can be made small compared to both λ and ξ. It is then safe to take both J_s and $|\psi|^2$ to be constant over the cross section, as is assumed in the theory. Some of the first careful experiments on samples of this sort were those of Hunt,[2] who worked on very narrow strips of a thin evaporated film. More recently, several groups have been working with tin "whiskers" only about 1 μm (micrometer) in diameter, generally composed of a single crystal and having smooth surfaces, which are nearly ideal for the purpose. Since these experiments have concentrated on the fluctuation effects giving rise to resistance even at currents below the J_c computed above, we shall defer further discussion of them until a later chapter.

For reasons of experimental simplicity, many other measurements of critical currents have been made on thin-film samples which are not narrow on the scale of λ or ξ. With these, the measured J_c is usually much less than (4-36) for a number of reasons. First, it is somewhat difficult to make films of uniform thickness and structure. More seriously, the electrodynamic equations cause the supercurrent to pile up at the edges of the film because the external magnetic flux density is greatest there as the flux lines circle the film strip. This effect makes the current density nonuniform, and also emphasizes the properties of the edges of the

[1] J. Bardeen, *Rev. Mod. Phys.* **34**, 667 (1962).
[2] T. K. Hunt, *Phys. Rev.* **151**, 325 (1966).

film, which generally are thinner and less perfect. This problem can be minimized in three ways: (1) One can simply make the strip narrow enough so that the product of the thickness d and the width w is less than λ^2; in this case, J_s will be nearly uniform even if $w > \lambda$. (2) One can use a ground-plane geometry, in which the film under study is deposited on a larger thick superconductor with only a thin insulating layer in between; in this geometry, the superconducting substrate forces the field lines to be parallel to the film, which in turn requires a uniform current density in the film. (3) One can use a cylindrical film, so that there are no edges, and symmetry guarantees a uniform current density provided a concentric current return is used. It is possible to reach critical currents within about 10 percent of the theoretical values by any of these techniques if enough care is taken.

A more microscopic test of the theory can be made by tunneling measurements, which allow a determination of the gap $\Delta(J)$ for $J < J_c$. According to (4-34), $|\psi|$, and hence Δ, should decrease as J^2 at first, going over to a more complicated variation at higher currents. Evidence for this decrease, based on an increase in the differential conductance at zero voltage, was first given by Levine;[1] much more definitive results, based on measuring the full spectral density of states, were subsequently obtained by Mitescu.[2]

4-5 FLUXOID QUANTIZATION AND THE LITTLE-PARKS EXPERIMENT

An ingenious experiment, in which $m^* v_s$ rather than the current is constrained by external conditions, and which clearly demonstrates that it is the fluxoid rather than the flux which is quantized, was performed by Little and Parks.[3] The experiment consists of measuring the resistive transition of a thin-walled superconducting cylinder in an axial magnetic field. From this, one can infer shifts $\Delta T_c(H)$ in the critical temperature depending on the magnetic flux enclosed by the cylinder. In this section we shall give an analysis[4] of this experiment in the framework of the Ginzburg-Landau theory.

4-5.1 The Fluxoid

In analyzing the state of a multiply connected superconductor in the presence of a magnetic field, F. London introduced the concept of the *fluxoid* Φ' associated with

[1] J. L. Levine, *Phys. Rev. Letters* **15**, 154 (1965).
[2] C. D. Mitescu, doctoral thesis, California Institute of Technology, Pasadena, 1966.
[3] W. A. Little and R. D. Parks, *Phys. Rev. Letters* **9**, 9 (1962); *Phys. Rev.* **133**, A97 (1964).
[4] M. Tinkham, *Phys. Rev.* **129**, 2413 (1963).

each hole (or normal region) passing through the superconductor. His definition was

$$\Phi' = \Phi + \left(\frac{4\pi}{c}\right)\oint\lambda^2\mathbf{J}_s \cdot d\mathbf{s} = \Phi + \left(\frac{m^*c}{e^*}\right)\oint\mathbf{v}_s \cdot d\mathbf{s} \qquad (4\text{-}42)$$

where

$$\Phi = \int \mathbf{h} \cdot d\mathbf{S} = \oint\mathbf{A} \cdot d\mathbf{s}$$

is the ordinary magnetic flux through the integration circuit. Since it is easily seen that $\Phi' = 0$ for any path which encloses no hole but only superconducting material, in which the first of the London equations [Eq. (3-1)] holds, it follows that Φ' has the same value for *any* path around a given hole. Similarly, if there is a time variation of $\mathbf{h}$, the change in $\mathbf{J}_s$ induced according to the second London equation is just enough to hold Φ' constant (unless, of course, the change is so violent as to drive the superconductor into the normal state). These two conservation laws imply that Φ' has a unique constant value for all contours enclosing any given hole. In fact, London argued[1] that the values of Φ' should be restricted to a discrete set, integral multiples of a fluxoid quantum hc/e^*. This can be seen simply by applying the Bohr-Sommerfeld quantum condition to (4-42), as follows:

$$\Phi' = \frac{c}{e^*}\oint\left(m^*\mathbf{v}_s + \frac{e^*\mathbf{A}}{c}\right) \cdot d\mathbf{s} = \frac{c}{e^*}\oint\mathbf{p} \cdot d\mathbf{s}$$

$$= n\frac{hc}{e^*} = n\Phi_0 \qquad (4\text{-}43)$$

Not anticipating the pairing theory of superconductivity, London presumed that e^* was e. We now know that $e^* = 2e$, so the fluxoid quantum has the value

$$\Phi_0 = \frac{hc}{2e} = 2.07 \times 10^{-7} \text{ gauss-cm}^2 \qquad (4\text{-}44)$$

This has been shown experimentally by measurements[2] of the flux trapped in hollow cylinders with walls sufficiently thick that the total flux and the fluxoid are indistinguishable, because $\mathbf{v}_s \to 0$ in (4-42) once the skin depth is passed.

It should not be thought that the exactness of the concept of fluxoid quantization is limited by the above argument, with its use of the semiclassical Bohr-Sommerfeld language and the inexact London equations. From the viewpoint of GL theory, it is based simply on the existence of a single-valued complex superconducting order parameter ψ. This requires that the phase φ must change by integral multiples of 2π in making a complete circuit, that is,

$$\oint\nabla\varphi \cdot d\mathbf{s} = 2\pi n \qquad (4\text{-}45)$$

---

[1] F. London, "Superfluids," vol. I, p. 152, Wiley, New York, 1950.
[2] B. S. Deaver and W. M. Fairbank, *Phys. Rev. Letters* **7**, 43 (1961); R. Doll and M. Näbauer, *Phys. Rev. Letters* **7**, 51 (1961).

which, with (4-9), leads directly to the results found above. We thus see that fluxoid quantization is the macroscopic analog of the quantization of angular momentum in an atomic system. Accordingly, it is not surprising that it provides a powerful tool for dealing with many problems of superconductors penetrated by magnetic flux.

4-5.2   The Little-Parks Experiment

Now let us apply this principle to the analysis of the Little-Parks experiment. Let R be the radius of the thin-walled cylinder and H be the applied field. [No distinction need be made between the applied field and the field inside the cylinder, since we are seeking the shifted $T_c(H)$, at which point $|\psi|^2 \to 0$, so $J_s \to 0$, and the fields are the same.] Then, using (4-42) with $\Phi = \pi R^2 H$ and $\Phi' = n\Phi_0$, the supercurrent velocity is fixed by

$$v_s = \frac{\hbar}{m^* R}\left(n - \frac{\Phi}{\Phi_0}\right) \qquad (4\text{-}46)$$

For the value of Φ imposed by a given H, the energy of the currents in the cylinder will be least for that integer n for which v_s is a minimum, and accordingly that choice of n will allow the system to remain superconducting at the highest possible temperature. With n chosen in this way, v_s will be a periodic function of Φ/Φ_0, as indicated in Fig. 4-4. Since v_s is now specified, we may apply (4-34) to find the reduction in $|\psi|^2$. In particular, the transition occurs when $|\psi|^2 = 0$, that is, when

$$\frac{1}{\xi^2} = \left(\frac{m^* v_s}{\hbar}\right)^2 = \frac{1}{R^2}\left(n - \frac{\Phi}{\Phi_0}\right)^2$$

Using the relations (4-24) to provide the proportionality constants between ξ^{-2} and $(1 - t)$, we find the fractional depression of T_c to be given by

$$\frac{\Delta T_c(H)}{T_c} = \begin{cases} 0.55\dfrac{\xi_0^2}{R^2}\left(n - \dfrac{\Phi}{\Phi_0}\right)^2 & \text{clean} \\[4mm] 0.73\dfrac{\xi_0 \ell}{R^2}\left(n - \dfrac{\Phi}{\Phi_0}\right)^2 & \text{dirty} \qquad (4\text{-}47) \end{cases}$$

The maximum depression of T_c occurs when $n - \Phi/\Phi_0 = \frac{1}{2}$. At that point $\Delta T_c/T_c$ reaches $0.14\xi_0^2/R^2$ and $0.18\xi_0 \ell/R^2$ for the clean and dirty cases, respectively. The samples actually used were typically dirty tin films evaporated around organic filaments $\sim 1~\mu$m in diameter. Taking typical values—$R = 7 \times 10^{-5}$ cm,

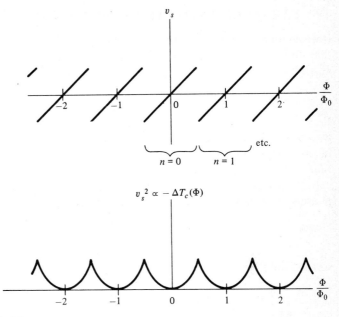

FIGURE 4-4
Variation of v_s and v_s^2 with flux threading the hollow cylinder in the Little-Parks experiment. The depression of T_c, and hence the increase in resistance in the actual experiment, is proportional to v_s^2, and thus displays the scalloped shape of the lower curve.

$\xi_0 = 2 \times 10^{-5}$ cm, $\ell = 10^{-6}$ cm—this leads to $\Delta T_c|_{max} \sim 0.8 \times 10^{-3} T_c \approx 3 \times 10^{-3}$ °K, which is readily measurable. For this diameter, the periodicity in field would be set by $\Phi_0 / \pi R^2 = 14$ G.

Although we have idealized the problem to finding $T_c(H)$, the actual experiment measures the periodic variation of the resistance of the film with changing H. This takes advantage of the finite breadth of the resistive transition to simplify the measurements, while $\Delta R(H)$ may be converted to $\Delta T_c(H)$ by using the measured dR/dT in the transition region. This is not entirely quantitative, however, since the shape of the resistive transition is observed to change with H, so that the inferred $\Delta T_c(H)$ depends to some extent on the choice of resistance level within the transition region. Another complication is an aperiodic quadratic shift of T_c with H which is sometimes observed and may be due to misalignment. Fortunately, more recent improvements of the experiment and analysis[1] have cleared up most of these problems and led to a rather satisfactory agreement with the theory.

[1] R. P. Groff and R. D. Parks, *Phys. Rev.* **176**, 567 (1968).

4-6 PARALLEL CRITICAL FIELD OF THIN FILMS

If we confine our attention initially to finding the critical field, and if we anticipate that thin films have second-order phase transitions, in which $|\psi|^2 \to 0$ and $\lambda_{\text{eff}}(H) \to \infty$, then we may neglect screening and write

$$A_y = \int_0^x h(x')\, dx' \approx Hx \qquad (4\text{-}48)$$

where H is the applied field. The latter form will also be a good approximation for films of thickness $d < \lambda$, even far from the transition. It will be noted this is the same "London gauge choice" we made in (3-19) in computing the weak-field value of λ_{eff} for films in the nonlocal electrodynamics. With this gauge choice, the phase φ as well as the amplitude $|\psi|$ is constant. Thus, from (4-9)

$$\mathbf{v}_s = -\frac{2e}{m^*c}\mathbf{A}$$

so the Gibbs free energy per unit area of film is

$$
\begin{aligned}
G &= \int_{-d/2}^{d/2} \left(f - \frac{hH}{4\pi} \right) dx \\
&= \int_{-d/2}^{d/2} \left[f_{n0} + \alpha|\psi|^2 + \frac{\beta}{2}|\psi|^4 + \tfrac{1}{2}m^*\left(\frac{2eHx}{m^*c}\right)^2 |\psi|^2 + \frac{(h-H)^2}{8\pi} - \frac{H^2}{8\pi} \right] dx \\
&= d\left[f_{n0} + \alpha|\psi|^2 + \frac{\beta}{2}|\psi|^4 - \frac{H^2}{8\pi} \right] + \frac{e^2 d^3 H^2}{6m^*c^2}|\psi|^2 + \int_{-d/2}^{d/2} \frac{(h-H)^2}{8\pi} dx
\end{aligned}
$$

$$(4\text{-}49)$$

Now, as mentioned above, in a thin film we can approximate h by H (as we have done in writing $A_y = Hx$), so the last term may be dropped.[1] Minimizing the remaining expression for G with respect to $|\psi|^2$ and using (4-11), we find

$$|\psi|^2 = \psi_\infty^2 \left(1 - \frac{d^2 H^2}{24\lambda^2 H_c^2} \right) \qquad (4\text{-}50)$$

where here λ refers to the appropriate λ_{eff} in zero field. Thus the film becomes normal, i.e., $|\psi|^2 \to 0$, when $H = H_{c\|}$, given by

$$H_{c\|} = 2\sqrt{6}\,\frac{H_c \lambda}{d} \qquad (4\text{-}51)$$

This parallel critical field can exceed the thermodynamic critical field H_c by a large factor if d/λ is small enough. The physical reason for this is simply that the thin

---

[1] It is easy to show that it is down by a factor of order $d^2 e^2 |\psi|^2/mc^2 \ll 1$ compared to the term proportional to d^3 which we keep.

film, being largely penetrated by the field, has little diamagnetic energy for a given applied field in comparison to an equal volume of a bulk superconductor.

Rewriting (4-50) in terms of $H_{c\|}$, it becomes

$$\frac{|\psi|^2}{\psi_\infty^2} = 1 - \frac{H^2}{H_{c\|}^2} \qquad (4\text{-}52)$$

Recalling the proportionality of ψ and Δ, we see that this predicts that the energy gap in a thin film will be depressed continuously to zero by increasing a parallel magnetic field up to $H_{c\|}$. This behavior has been confirmed qualitatively by electron tunneling experiments.[1] However, later work[2] has shown that the BCS excitation spectrum is progressively smeared out by the field, until at $H \approx 0.95H_{c\|}$ the spectrum becomes gapless, although the film stays superconducting as measured by its resistance. Such "gapless" superconductivity, first proposed by Abrikosov and Gor'kov,[3] turns out to be quite characteristic of superconductors subjected to time-reversal noninvariant perturbations (typically magnetic), but we shall defer detailed discussion of this until Sec. 8-2.

4-6.1   Thicker Films

What if the film is not extremely thin? So long as it is thin enough to have a second-order phase transition, (4-51) remains *exactly*[4] correct, because at $H_{c\|}$, $|\psi|^2 \to 0$, so $\lambda_{\text{eff}}(H) \to \infty$, and $d/\lambda_{\text{eff}}(H) \to 0$. The question then is, how thin must the film be to show a second-order transition at $H_{c\|}$. To answer this question, the full expression for the free energy must be used, allowing for the fact that screening currents will make $h < H$ inside the film.

So long as $d \ll \xi$, we can retain the approximation that $|\psi|$ is constant over the film. Then h is governed by the simple London equations with a field-dependent λ_{eff} given by

$$\lambda_{\text{eff}}^2(H) = \frac{16\pi e^2 |\psi|^2}{m^* c^2} \qquad (4\text{-}53)$$

[1] R. Meservy and D. H. Douglass, Jr., *Phys. Rev.* **135**, A24 (1964).

[2] J. L. Levine, *Phys. Rev.* **155**, 373 (1967); J. Millstein and M. Tinkham, *Phys. Rev.* **158**, 325 (1967).

[3] A. A. Abrikosov and L. P. Gor'kov, *Zh. Eksperim. i Teor. Fiz.* **39**, 1781 (1960) [*Soviet Phys.— JETP* **12**, 1243 (1961)].

[4] Of course, there will be small corrections if d/ξ is not very small, because then $|\psi|$ will not be exactly constant across the film thickness, as has been assumed. A variational calculation shows that this correction increases $H_{c\|}$ above (4-51) by a fractional amount of order $d^2/100\xi^2$. This never exceeds 3 percent before the film is thick enough $[d \approx 1.8\xi(T)]$ to have an entirely different solution in which $|\psi|$ is maximum near a surface. This case will be treated later in Sec. 4-10.2.

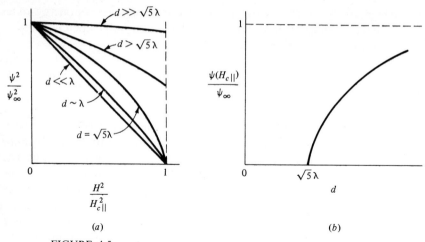

(a)   (b)

FIGURE 4-5

Dependence of ψ on magnetic field for various film thicknesses. The size of the discontinuity of ψ at the first-order transition for thickness $d > \sqrt{5}\,\lambda$ is shown in (b). It is assumed that $d \ll \xi(T)$ throughout.

Thus the usual symmetric combination of exponential solutions, satisfying the boundary conditions $h = H$ at both surfaces, will hold, so that

$$h = H\,\frac{\cosh x/\lambda_{\text{eff}}(H)}{\cosh d/2\lambda_{\text{eff}}(H)} \qquad (4\text{-}54a)$$

The corresponding vector potential is

$$A_y = H\lambda_{\text{eff}}(H)\,\frac{\sinh x/\lambda_{\text{eff}}(H)}{\cosh d/2\lambda_{\text{eff}}(H)} \qquad (4\text{-}54b)$$

When these are used in the expression for G, and the condition for a minimum is computed, the result is a somewhat awkward transcendental equation. After a little manipulation,[1] it emerges that the maximum thickness for which a second-order transition occurs is

$$d_{\text{max, 2d–order}} = \sqrt{5}\lambda \qquad (4\text{-}55)$$

where this λ is again the value of $\lambda_{\text{eff}}(H)$ at zero field. Below this thickness, $H_{c\parallel}$ is rigorously given by (4-51), but the decrease of $|\psi|^2$ to zero as H approaches $H_{c\parallel}$ is not in general as simple as (4-52). For $d > \sqrt{5}\,\lambda$, the transition becomes first

[1] For details, see, for example, P. G. de Gennes, "Superconductivity of Metals and Alloys," p. 189, W. A. Benjamin, New York, 1966. These results were obtained first in the original paper of Ginzburg and Landau.

order, with a discontinuous drop in $|\psi|^2$ to zero, as in the transition of a bulk sample in a magnetic field. However, if d is not much larger than $\sqrt{5}\lambda$, $|\psi|^2$ decreases substantially before the transition occurs. These results are shown schematically in Fig. 4-5.

4-7 THE LINEARIZED GL EQUATION

In the examples treated above, we have always restricted our consideration to thin films or wires in which $d \ll \xi(T)$, so that $|\psi|$ did not vary appreciably. Evidently this restriction excludes most of the interesting cases, such as bulk samples and films in fields other than those parallel to the surface. We now wish to treat such cases. However, to avoid facing immediately the full complications of solving a pair of coupled nonlinear partial differential equations, we first study the solutions of the *linearized* GL equation, obtained by dropping the term $\beta|\psi|^2\psi$ in (4-13), which corresponds to dropping $\frac{1}{2}\beta|\psi|^4$ in (4-1). These omissions will be justified only if $|\psi|^2 \ll \psi_\infty^2 = -\alpha/\beta$, because when $\psi \approx \psi_\infty$, the term in β is of the same order of magnitude as the one in α which we retain. Thus, this linearized theory will be appropriate only when the magnetic field has reduced ψ to a value much smaller than ψ_∞. Using our definition (4-17) relating α to ξ, we can write the linearized form of the differential equation (4-13) as

$$\left(\frac{\nabla}{i} - \frac{2\pi\mathbf{A}}{\Phi_0}\right)^2 \psi = -\frac{2m^*\alpha}{\hbar^2}\psi \equiv \frac{\psi}{\xi^2(T)} \qquad (4\text{-}56)$$

A further essential simplification arises from the fact that in (4-56) $\mathbf{A} = \mathbf{A}_{\text{ext}}$, since all screening effects due to supercurrents are proportional to $|\psi|^2$, and hence lead to higher-order terms which are dropped in the linearized approximation. Thus, in this approximation, the second GL equation (4-14), giving the current, is decoupled from the first, which governs ψ, leading to great mathematical simplification.

We note that (4-56) is identical with the Schrödinger equation for a free particle of mass m^* and charge $e^* = 2e$ in a magnetic field $\mathbf{h} = \text{curl } \mathbf{A}$, with $-\alpha = |\alpha|$ playing the role of the energy eigenvalue. This property allows various solutions and methods, familiar from quantum mechanics, to be applied directly to superconductivity. In particular, one may determine the fields at which solutions of the linearized GL equation exist, and hence at which solutions to the full nonlinear GL equations are possible with infinitesimal amplitude, by simply equating $1/\xi^2(T)$ with the field-dependent eigenvalues of the operator on the left in (4-56). The field values determined in this way correspond to the critical fields for second-order phase transitions, or, if the transition is of first order, to the nucleation field which sets a limit to the extent of supercooling. We now consider some important illustrations of this technique.

4-8 NUCLEATION IN BULK SAMPLES: H_{c2}

Let us first solve the problem of the nucleation of superconductivity in a bulk sample in the presence of a field **H** along the z axis. A convenient gauge choice is

$$A_y = Hx$$

The origin of coordinates is immaterial, since we consider an infinite sample. Expanding the left member of (4-56), we have

$$\left[-\nabla^2 + \frac{4\pi i}{\Phi_0} Hx \frac{\partial}{\partial y} + \left(\frac{2\pi H}{\Phi_0} \right)^2 x^2 \right] \psi = \frac{1}{\xi^2} \psi \qquad (4\text{-}57)$$

Since the effective potential depends only on x, it is reasonable to look for a solution of the form

$$\psi = e^{ik_y y} e^{ik_z z} f(x) \qquad (4\text{-}58)$$

Substituting this into (4-57) and rearranging terms, we find

$$-f''(x) + \left(\frac{2\pi H}{\Phi_0} \right)^2 (x - x_0)^2 f = \left(\frac{1}{\xi^2} - k_z^2 \right) f \qquad (4\text{-}59)$$

where

$$x_0 = \frac{k_y \Phi_0}{2\pi H} \qquad (4\text{-}59a)$$

Thus, inclusion of the factor $e^{ik_y y}$ only shifts the location of the minimum of the effective potential. This is unimportant for the present, but it will become important when we deal with superconductivity near surfaces of finite samples, and when we construct a space-filling solution rather than a localized one.

We can obtain the solutions to (4-59) immediately by noting that (after multiplying by $\hbar^2/2m^*$) it is the Schrödinger equation for a particle of mass m^* bound in a harmonic oscillator potential with force constant $(2\pi\hbar H/\Phi_0)^2/m^*$. This problem is formally the same as that of finding the quantized states of a normal charged particle in a magnetic field, which leads to the so-called Landau levels, separated by the cyclotron energy $\hbar\omega_c$. The resulting harmonic oscillator eigenvalues are

$$\epsilon_n = \left(n + \frac{1}{2} \right) \hbar\omega_c = \left(n + \frac{1}{2} \right) \hbar \left(\frac{2eH}{m^* c} \right)$$

In view of (4-59), these are to be equated to $(\hbar^2/2m^*)(\xi^{-2} - k_z^2)$. Thus,

$$H = \frac{\Phi_0}{2\pi(2n + 1)} \left(\frac{1}{\xi^2} - k_z^2 \right) \qquad (4\text{-}60)$$

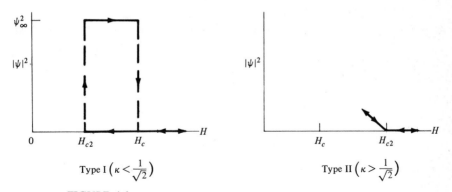

FIGURE 4-6
Contrast of behavior of order parameter at H_{c2} in type I and type II superconductors. Note hysteretic behavior with type I and reversible behavior with type II.

Evidently this has its highest value if $k_z = 0$ and $n = 0$. The corresponding value, defined as H_{c2}, is

$$H_{c2} = \frac{\Phi_0}{2\pi\xi^2(T)}$$

This is the highest field at which superconductivity can nucleate in the interior of a large sample in a decreasing external field. As may be verified by substitution in (4-59), the corresponding eigenfunction is

$$f(x) = \exp\left[-\frac{(x - x_0)^2}{2\xi^2}\right] \qquad (4\text{-}61)$$

The relation of H_{c2} to the thermodynamic critical field H_c is clarified if we reexpress H_{c2} in terms of H_c using (4-20) and (4-27). In this way, we arrive at the three equivalent expressions for H_{c2}

$$H_{c2} = \frac{\Phi_0}{2\pi\xi^2} = \frac{4\pi\lambda^2 H_c^2}{\Phi_0} = \sqrt{2}\,\kappa H_c \qquad (4\text{-}62)$$

The third form makes it clear that the value $\kappa = 1/\sqrt{2}$ does indeed separate the materials for which $H_{c2} > H_c$ (type II superconductors) from those for which $H_{c2} < H_c$ (type I). Because of the significance of H_{c2} as a nucleation field, these inequalities imply that in a decreasing field, type II superconductors become superconducting in a second-order phase transition (with $|\psi|^2$ starting up continuously from zero) at $H_{c2} > H_c$. On the other hand, type I superconductors "supercool," remaining normal even below H_c, ideally until $H_{c2} < H_c$ is reached. At this point nucleation occurs, followed by a discontinuous and irreversible jump of $|\psi|^2$ to ψ_∞^2. These features are illustrated in Fig. 4-6. In practice, nucleation at sample defects usually limits the amount of supercooling actually observed in type I superconductors to less than the theoretical limit set by H_{c2}. Nonetheless rather ideal and tiny samples may remain normal all the way to the theoretical limit.[1] For example, Feder and McLachlan[2] found $\kappa = 0.062$ for indium by observing supercooling down to $H_{c2} \approx 0.09\,H_c$.

4-9   NUCLEATION AT SURFACES: H_{c3}

Since real superconductors are finite in size, behavior near the surfaces must be considered. In fact, Saint-James and de Gennes[3] showed that superconductivity can nucleate at a metal-insulator interface in a parallel field H_{c3} higher by a factor of 1.695 than H_{c2}. For field values between H_{c2} and H_{c3} there is a superconducting surface sheath of thickness $\sim \xi(T)$, while $\psi \to 0$ in the interior. Let us see how this comes about.

At an insulating surface, the boundary condition on ψ is

$$\left(\frac{\nabla}{i} - \frac{2\pi \mathbf{A}}{\Phi_0}\right)\psi\bigg|_n = 0 \qquad (4\text{-}63)$$

For $\mathbf{H} \parallel \hat{\mathbf{z}}$ normal to the surface, this is satisfied by (4-58) for $k_z = 0$ (the case relevant for H_{c2}), since with our gauge choice $\mathbf{A}$ is along $\hat{\mathbf{y}}$, which is in the plane of the surface. Thus H_{c2} as found above would correctly give the nucleation field near a surface *normal* to the field.

Now consider a surface *parallel* to the field, the yz plane for example. In our chosen gauge, $A_x = A_n = 0$, so the boundary condition becomes simply

$$\frac{\partial \psi}{\partial x}\bigg|_{\text{surface}} = \frac{df}{dx}\bigg|_{\text{surface}} = 0 \qquad (4\text{-}64)$$

Our eigenfunction (4-61) satisfies this if $x_0 = 0$ or ∞, in both cases with the eigenvalue corresponding to H_{c2}. An eigenfunction with lower eigenvalue, still satisfying (4-64), can be constructed if k_y is chosen so as to put x_0, the minimum of the effective potential, inside the surface by a distance of the order of ξ. This can be seen to be the case by a qualitative argument: Imagine the potential well about x_0 to be extended by a mirror image outside the surface, as shown in Fig. 4-7b, thus

---

[1] To avoid nucleation at the surface at $H_{c3} > H_{c2}$, as treated in the next section, the sample surface must be plated with a normal metal.

[2] J. Feder and D. S. McLachlan, *Phys. Rev.* **177**, 763 (1969).

[3] D. Saint-James and P. G. de Gennes, *Phys. Letters* **7**, 306 (1963).

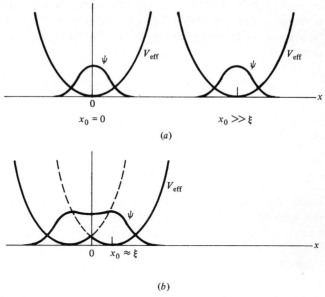

FIGURE 4-7
(a) Surface and interior nucleation at H_{c2}. (b) Surface nucleation at H_{c3}.

forming a potential symmetric about the surface. Now, the lowest eigenfunction of a symmetric potential is itself symmetric, hence has $df/dx = 0$ at the surface, satisfying the boundary condition. [Of course, the half of $f(x)$ outside the sample surface has no physical significance and is discarded.] It is clear by inspection that this new "surface" eigenfunction must have a lower eigenvalue than the "interior" ones, because it arises from a potential curve that is lower and broader than the simple parabola about x_0. The exact solution shows that this eigenvalue is lower by a factor of 0.59, leading to

$$H_{c3} = \frac{1}{0.59} H_{c2} = 1.695 H_{c2} = 1.695(\sqrt{2}\,\kappa\, H_c) \qquad (4\text{-}65)$$

This exact result was obtained using the tabulated Weber functions. However, a very simple variational approach (suggested by C. Kittel) gives quite a good approximation and illustrates the usefulness of variational methods in working with the GL theory. We outline the calculation here, leaving the details as an exercise. Motivated by (4-61), we take our trial function to be

$$\psi = f(x)e^{ik_y y} = e^{-ax^2}e^{ik_y y} \qquad (4\text{-}66)$$

With x measured from the sample surface, this function automatically satisfies the boundary condition (4-64). The parameters a and k_y are then determined variationally so as to minimize the Gibbs free energy per unit surface area. In the linearized approximation, this can be written as

$$G - G_n = \frac{\hbar^2}{2m^*} \int_0^\infty \left[-\frac{1}{\xi^2} |\psi|^2 + \left| \left(\frac{\nabla}{i} - \frac{2\pi}{\Phi_0} \mathbf{A} \right) \psi \right|^2 \right] dx \qquad (4\text{-}67)$$

Substituting (4-66) into this expression and differentiating under the integral sign, one is led to

$$x_0 = \frac{k_y \Phi_0}{2\pi H} = (2\pi a)^{-1/2} \qquad (4\text{-}68)$$

Using this value of x_0, $G - G_n$ becomes a function of a, H, and $\xi(T)$. Minimizing this with respect to a determines a as a function of H for given ξ. However, the linearized theory is valid only at the transition, where $G = G_n$. This gives a second condition, allowing the determination of the critical field as well as the optimum value of a at that field. The final results are

$$a = \frac{1}{2\xi^2} \qquad x_0 = \frac{\xi}{\sqrt{\pi}}$$

$$H_{c3} \approx \left(\frac{\pi}{\pi - 2} \right)^{1/2} \frac{\Phi_0}{2\pi\xi^2} = 1.66 H_{c2} \qquad (4\text{-}69)$$

This value of H_{c3} is only 2 percent below the exact result (4-65). A two-term variational function of the form $f(x) = (1 + cx^2)e^{-ax^2}$ leads to essentially perfect agreement with the exact result.

Our conclusion is that in a magnetic field parallel to the surface, superconductivity will nucleate in a surface layer of thickness $\sim \xi$ at a field 70 percent higher than that at which nucleation occurs in the volume of the material. This means that a sample may be able to carry a surface supercurrent over a wide range of fields in which there is no volume superconductivity as measured by magnetization, for example. As is readily imagined, the theoretical discovery of surface superconductivity by Saint-James and de Gennes provided a rational explanation for great amounts of experimental data on the persistence of superconductivity at high fields which had previously been dismissed as due to inhomogeneous samples.

An interesting consequence of these results is that H_{c3}, not H_{c2}, should limit the range of supercooling of a type I superconductor, since the surface sheath of superconductivity will serve to initiate the transformation of the interior. In the experiments of Feder and McLachlan, cited above, both H_{c3} and H_{c2} were measured as supercooling fields by the expedient of plating some of the samples with a normal metal. The plated samples supercooled to H_{c2}, because the normal plating changes the boundary condition on ψ so that (4-64) is replaced by the less favorable condition $d\psi/dx = -\psi/b$. In more physical terms, surface superconductivity is inhibited by the normal metal overlay because any pairs formed at the surface tend to diffuse into the normal metal and be destroyed. In other words, the contact with normal metal serves as a pair-breaking mechanism which locally augments the effect of the magnetic field, thus suppressing the surface superconductivity.

For materials with κ between $1/1.695\sqrt{2}$ and $1/\sqrt{2}$, that is, in the range $0.42 < \kappa < 0.707$, we have the inequality $H_{c2} < H_c < H_{c3}$. Thus, the type I superconductors with κ in this range should not show any supercooling at H_c, despite the fact that the volume of the sample makes a first-order transition there. Such superconductors are sometimes referred to as "type $1\frac{1}{2}$ superconductors," a somewhat questionable classification since its validity depends on the nature of the boundary condition on ψ at the sample surface, not on the intrinsic properties of the bulk of the material.

4-10 NUCLEATION IN FILMS AND FOILS

In our discussion of the surface nucleation field H_{c3}, we tacitly assumed that the medium was semi-infinite, allowing us to ignore all surfaces except the one of direct concern. We now consider the nucleation of superconductivity in a film, where both surfaces must be considered. First we discuss the angular dependence of the critical field for a film thin enough $[d \ll \xi(T)]$ that we may treat $|\psi|$ as constant through the thickness of the film. We then consider films of intermediate thickness $d \approx \xi$, not only for their intrinsic interest, but also because they provide a simple introduction to the vortex state of bulk samples.

4-10.1 Angular Dependence of the Critical Field of Thin Films

We noted above that H_{c2} would be the critical field for nucleation of superconductivity near a surface normal to the magnetic field, since the corresponding eigenfunction has $\partial\psi/\partial z = 0$ everywhere and hence automatically satisfies the boundary condition of zero normal derivative. For the same reason, H_{c2} will also give the value of $H_{c\perp}$, the perpendicular critical field of a thin film. It is only necessary to insert in (4-62) the appropriate effective value of κ, which will depend on the thickness d if the film is very thin. In fact, it is often more convenient to think in terms of the thickness dependence of λ_{eff}, as was analyzed in Sec. 3-2, and use the intermediate form in (4-62).

The other limiting case, the parallel critical field, was worked out for thin films in Sec. 4-6. Since these two limiting values are very different in the case of thin films, $H_{c\parallel}$ being much greater than $H_{c\perp}$, it is of interest to work out the angular dependence which interpolates between these limits. In first treating this problem, Tinkham[1] used a simple physical argument based on fluxoid quantization to obtain the answer that $H_c(\theta)$ is determined implicitly by the relation

$$\left| \frac{H_c(\theta) \sin \theta}{H_{c\perp}} \right| + \left(\frac{H_c(\theta) \cos \theta}{H_{c\parallel}} \right)^2 = 1 \qquad (4\text{-}70)$$

Note that this implies a cusp in $H_c(\theta)$ at $\theta = 0$, since

$$\left| \frac{dH_c}{d\theta} \right|_{\theta=0} = \frac{H_{c\parallel}^2}{2H_{c\perp}} > 0 \qquad (4\text{-}70a)$$

These results were subsequently[2] confirmed as valid thin-film limits by using the linearized GL equation. We now outline this calculation.

We choose a coordinate system in which x is measured normal to the film from its midplane. The magnetic field is chosen to lie in the xz plane at an angle θ from the plane of the film. For convenience, the vector potential is chosen to have only a y component, given by

$$A_y = H(x \cos \theta - z \sin \theta) \qquad (4\text{-}71)$$

When this is inserted in (4-56), the resulting partial differential equation is difficult to solve with the appropriate boundary condition (4-63) because it does not separate in these coordinates. We can safely take ψ to be independent of y, however, since y does not enter the differential equation. Moreover, if $d \ll \xi$, we can take ψ to be independent of x, as we did in Sec. 4-6, which automatically satisfies the boundary condition of zero normal derivative at the surface. Thus, in the limit $d \to 0$, we seek a function of a single variable $\psi(z)$. This problem may be approached variationally, using Eq. (4-67) with suitably changed integration limits. Applying the calculus of variations to the resulting expression, we are led to the ordinary differential equation

$$-\frac{d^2\psi}{dz^2} + \left(\frac{2\pi H \sin \theta}{\Phi_0} \right)^2 z^2 \psi = \left[\frac{1}{\xi^2} - \left(\frac{\pi H d \cos \theta}{\sqrt{3}\,\Phi_0} \right)^2 \right] \psi \qquad (4\text{-}72)$$

[This could have been obtained directly from the general form Eq. (4-56), with $\mathbf{A}$ given by Eq. (4-71), by taking $\partial\psi/\partial x = \partial\psi/\partial y = 0$ and making the plausible replacement of x and x^2 by the average values $\langle x \rangle = 0$ and $\langle x^2 \rangle = d^2/12$.] Since Eq. (4-72) has the same structure as Eq. (4-59), we can take over the solution found there. Thus, the eigenfunction has the form

$$\psi \propto \left(\exp \left[\frac{-\pi H z^2 \sin \theta}{\Phi_0} \right] \right) \qquad (4\text{-}73)$$

and by equating the corresponding eigenvalue of the operator on the left side of

[1] M. Tinkham, *Phys. Rev.* **129**, 2413 (1963).
[2] M. Tinkham, *Conf. on the Phys. of Type-II Superconductivity*, Cleveland, Ohio (1964, unpublished); F. E. Harper and M. Tinkham, *Phys. Rev.* **172**, 441 (1968).

(4-72) with the coefficient on the right, we are led to Eq. (4-70). In this way, Eq. (4-70) is established as a rigorous result in the limit of a very thin film.

The quality of the approximation may be tested by considering a more general variational trial function, one allowing variation in the x direction. In this way, one may estimate that corrections to $H_{c\parallel}$ are of order $d^2/100\xi^2$, whereas corrections to $\left| dH_c(\theta)/d\theta \right|_{\theta=0}$ are of order $d^2/5\xi^2$. Thus, so long as $d/\xi \ll 1$, the simple results for the limiting values and for the interpolation for intermediate angles should be quite accurate. A somewhat more accurate form for the case when d/ξ is not too small has been given by Yamafuji et al.[1] Saint-James[2] has carried out exact computer solutions of the initial slope at $\theta = 0$. His results follow our simple analytic approximation (4-70a) quite well for $d < \xi$, but they reveal a singular change in behavior at the critical thickness $d_c \approx 1.8\xi$, at which surface solutions become favored. We now address ourselves to considering nucleation in films of such intermediate thicknesses. For simplicity, we shall restrict our attention to the case of a film in a parallel magnetic field.

4-10.2 Nucleation in Films of Intermediate Thickness

We start with $d \ll \xi$ and consider what happens as d increases. First, we must relax the approximation that ψ is independent of x. As remarked above, one way to do this is to use a variational approach. For example, we might take an x-dependence

$$1 + c \cos\frac{2\pi x}{d}$$

which satisfies the boundary condition $d\psi/dx = 0$ at $x = \pm d/2$, and choose c so as to minimize the free energy. The result of doing this is that $c > 0$, that is, ψ tends to decrease near the surfaces of the film, where $\frac{1}{2}m^*v_s^2$ is large, as is obvious on qualitative grounds. As remarked earlier, this improved ψ leads to a slightly higher critical field, namely,

$$H_{c\parallel} = \frac{2\sqrt{6}\,H_c\lambda}{d}\left(1 + \frac{9d^2}{\pi^6\xi^2}\right) \qquad (4\text{-}74)$$

This correction term never exceeds 3 percent, because when $d > d_c \approx 1.8\xi$, this symmetrical solution is superseded by the surface solution of Saint-James and de Gennes, suitably modified by the presence of two surfaces in close proximity. Let us now see how this comes about.

In our analysis of nucleation at a single surface, we found it optimum to choose k_y (or x_0) so that the minimum of the effective potential was located a distance $\sim \xi$ behind the surface. Evidently a qualitative change in the nature of

[1] K. Yamafuji, T. Kawashima, and F. Irie, *Phys. Letters* **20**, 123 (1966).
[2] D. Saint-James, *Phys. Letters* **16**, 218 (1965).

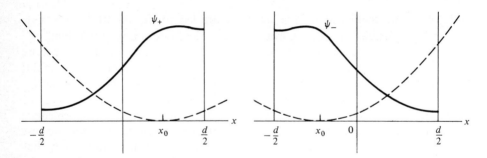

FIGURE 4-8
The two equivalent asymmetric solutions for nucleation in a film of intermediate thickness. The dashed curves centered at $\pm x_0$ represent the effective potential due to the magnetic field.

the solutions at the two surfaces must be expected when they are close enough together to cause these two minima to come together, i.e., at some critical thickness $d_c \sim 2\xi$. For $d < d_c$, the lowest eigenvalue (highest critical field) is obtained for $k_y = x_0 = 0$, so that the minimum of the effective potential is in the midplane of the film. These are the symmetric solutions of the previous paragraph. However, for $d > d_c$, the optimum place for x_0 shifts away from the center so as to stay an optimum distance behind one of the surfaces. It is difficult to predict the details of the changeover at d_c without numerical computation. The results of such work[1] are that $x_0 = k_y = 0$ below $d_c = 1.81\xi$, and that above d_c they rise from zero with infinite initial slope.

It is evident that when $x_0 \neq 0$, there must be two equivalent positions for the minimum, at $x = \pm |x_0|$ with $k_y = \pm |k_y|$, both having eigenfunctions of the linearized GL equation with exactly the same eigenvalue. These are shown schematically in Fig. 4-8. Either may be used for calculating the critical field, as was done by Saint-James and de Gennes for arbitrary values of d/ξ. However, as soon as one goes slightly below the nucleation field H_{c3} of the film, so that ψ becomes finite, then the nonlinear term of the full GL equations comes into play. This has the effect of resolving degeneracies in favor of extended solutions, since the $\beta|\psi|^4$ term penalizes wavefunctions with ψ peaked up in a small volume. Thus we must expect a linear combination of the two asymmetric solutions, with equal weights, such as

$$\psi = \psi_+ + \psi_- = e^{ik_y y}f(x) + e^{-ik_y y}f(-x)$$
$$= \cos k_y y[f(x) + f(-x)] + i \sin k_y y[f(x) - f(-x)] \qquad (4\text{-}75)$$

[1] D. Saint-James, *Phys. Letters* **16**, 218 (1965); H. J. Fink, *Phys. Rev.* **177**, 732 (1969); H. A. Schultens, *Z. Phys.* **232**, 430 (1970).

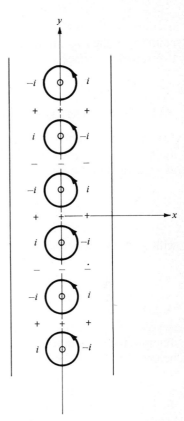

FIGURE 4-9
Vortex pattern in superconducting film of intermediate thickness set up by superposition of the two asymmetric solutions of Fig. 4-8. Notations $\pm$, $\pm i$ denote phase factor $e^{i\varphi}$ of ψ. Arrows indicate $\nabla\varphi$, to which J_s is proportional.

Because of the "interference" of the two solutions, there are nodes along the midplane ($x = 0$) whenever $\cos k_y y = 0$, that is, at intervals

$$\Delta y = \frac{\pi}{k_y} = \frac{\Phi_0}{2x_0 H} \approx \frac{\Phi_0}{H(d - d_c)} \qquad (4\text{-}76)$$

The approximate equality would hold if $|x_0|$ were given by $d/2 - d_c/2$, so that the potential minima stayed a constant distance behind the surfaces for $d > d_c$, as is approximately the case. If one works out the phase of (4-75), one finds that it varies through 2π in a circuit around each of these nodes, corresponding to vortices of current as shown in Fig. 4-9. Thus, by superimposing two essentially one-dimensional solutions, we have generated a two-dimensional solution with currents circulating around nodes of $|\psi|$.

As the slab gets thicker, the two surface solutions pull further apart, and the vortex spacing Δy decreases. According to the numerical calculations, after a certain point is reached, the vortex pattern becomes more complex. However, the overlap of the two surface waves $\psi_\pm$ also decreases. Eventually the energetic coupling of the two surface solutions becomes negligible compared to kT, and the two separate surface solutions become independent, as treated earlier.

4-11 THE ABRIKOSOV VORTEX STATE AT H_{c2}

Just as there are two degenerate surface solutions to the linearized GL equations at H_{c3} for a sufficiently thick slab, there are an infinite number of interior solutions at H_{c2}, each of the form

$$\psi_k = e^{iky}f(x) = \exp(iky) \exp\left[-\frac{(x - x_k)^2}{2\xi^2}\right] \qquad (4\text{-}77)$$

with k denoting k_y and
$$x_k = \frac{k\Phi_0}{2\pi H} \qquad (4\text{-}77a)$$

as found above. Each of these describes a gaussian slice of superconductivity at the plane $x = x_k$. All ψ_k are orthogonal because of the different e^{iky} factors. Each of these solutions is equally valid exactly at H_{c2}, and all give the same H_{c2}. Just as in the case of the thin film, however, as soon as H is reduced below H_{c2} by any finite amount, the minimum free-energy solution of the nonlinear equations will have to be one that fills the entire sample and that does so in such a way as to minimize the $\beta|\psi|^4$ term as well as the magnetic and kinetic-energy terms.

The two solutions for the film case automatically set up a periodicity in the y direction. Since qualitatively we expect a crystalline array of vortices to have lower energy than a random one, we want to enforce periodicity also in the interior solution. This can be done simply by restricting the values of k in (4-77) to a discrete set

$$k_n = nq \qquad (4\text{-}78)$$

in which case there will be a periodicity in y with period

$$\Delta y = \frac{2\pi}{q} \qquad (4\text{-}79)$$

This restriction also automatically facilitates a periodicity in x through (4-77a), since the gaussian solutions are located at

$$x_n = \frac{k_n\Phi_0}{2\pi H} = \frac{nq\Phi_0}{2\pi H} \qquad (4\text{-}80)$$

Thus, if all ψ_n enter with equal weight, there will be periodicity in the x direction with period

$$\Delta x = \frac{q\Phi_0}{2\pi H} = \frac{\Phi_0}{H\,\Delta y} \qquad (4\text{-}81)$$

From (4-81) we see at once that

$$H\Delta x\,\Delta y = \Phi_0 \qquad (4\text{-}82)$$

so that each unit cell of the periodic array carries one quantum of flux. Although we have shown this only for the case when $H = H_{c2}$, it is in fact true also below H_{c2} if we replace H by B. This can be seen simply from the fluxoid quantization discussion, since by symmetry, $\mathbf{J}_s = 0$ on the boundary between two identical cells of the vortex pattern.

More generally, we may consider a function

$$\psi_L = \sum_n C_n \psi_n = \sum_n C_n \exp\,(inqy)\exp\left[-\frac{(x-x_n)^2}{2\xi^2}\right] \qquad (4\text{-}83)$$

This is a general solution to the *linearized* GL equation at H_{c2}, periodic in y by construction. It will also be periodic in x if the C_n are periodic functions of n, such that $C_{n+\nu} = C_n$, for some ν. For example, the square lattice of Abrikosov arises if all C_n are equal, so that $\nu = 1$. This is the case considered above in the qualitative discussion leading to (4-81). On the other hand, a triangular lattice results if $\nu = 2$ and $C_1 = iC_0$.

All solutions of the form (4-83) are possible at H_{c2}. To determine which one should actually be observed, it is necessary to bring in the nonlinear terms and do some numerical calculations. As noted by Abrikosov, the parameter determining the relative favorability of various possible solutions is

$$\beta_A \equiv \frac{\langle \psi_L^4 \rangle}{\langle \psi_L^2 \rangle^2} \qquad (4\text{-}84)$$

This parameter is obviously independent of the normalization of ψ. It has the value unity if ψ is constant, and becomes increasingly large for functions which are more and more peaked up and localized. For example, if ψ were approximately constant over a localized fraction f of the volume, and approximately zero everywhere else, $\beta_A \approx f^{-1} \gg 1$.

An indication of why this quantity enters can be obtained as follows. First, imagine that ψ is forced to vary in space to satisfy external conditions, but that the variation in space is so slow that gradient and current terms in the energy may be neglected. Then the free energy may be approximated by the two terms in (4-2). To allow the form and the amplitude of ψ to be adjusted separately, we write $\psi(\mathbf{r}) = c\chi(\mathbf{r})$. Inserting this in (4-2) and minimizing with respect to c^2, we find $c^2 = -(\alpha/\beta)\langle \chi^2 \rangle/\langle \chi^4 \rangle$. When this is put back into (4-2), we find

$$\langle f_s - f_n \rangle = -\frac{\alpha^2}{2\beta}\frac{\langle \chi^2 \rangle^2}{\langle \chi^4 \rangle} = -\frac{\alpha^2}{2\beta}\beta_A^{-1} \qquad (4\text{-}85)$$

If χ is constant, so $\beta_A = 1$, this reduces to the usual condensation energy (4-4). If χ is not constant, $\beta_A > 1$, and the more β_A increases, the less favorable is the energy.

Again taking the extreme example of a localized solution filling only a fraction f of the sample volume, for which $\beta_A \approx f^{-1}$, we see that the condensation energy is only a fraction f of what would be obtained with a space-filling solution, for which $\beta_A \approx 1$. This result is, of course, intuitively very plausible, and it shows that any form of space-filling solution of the linearized equation will be clearly favored over any localized one, when the quartic terms in the free energy are taken into consideration.

A more realistic version of Eq. (4-85) can be derived by including in the free energy the term in Eq. (4-7) arising from the gradient of $|\psi|$, but still excluding currents and vector potentials, which require more complex self-consistent solutions. If that is done, and the calculation outlined above is repeated, an additional factor appears, so that Eq. (4-85) is replaced by

$$\langle f_s - f_n \rangle = -\left(\frac{\alpha^2}{2\beta}\right)\beta_A^{-1}\left[1 - \xi^2\frac{\langle|\nabla\chi|^2\rangle}{\langle\chi^2\rangle}\right]^2 \qquad (4\text{-}86)$$

This extra factor goes to zero at a second-order phase transition point, where χ satisfies the linearized GL equation. Thus, we see that β_A^{-1} still measures the effectiveness of a wavefunction with respect to minimizing the effect of the nonlinear terms, and the condensation energy still increases quadratically with temperature below the point of the second-order phase transition. That transition point is shifted, however, from T_c (where $\alpha \to 0$) to some lower temperature at which nucleation can occur in the presence of gradients. This temperature is determined by the vanishing of the final factor in Eq. (4-86), that is, by

$$\xi^2(T) = \frac{\langle\chi^2\rangle}{\langle|\nabla\chi|^2\rangle} \qquad (4\text{-}87)$$

Although we have not proved it here, these qualitative features of Eq. (4-86) carry through to the cases of interest here, in which magnetic field and current terms in the energy play a critical role.

Returning now to the optimization of (4-83), we see that it is equivalent to finding the set of C_n for which β_A is the smallest, since the final factor in (4-86) is the same for any linear combination of solutions with the same H_{c2}. Numerical calculations show that for the square lattice of Abrikosov, $\beta_A = 1.18$, while for the triangular lattice mentioned above (with $iC_{2n+1} = C_{2n} = \text{const}$), $\beta_A = 1.16$.[1] Considering this small difference, it is understandable that a numerical error could have led Abrikosov originally to conclude that the square array was more stable. Later work by Kleiner, Roth, and Autler[2] rectified this error, and showed that the

---

[1] Some feeling for these numbers can be gained by noting that they correspond to cells in which "normal cores" with $\psi = 0$ occupy some 15 percent of the cell area, with $\psi \approx \text{const}$ over the rest. In reality, ψ goes *smoothly* to zero at the center of each vortex.

[2] W. H. Kleiner, L. M. Roth, and S. H. Autler, *Phys. Rev.* **133**, A1226 (1964).

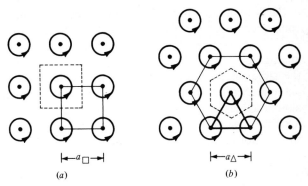

FIGURE 4-10
Schematic diagram of square and triangular vortex arrays. The dashed lines outline
the basic unit cell.

triangular array had in fact the most favorable value of β_A of all possible periodic
solutions.

It is interesting that this result agrees with that of a simple argument based
on the fact that the triangular array is a "closed-packed" one, in which each
vortex is surrounded by a hexagonal array of other vortices. (See Fig. 4-10.) In this
array, the nearest-neighbor distance is

$$a_\triangle = \left(\frac{4}{3}\right)^{1/4}\left(\frac{\Phi_0}{B}\right)^{1/2} = 1.075\left(\frac{\Phi_0}{B}\right)^{1/2} \qquad (4\text{-}88a)$$

whereas for the four neighbors in a square array

$$a_\square = \left(\frac{\Phi_0}{B}\right)^{1/2} \qquad (4\text{-}88b)$$

Thus, for a given flux density, $a_\triangle > a_\square$. Taking account of the mutual repulsion
of the vortices, it is reasonable that the structure with greatest separation of
nearest neighbors would be favored.

In general, experiments[1] confirm the triangular array, but in some materials,
symmetries of the underlying crystal structure appear to dominate over the small
theoretical energy difference for a structureless medium, leading to the observation of square or even rectangular arrays. Also, defects in the material may introduce sufficient inhomogeneity to destroy the regular array entirely, leading to
observation of a "liquidlike" rather than a crystalline array of vortices.

Since the detailed nature of the vortex array is unimportant for most applications, we shall not carry through the details of the Abrikosov solution. Rather,

[1] U. Essmann and H. Träuble, *Phys. Letters* **24A**, 526 (1967); A. Seeger, *Comments Solid State Phys.* **3**, 97 (1970); U. Essmann, *Phys. Letters* **41A**, 477 (1972).

in the next chapter, we shall outline the principal results and discuss their physical consequences largely with the simplification that the unit cell of the vortex array may be approximated by a circular one of the same area. This simplification is analogous to that of replacing the actual Wigner-Seitz cell by a spherical one of the same volume in the calculation of the electronic energy bands in solids. Typical numerical results such as β_A differ about as much between circular and hexagonal cells as between hexagonal and square, i.e., by a few percent. Of course, H_{c2} itself must be exactly the same for *any* solution of the linearized GL equation in the interior of an infinite medium.

5

MAGNETIC PROPERTIES OF TYPE II SUPERCONDUCTORS

In the previous chapter we found that superconductors with $\kappa > 1/\sqrt{2}$, that is, type II superconductors, have solutions of the GL equations with $|\psi| > 0$ until fields $H_{c2} > H_c$ are reached. In particular, the Abrikosov solution (4-83) corresponded to a regular array of vortices of current surrounding nodal lines of ψ. Each unit cell of the array carried total flux equal to $\Phi_0 = hc/2e$.

It is the objective of the present chapter to investigate the behavior of type II superconductors over the entire field range from zero to H_{c2}. This will allow us to find H_{c1}, the field at which flux first penetrates (in an ideally long, thin sample of zero demagnetizing factor), and the entire magnetization curve which describes the increase of B from zero at H_{c1} to H at H_{c2}. We shall then consider the effect of the Lorentz force due to a transport current on the flux in a type II superconductor, and show how it can lead to electrical resistance associated with flux creep or flow. Finally, we examine the design conflict between thermal stability and low ac loss in practical magnets.

5-1 BEHAVIOR NEAR H_{c1}: THE STRUCTURE OF AN ISOLATED VORTEX

When the first flux enters a type II superconductor, it is carried within an array of vortices sparsely distributed through the material. So long as the separation is large compared to λ, there will be negligible overlap or interaction of the vortices, so each can be treated in isolation. Given the axial symmetry of the situation, the problem reduces to finding a self-consistent solution of the GL equations for $\psi(r)$ and $h(r)$. From this, one can calculate the extra free energy ϵ_1 per unit length of the line. This determines H_{c1} in the following manner. By definition, when $H = H_{c1}$ the Gibbs free energy must have the same value whether the first vortex is in or out of the sample. Thus, at H_{c1}

$$G_s \bigg|_{\text{no flux}} = G_s \bigg|_{\text{first vortex}}$$

or, since $G = F - (H/4\pi) \int h \, d\mathbf{r}$,

$$F_s = F_s + \epsilon_1 L - \frac{H_{c1} \int h \, d\mathbf{r}}{4\pi}$$

$$= F_s + \epsilon_1 L - \frac{H_{c1} \Phi_0 L}{4\pi}$$

where L is the length of the vortex line in the sample. Thus,

$$H_{c1} = \frac{4\pi\epsilon_1}{\Phi_0} \qquad (5\text{-}1)$$

The calculation of ψ, h, and ϵ_1 for arbitrary κ unfortunately requires numerical solution of the GL equations. Thus, considerable attention has been given to the extreme type II limit, in which $\kappa = \lambda/\xi \gg 1$, because useful analytic results can be obtained. The simplification results because ψ can rise from zero to a limiting value (which will be ψ_∞ if we are dealing with an isolated vortex) within a "core" region of radius $\sim \xi$. Thus, over most of the vortex (of radius $\sim \lambda \gg \xi$) the superconductor will act like an ordinary London superconductor.

Before making this restrictive assumption, however, let us first see how far we can go in approaching the full solution of the nonlinear GL equations (4-13) and (4-14). It is convenient to introduce a vortex wavefunction of the form

$$\psi = \psi_\infty f(r)e^{i\theta} \qquad (5\text{-}2)$$

which builds in the axial symmetry and the fact that the phase of ψ varies by 2π in making a complete circuit, corresponding to there being a single flux quantum associated with the vortex. This phase choice for ψ fixes the gauge choice[1] for **A** so that

$$\mathbf{A} = A(r)\hat{\theta} \qquad (5\text{-}3)$$

with

$$A(r) = \left(\frac{1}{r}\right) \int_0^r r'h(r')\, dr' \qquad (5\text{-}3a)$$

Near the center of the vortex, this becomes

$$A(r) = \frac{h(0)r}{2} \qquad (5\text{-}3b)$$

whereas far from the center of an isolated vortex, it becomes

$$A_\infty = \frac{\Phi_0}{2\pi r} \qquad (5\text{-}3c)$$

since the total flux contained is $\oint \mathbf{A} \cdot d\mathbf{s} = 2\pi r A_\infty = \Phi_0$.

When (5-2) is substituted in the GL equation (4-13), we find, after simplifying, that f satisfies the equation

$$f - f^3 - \xi^2 \left[\left(\frac{1}{r} - \frac{2\pi A}{\Phi_0} \right)^2 f - \frac{1}{r}\frac{d}{dr}\left(r\frac{df}{dr} \right) \right] = 0 \qquad (5\text{-}4)$$

The current has only a θ component, and from (4-14) it is

$$J = -\frac{c}{4\pi}\frac{dh(r)}{dr} = -\frac{c}{4\pi}\frac{d}{dr}\left[\frac{1}{r}\frac{d}{dr}(rA) \right] = \frac{e^*\hbar}{m^*}\psi_\infty^2 f^2\left(\frac{1}{r} - \frac{2\pi A}{\Phi_0} \right) \qquad (5\text{-}5)$$

The problem is now to find simultaneous solutions of these two nonlinear differential equations for $f(r)$ and $A(r)$. Since this requires numerical methods, in general, we shall examine certain limiting cases in which progress can be made analytically.

First, let us look right at the center of the vortex, as $r \to 0$. Using (5-3b), (5-4) becomes

$$f - f^3 - \xi^2 \left[\left(\frac{1}{r} - \frac{\pi h(0)r}{\Phi_0} \right)^2 f - \frac{1}{r}\frac{d}{dr}\left(r\frac{df}{dr} \right) \right] = 0 \qquad (5\text{-}6)$$

---

[1] Note that this gauge choice is quite different from the London gauge, in which ψ is real, so $J \propto A$, and A vanishes exponentially with distance from the vortex. In the London gauge, our $A(r)$ is replaced by

$$A'(r) = A(r) - A_\infty = A(r) - \frac{\Phi_0}{2\pi r}$$

Since curl $(\hat{\theta}/r) = 0$ (except along the line $r = 0$), A and A' give the same $h(r)$, $J(r)$, and $|\psi(r)|$. We shall use $A(r)$ because it is nonsingular, but equivalent results could be obtained with either choice.

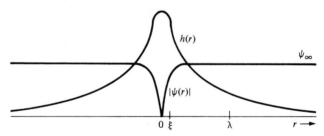

FIGURE 5-1
Structure of an isolated Abrikosov vortex in a material with $\kappa \approx 8$. The maximum value of $h(r)$ is approximately $2H_{c1}$.

Let us assume that for $r \approx 0$ the solution starts as

$$f = cr^n \qquad n \geq 0$$

Then (5-6) becomes

$$cr^n - c^3 r^{3n} - \xi^2 \left[\left(\frac{1}{r} - \frac{\pi h(0)r}{\Phi_0} \right)^2 cr^n - n^2 cr^{n-2} \right] = 0$$

As $r \to 0$, the leading term is proportional to $r^{n-2}(1 - n^2)$. For this to vanish, $n = 1$, and f must start out proportional to r at the origin. (It can readily be shown in general that if the vortex contains m quanta of flux, $\psi \sim e^{im\theta}$, and $f \sim r^m$, as $r \to 0$.) From the structure of (5-6), we can see that only odd powers of r now enter in the expansion of f. Working out the coefficient of the next term, we find

$$f \approx cr \left\{ 1 - \frac{r^2}{8\xi^2} \left[1 + \frac{h(0)}{H_{c2}} \right] \right\} \qquad (5\text{-}7)$$

which shows that the rise of $f(r)$ starts to saturate at $r \approx 2\xi$, as might be expected. To get the normalization constant c, we must go further to bring the f^3 term into play. [It plays no role in the terms $\sim f$ which give (5-7).] However, it is clear that c must be $\sim 1/2\xi$ for isolated vortices, so that the series (5-7) will join on to the distant solution, where $f \to 1$. A reasonable approximation to f over the entire range is

$$f \approx \tanh \frac{vr}{\xi} \qquad (5\text{-}8)$$

where v is a constant ~ 1. This dependence is sketched in Fig. 5-1.

5-1.1 The High-κ Approximation

Because f rises almost to unity in a distance $\sim \xi$, one can make a very convenient approximation when $\lambda \gg \xi$, or $\kappa \gg 1$. Namely, over all except a core region of radius $\sim \xi$, one can treat f as constant $\simeq 1$, in which case the London equations govern the fields and currents. Thus, outside the core

$$\frac{4\pi\lambda^2}{c} \text{curl } \mathbf{J}_s + \mathbf{h} = 0 \qquad (5\text{-}9)$$

If this relation held everywhere, the fluxoid for any path would be zero. We correct this by inserting a term to take account of the presence of the core, so that (5-9) becomes

$$\frac{4\pi\lambda^2}{c} \text{curl } \mathbf{J}_s + \mathbf{h} = \hat{\mathbf{z}}\Phi_0\delta_2(\mathbf{r}) \qquad (5\text{-}10)$$

where $\hat{\mathbf{z}}$ is a unit vector along the vortex and $\delta_2(\mathbf{r})$ is a two-dimensional δ function at the location of the core. Note that (5-10) can be obtained directly by taking the curl of (5-5). Combining (5-10) with the Maxwell equation

$$\text{curl } \mathbf{h} = \frac{4\pi}{c}\mathbf{J} \qquad (5\text{-}11)$$

we obtain

$$\lambda^2 \text{ curl curl } \mathbf{h} + \mathbf{h} = \hat{\mathbf{z}}\Phi_0\delta_2(\mathbf{r}) \qquad (5\text{-}12)$$

Since div $\mathbf{h} = 0$, this can be written

$$\nabla^2\mathbf{h} - \frac{\mathbf{h}}{\lambda^2} = -\frac{\Phi_0}{\lambda^2}\hat{\mathbf{z}}\delta_2(\mathbf{r}) \qquad (5\text{-}13)$$

This equation has the exact solution

$$h(r) = \frac{\Phi_0}{2\pi\lambda^2} K_0\left(\frac{r}{\lambda}\right) \qquad (5\text{-}14)$$

where K_0 is a zero-order Hankel function of imaginary argument. Qualitatively, $K_0(r/\lambda)$ cuts off as $e^{-r/\lambda}$ at large distances and diverges logarithmically as $\ln(\lambda/r)$ as $r \to 0$. Of course, in reality this divergence is cut off at $r \sim \xi$, where $|\psi|^2$ starts dropping to zero. Thus, $h(r)$ is actually regular at the center of the vortex, as shown in Fig. 5-1. More precisely, the two limiting forms of (5-14) are known to be

$$h(r) \to \frac{\Phi_0}{2\pi\lambda^2}\left(\frac{\pi}{2}\frac{\lambda}{r}\right)^{1/2} e^{-r/\lambda} \qquad r \to \infty \qquad (5\text{-}14a)$$

$$h(r) \approx \frac{\Phi_0}{2\pi\lambda^2}\left[\ln\frac{\lambda}{r} + 0.12\right] \qquad \xi \ll r \ll \lambda \qquad (5\text{-}14b)$$

These two limiting forms can be checked directly by considering the problem in the appropriate limits. For example, the logarithmic behavior of (5-14b) results from $J \propto v_s = \hbar/m^*r$, which follows from fluxoid quantization so long as $r \ll \lambda$, so that the flux enclosed in a circle of radius r is much less than Φ_0.

5-1.2 Vortex Line Energy

Now let us find the line tension, or free energy per unit length, ϵ_1. Neglecting the core, we have only the contributions from the field energy and the kinetic energy of the currents

$$\epsilon_1 = \frac{1}{8\pi} \int (h^2 + \lambda^2 |\operatorname{curl} \mathbf{h}|^2)\, dS \qquad (5\text{-}15)$$

This can be transformed by a vector identity to

$$\epsilon_1 = \frac{1}{8\pi} \int (\mathbf{h} + \lambda^2 \operatorname{curl} \operatorname{curl} \mathbf{h}) \cdot \mathbf{h}\, dS + \frac{\lambda^2}{8\pi} \oint (\mathbf{h} \times \operatorname{curl} \mathbf{h}) \cdot d\mathbf{s}$$

$$= \frac{1}{8\pi} \int |\mathbf{h}| \Phi_0\, \delta_2(\mathbf{r})\, dS + \frac{\lambda^2}{8\pi} \oint (\mathbf{h} \times \operatorname{curl} \mathbf{h}) \cdot d\mathbf{s} \qquad (5\text{-}15a)$$

where the line integrals are around the inner and outer perimeter of the integration area. Since the integration excludes the core, the first term contributes nothing. The second term goes to zero at infinity, but gives a finite contribution in encircling the core, namely

$$\epsilon_1 = \frac{\lambda^2}{8\pi} \left[h \frac{dh}{dr}\, 2\pi r \right]_\xi$$

Using Eq. (5-14b), $dh/dr = \Phi_0/2\pi\lambda^2 r$, so this reduces to

$$\epsilon_1 = \frac{\Phi_0}{8\pi} h(\xi) \approx \frac{\Phi_0}{8\pi} h(0) \qquad (5\text{-}16)$$

where $h(\xi) \approx h(0)$ because $f \to 0$ and hence $J_s \to 0$ in the core.* Using (5-14b) again, but dropping the 0.12 as not significant in view of the approximation made in imposing a cutoff at ξ,

$$\epsilon_1 \approx \left(\frac{\Phi_0}{4\pi\lambda} \right)^2 \ln \kappa \qquad (5\text{-}17)$$

Since this depends only logarithmically on the core size, the result should be quite reliable despite the crude treatment of the core.

The magnitude can be reexpressed in more physical terms using the relation (4-20), namely, $\Phi_0 = 2\sqrt{2}\,\pi\lambda\xi H_c$. Thus,

$$\epsilon_1 = \frac{H_c^2}{8\pi} 4\pi\xi^2 \ln \kappa \qquad (5\text{-}17a)$$

This shows that the line energy is of the same order of magnitude as the condensation energy lost in the core, but it is larger by a factor of order $4 \ln \kappa$. Thus, for $\kappa \gg 1$, errors in handling the core should be unimportant. Similarly, we can see

---

* It is interesting to note that Eq. (5-16) would also arise [but from the *first* term of (5-15a)] if the core were *not* excluded from the integration.

that ϵ_1 is of the order of the total field energy $h^2/8\pi$, since $h \approx \Phi_0/\pi\lambda^2$, and this is integrated over an area $\sim \pi\lambda^2$, giving an energy $\Phi_0^2/8\pi^2\lambda^2$. This estimate differs from (5-17) only by the absence of a factor of $\frac{1}{2}\ln\kappa \approx 1$.

Now that we have evaluated the line tension ϵ_1, we can substitute back into (5-1) to get H_{c1}, the field at which flux first penetrates. This is

$$H_{c1} = \frac{4\pi}{\Phi_0}\epsilon_1 \approx \tfrac{1}{2}h(0) \approx \frac{\Phi_0}{4\pi\lambda^2}\ln\kappa = \frac{H_c}{\sqrt{2}\,\kappa}\ln\kappa \qquad (5\text{-}18)$$

Thus, apart from the $\ln\kappa$ term, $H_c/H_{c1} = H_{c2}/H_c = \sqrt{2}\,\kappa$, so that H_c is approximately the geometric mean of H_{c1} and H_{c2}.

5-2   INTERACTION BETWEEN VORTEX LINES

If we continue to make the approximation $\kappa \gg 1$, it is easy to treat the interaction energy between two vortices, since in this approximation the medium is linear and we may use superposition. Thus, the field is given by

$$\mathbf{h}(\mathbf{r}) = \mathbf{h}_1(\mathbf{r}) + \mathbf{h}_2(\mathbf{r})$$
$$= [h(|\mathbf{r} - \mathbf{r}_1|) + h(|\mathbf{r} - \mathbf{r}_2|)]\hat{\mathbf{z}} \qquad (5\text{-}19)$$

where $\mathbf{r}_1$ and $\mathbf{r}_2$ specify the positions of the cores of the two vortex lines and $h(r)$ is given by (5-14). The energy may be calculated by substituting this into (5-15) and using vector transformations as we did in reaching (5-16). The result for the total increase in free energy per unit length can be written

$$\Delta F = \frac{\Phi_0}{8\pi}[h_1(\mathbf{r}_1) + h_1(\mathbf{r}_2) + h_2(\mathbf{r}_1) + h_2(\mathbf{r}_2)]$$
$$= 2\left[\frac{\Phi_0}{8\pi}h_1(\mathbf{r}_1)\right] + \frac{\Phi_0}{4\pi}h_1(\mathbf{r}_2)$$

by symmetry. The first term is just the sum of the two individual line energies. The second term is the interaction energy we were looking for

$$F_{12} = \frac{\Phi_0 h_1(\mathbf{r}_2)}{4\pi} = \frac{\Phi_0^2}{8\pi^2\lambda^2}K_0\left(\frac{r_{12}}{\lambda}\right) \qquad (5\text{-}20)$$

As noted above, this falls off as $r_{12}^{-1/2}e^{-r_{12}/\lambda}$ at large distances and varies logarithmically at small distances. The interaction is *repulsive* for the usual case, in which the flux has the same sense in both vortices.

We may compute the *force* arising from this interaction by taking a derivative of F_{12}. For example, the force on line 2 in the x direction is

$$f_{2x} = -\frac{\partial F_{12}}{\partial x_2} = -\frac{\Phi_0}{4\pi}\frac{\partial h_1(\mathbf{r}_2)}{\partial x_2} = \frac{\Phi_0}{c}J_{1y}(\mathbf{r}_2) \qquad (5\text{-}21)$$

using the Maxwell equation: curl $\mathbf{h} = 4\pi\mathbf{J}/c$. Putting this back into vector form, the force per unit length on vortex 2 is

$$\mathbf{f}_2 = \mathbf{J}_1(\mathbf{r}_2) \times \frac{\mathbf{\Phi}_0}{c} \qquad (5\text{-}22)$$

where the direction of $\mathbf{\Phi}_0$ is parallel to the flux density. Making the obvious generalization to an arbitrary array,

$$\mathbf{f} = \mathbf{J}_s \times \frac{\mathbf{\Phi}_0}{c} \qquad (5\text{-}23)$$

where now $\mathbf{J}_s$ represents the total supercurrent density due to all other vortices (and even including any net transport current) at the location of the core of the vortex in question.

An implication of (5-23) is that a vortex line can be in static equilibrium at any given position only if the superfluid velocity from all other sources is zero there. This can be accomplished if each vortex is surrounded by a symmetrical array, as in the square or triangular arrays discussed above. However, it turns out that the square array has only *unstable* equilibrium, so that small displacements tend to grow. On the other hand, the triangular array is stable,[1] as is reasonable since it has the lowest energy. Further, this result gives warning that even the triangular array will feel a force transverse to any transport current, so that the vortices will move unless "pinned" in place by inhomogeneities in the medium. Since flux motion causes energy dissipation and induces a longitudinal "resistive" voltage, this situation is crucial in determining the usefulness of type II superconductors in the construction of high-field superconducting solenoids, where strong currents and fields inevitably must coexist. We shall return to this point a little later.

5-3 MAGNETIZATION CURVES

Now that we have worked out the energy of a single vortex and the interaction energy between two vortices, we can work out the magnetization curve from the first flux penetration up to the vicinity of H_{c2}. One can distinguish three regimes between H_{c1} and H_{c2}:

1 Very near H_{c1}, $\Phi_0/B \gg \lambda^2$, and the vortices are separated by more than λ. In this case only a few nearest neighbors are important.
2 For moderate values of B such that $\xi^2 \ll \Phi_0/B \ll \lambda^2$, many vortices are within interaction range of any given one, so that more elaborate summing procedures are required. However, it is still a good approximation to neglect details of the core.

---

[1] See, for example, A. L. Fetter, P. C. Hohenberg, and P. Pincus, *Phys. Rev.* **147**, 140 (1966).

3   Near H_{c2}, $\xi^2 \approx \Phi_0/B$, so that the cores are almost overlapping. This requires more detailed treatment of the core. Our simple superposition technique is no longer accurate, but the Abrikosov solution (4-83) to the linearized GL equation at H_{c2} is a helpful approximation.

In the first two of these regimes, we can write the increase in Gibbs function per unit volume as

$$G - G_{s0} = \frac{B}{\Phi_0}\epsilon_1 + \sum_{i>j} F_{ij} - \frac{BH}{4\pi} \qquad (5\text{-}24)$$

where we have used the fact that B/Φ_0 gives the number of vortices per unit area normal to the field. For fixed H, B should then take on the value which minimizes G. Since $\sum_{i>j} F_{ij}$ is positive and increases as B increases, we see at once that for $H < H_{c1} = 4\pi\epsilon_1/\Phi_0$, the minimum of G occurs when $B = 0$, in agreement with our earlier analysis. If H is slightly above H_{c1}, flux will enter until limited by the increase of the interaction energy between the vortices. Formally,

$$\frac{\partial G}{\partial B} = 0 = \frac{\epsilon_1}{\Phi_0} - \frac{H}{4\pi} + \frac{\partial}{\partial B}\sum_{i>j}F_{ij}$$

so B is determined by

$$\frac{\partial}{\partial B}\sum_{i>j}F_{ij} = \frac{H - H_{c1}}{4\pi} \qquad (5\text{-}25)$$

5-3.1   Low Flux Density

To start, let us consider the first regime and assume a regular array of vortices such that each vortex has z nearest neighbors at a distance $a = c(\Phi_0/B)^{1/2}$. For example, from Eq. (4-88), $c_\square = 1$ for the square array ($z_\square = 4$) and $c_\triangle = 1.075$ for the triangular array ($z_\triangle = 6$). Given the exponential decrease of F_{ij}, we neglect all contributions except those of the nearest neighbors. Then, using Eq. (5-20)

$$\sum_{i>j} F_{ij} = \left(\frac{B}{\Phi_0}\right)\frac{z}{2}\frac{\Phi_0^2}{8\pi^2\lambda^2}K_0\left(\frac{a}{\lambda}\right)$$

$$\approx \frac{Bz\Phi_0}{16\pi^2\lambda^2}\left(\frac{\pi\,\lambda}{2\,a}\right)^{1/2}e^{-a/\lambda} \qquad (5\text{-}26)$$

Because this varies exponentially with a but only linearly with z, we can see that the triangular array will surely be lower in energy if a is large enough (i.e., if B is small enough).

Given Eq. (5-26), it is straightforward to take the derivative with respect to B (including the change occurring via the change in a with B) and insert the result in Eq. (5-25) to find $B(H)$. In view of the dominance of the exponential variation of F_{ij} with $a/\lambda = c(\Phi_0/B)^{1/2}/\lambda$, we might anticipate that $(H - H_{c1})$ would vary roughly as $e^{-a/\lambda}$, so that B would vary inversely as the square of the logarithm of

$H - H_{c1}$. This turns out to be the case, the leading term at H_{c1} being

$$B = \frac{2\Phi_0}{\sqrt{3}\,\lambda^2}\left\{\ln\left[\frac{3\Phi_0}{4\pi\lambda^2(H - H_{c1})}\right]\right\}^{-2} \quad (5\text{-}27)$$

Note that B is continuous at H_{c1}, corresponding to a second-order phase transition, but it rises there with infinite initial slope. The physical reason for this steep rise is that, once $H > H_{c1}$, so that the first vortex can enter, there is little to inhibit a large number of vortices from entering, until they get within a distance $\sim \lambda$ of each other. Since $H_{c1} \approx (\Phi_0/4\pi\lambda^2)\ln\kappa$, the mutual repulsion will not become strong until B has risen from zero to some significant fraction of H_{c1}. In fact, experimental data[1] show that in some systems B rises a finite amount discontinuously, exactly at H_{c1}, corresponding to a first-order phase transition.

Since such a first-order transition is contrary to the predictions of ordinary GL theory, as reviewed here, it has been the subject of considerable recent work. The basic qualitative requirement is that at some intermediate distance the vortices have a lower energy than at infinite separation, although they repel at still smaller separations. It seems plausible that such a situation might arise from the eventual field reversal which accompanies screening in the true nonlocal electrodynamics of superconductors in which κ is not too large. [See the discussion preceding (3-8).] This effect is completely missing in the GL theory, which uses local electrodynamics. In fact, detailed calculations of Eilenberger and Büttner,[2] using the microscopic theory, showed the existence of a damped oscillatory component in the variation of $\Delta(r)$ and $A(r)$ if $\kappa \leq 1.7$. Although this behavior is suggestive of an explanation of the locally attractive potential, this calculation has been criticized by Cleary.[3] A very thorough alternate discussion has been given more recently by Jacobs,[4] using somewhat different techniques. It appears from these various approaches that the more microscopic theories will be able to account for the conditions of κ, temperature, and mean free path under which the discontinuous entry of a finite flux density at H_{c1} is observed. Since questions remain open at the time of this writing, however, we shall not go further into the topic here.

5-3.2   Intermediate Flux Densities

For flux densities in the range $\Phi_0/\lambda^2 \lesssim B \ll \Phi_0/\xi^2$, which corresponds to $H_{c1} \lesssim B \ll H_{c2}$, we can continue to use our modified London equation to treat the electrodynamics, but we must carefully include the interaction with many neighbors. This is most conveniently done by Fourier analysis of the local flux

[1] U. Kumpf, *Phys. Stat. Sol.* **44**, 829 (1971); **52**, 653 (1972); J. Auer and H. Ullmaier, *Phys. Rev.* **B7**, 136 (1973).
[2] G. Eilenberger and H. Büttner, *Z. Phys.* **224**, 335 (1969).
[3] R. M. Cleary, *Phys. Rev. Letters* **24**, 940 (1970).
[4] A. E. Jacobs, *Phys. Rev.* **B4**, 3016, 3022, and 3029 (1971); *J. Low Temp. Phys.* **10**, 137 (1973).

density $h(x, y)$ in a plane perpendicular to the field. Since the vortex array is periodic, a Fourier series is used:

$$h_z(\mathbf{r}) = \sum_Q h_Q e^{i\mathbf{Q} \cdot \mathbf{r}} \qquad (5\text{-}28)$$

where the $\mathbf{Q}$'s run over the two-dimensional reciprocal lattice of the array. For example, in a square array, the $\mathbf{Q}$'s are of the form

$$\mathbf{Q}_{mn} = \frac{2\pi}{a_\square} (m\hat{\mathbf{x}} + n\hat{\mathbf{y}}) \qquad (5\text{-}29)$$

In the triangular array, the primitive translations are not orthogonal, and the situation is a bit more complicated. If we take the translations in coordinate space to be

$$\mathbf{a}_1 = a_\triangle \hat{\mathbf{x}} \qquad \mathbf{a}_2 = \frac{a_\triangle}{2} (\hat{\mathbf{x}} + \sqrt{3}\,\hat{\mathbf{y}}) \qquad (5\text{-}30)$$

then the $\mathbf{Q}$'s are linear combinations of integral multiples of

$$\mathbf{Q}_1 = \frac{2\pi}{a_\triangle}\left(\hat{\mathbf{x}} - \frac{\hat{\mathbf{y}}}{\sqrt{3}}\right) \qquad \mathbf{Q}_2 = \frac{2\pi}{a_\triangle}\frac{2}{\sqrt{3}}\hat{\mathbf{y}} \qquad (5\text{-}31)$$

which have the required property that $\mathbf{a}_i \cdot \mathbf{Q}_j = 2\pi\delta_{ij}$. For the moment, though, we retain the generality of an arbitrary periodic array.

   We now determine the coefficients h_Q by requiring that $h(x, y)$ satisfy the modified London equation (5-12)

$$\mathbf{h} + \lambda^2 \operatorname{curl} \operatorname{curl} \mathbf{h} = \Phi_0\, \hat{\mathbf{z}} \sum_i \delta_2(\mathbf{r} - \mathbf{r}_i) \qquad (5\text{-}32)$$

where the sum over $\mathbf{r}_i$ runs over the locations of the various vortices. Inserting the series (5-28), this becomes

$$\sum_Q (h_Q + \lambda^2 Q^2 h_Q) e^{i\mathbf{Q} \cdot \mathbf{r}} = B \sum_Q e^{i\mathbf{Q} \cdot \mathbf{r}} \qquad (5\text{-}33)$$

In obtaining the right member, we have taken the origin of coordinates at the center of a vortex, and have used the fact that the area of the unit cell is Φ_0/B. Solving for h_Q, we find

$$h_Q = \frac{B}{1 + \lambda^2 Q^2} \qquad (5\text{-}34)$$

so that

$$h_z(\mathbf{r}) = B \sum_Q \frac{e^{i\mathbf{Q} \cdot \mathbf{r}}}{1 + \lambda^2 Q^2} \qquad (5\text{-}35)$$

   Given $h(\mathbf{r})$, we can calculate the increase in free energy per unit length (still neglecting core effects) by

$$F - F_{s0} = \frac{1}{8\pi} \int [h^2 + \lambda^2 |\operatorname{curl} \mathbf{h}|^2]\, dS$$

Using the same vector transformations as were used to reach Eq. (5-16) or Eq. (5-20), this can be reduced to

$$F - F_{s0} = \frac{\Phi_0}{8\pi} \sum_i h(\mathbf{r}_i)$$

where the $h(\mathbf{r}_i)$ are the total fields at the cores of the vortices, i.e., self-field plus fields due to other vortices. Since all $h(\mathbf{r}_i) = h(0)$, and since there are B/Φ_0 vortices per unit area, we can write the increase in F per unit volume as

$$F - F_{s0} = \frac{Bh(0)}{8\pi} = \frac{B^2}{8\pi} \sum_Q \frac{1}{1 + \lambda^2 Q^2} \qquad (5\text{-}36)$$

using Eq. (5-35). As it stands, this sum is not convergent, since the number of $\mathbf{Q}$ values between Q and $Q + \delta Q$ varies as $Q\, \delta Q$. Thus, a cutoff is required at high Q. It is reasonable to take this as $Q_{max} \approx 1/\xi$, since Fourier components higher than $1/\xi$ come dominantly from the spurious logarithmic singularity of $h(r)$ as $r \to 0$ which results from use of Eq. (5-32) in the core region where $|\psi|^2$ is going to zero.

Without actually evaluating the sum in Eq. (5-36), which depends on the lattice structure of the vortices, let us use it formally to define the Gibbs free energy $G = F - BH/4\pi$. In this case, the equilibrium value of B is found by setting $\partial G/\partial B = 0$. This leads to the general relation

$$H = \frac{1}{2}\left[h(0) + B\frac{dh(0)}{dB} \right] \qquad (5\text{-}37)$$

If $B = 0$, this reduces to $\frac{1}{2}h(0)$, which should be H_{c1}; comparing with Eq. (5-18), we see that this is correct. For arbitrary values of B, both terms play a role, and they may be computed by summing the series. Since the term with $\mathbf{Q} = 0$ will dominate when the vortices are highly overlapping and the field almost uniform, it is useful to take this term out explicitly. When this is done, Eq. (5-37) can be written as

$$H = B\left\{ 1 + \frac{1}{2} \sum_Q{}' [(1 + \lambda^2 Q^2)^{-1} + (1 + \lambda^2 Q^2)^{-2}] \right\} \qquad (5\text{-}38)$$

where the summation is now over $\mathbf{Q} > 0$. Once $B \gg H_{c1}$, so that the vortices are highly overlapping, the second term under the summation sign becomes negligible compared to the first, and may be dropped. In the same spirit, we may make an integral approximation to the first sum, integrating from Q_{min} to Q_{max}, with weighting factor $(\Phi_0/2\pi B)|\mathbf{Q}| d|\mathbf{Q}|$. In this, $Q_{min}^2 \approx 4\pi B/\Phi_0$, to take account of the omitted term in which $Q = 0$, and $Q_{max} \approx 1/\xi$. Proceeding in this way, and neglecting numerical factors, (5-38) can be roughly approximated

$$H \approx B + H_{c1} \frac{\ln (H_{c2}/B)}{\ln \kappa} \qquad (5\text{-}39)$$

which gives a reasonable description of the data well above H_{c1}.

5-3.3 Regime Near H_{c2}

Near H_{c2}, the vortices are packed so tightly that the cores fill much of the volume. Thus, it becomes essential to abandon the simple approach of the modified London theory and go back to the full GL equations. Fortunately, the small value of ψ/ψ_∞ as one approaches H_{c2} makes it possible to use an expansion scheme starting from the Abrikosov solution for the linearized problem (4-83). It turns out to be a good approximation to retain the form of the linearized solution, simply scaling the periodicity to correspond to $B(H)$ rather than to H_{c2}, in which case $\psi(\mathbf{r})$ can be characterized by a simple amplitude parameter, for example, $\langle\psi^2\rangle$. Although we shall not go through the calculations here, Abrikosov was able to show that the local flux density was less than the applied field by an amount proportional to the local value of $|\psi(\mathbf{r})|^2$. As a result $M = (B - H)/4\pi$ is proportional to $\langle\psi^2(\mathbf{r})\rangle$. Since $\langle\psi^2(\mathbf{r})\rangle$ goes to zero linearly with $(H_{c2} - H)$ in the second-order phase transition at H_{c2}, this leads to M vanishing in the same way. The explicit result is

$$B = H + 4\pi M = H - \frac{H_{c2} - H}{(2\kappa^2 - 1)\beta_A} \tag{5-40}$$

where $\beta_A = \langle\psi^4\rangle/\langle\psi^2\rangle^2 \geq 1$ is the parameter introduced in (4-84) to characterize the inconstancy of ψ over space. As noted there, the parameter β_A is independent of the amplitude of ψ, depending only on the configuration of the particular solution of the linearized GL equations one is using. The most important case is the triangular lattice, for which $\beta_A = 1.16$; for the square lattice, it is 1.18. Note that our approximation (5-39) agrees with this result in the limit $B \approx H \approx H_{c2}$ if $2\kappa^2 \gg 1$ and if we take $\beta_A \approx 1$.

In view of the thermodynamic relation

$$\left(\frac{\partial G}{\partial H}\right)_T = -\frac{B}{4\pi} \tag{5-41}$$

the lattice with the lowest value of β_A is most stable. We can see this by integrating down from the normal state at H_{c2}, where $G_s(H_{c2}) = G_n(H_{c2})$ in all cases. Then

$$G_s(H) = G_n(H_{c2}) + \frac{1}{4\pi}\int_H^{H_{c2}} B(H)\,dH$$

$$= G_n(H_{c2}) + \frac{H_{c2}^2 - H^2}{8\pi} - \frac{1}{8\pi}\frac{(H_{c2} - H)^2}{(2\kappa^2 - 1)\beta_A} \tag{5-42}$$

which makes it clear that the lower the value of β_A, the lower the value of $G_s(H < H_{c2})$, as anticipated in (4-86). Thus, the triangular lattice is more stable than the square lattice near H_{c2} as well as near H_{c1}.

An important feature of (5-40) is the presence of a factor $(2\kappa^2 - 1)$ in the denominator. This factor goes to zero at $\kappa = 1/\sqrt{2}$, which is exactly the criterion for distinguishing type I and II superconductors. This is reasonable, since a type I superconductor undergoes a first-order transition at H_c, in which $4\pi |M|$ rises discontinuously from 0 to H_c. The qualitative change in the shape of the magnetization curve with the value of κ is sketched in Fig. 5-2. Despite these changes in shape, however, the area under the curve is in *all* cases given by the condensation energy $H_c^2/8\pi$. This may be shown by integrating (5-41), noting that $F = G$ when $H = 0$, and that $G_s = G_n$ when $H > H_{c2}$. Thus, whatever the value of κ, we can write

$$-\int M \, dH = F_n(0) - F_s(0) = \frac{H_c^2}{8\pi} \qquad (5\text{-}43)$$

Given the various relations derived or stated above, κ values can be determined in several ways from the experimental magnetization curves. Very near T_c, where the simple GL equations can be rigorously justified by microscopic theory, one expects all these determinations to be consistent, but at lower temperatures there may be small discrepancies. Maki[1] first treated these deviations theoretically from a microscopic point of view; he introduced the notations κ_1, κ_2, and κ_3 for the values determined by fitting data to the three relations

$$H_{c2} = \sqrt{2}\,\kappa_1 H_c \qquad (5\text{-}44a)$$

$$4\pi \left.\frac{dM}{dH}\right|_{H_{c2}} = (2\kappa_2^2 - 1)^{-1}\beta_A^{-1} \qquad (5\text{-}44b)$$

$$H_{c1} = H_c \frac{\ln \kappa_3}{\sqrt{2}\,\kappa_3} \qquad (5\text{-}44c)$$

More properly, (5-44c) should be replaced by the exact numerical relation between H_{c1}/H_c and κ, as first worked out by Harden and Arp,[2] to which (5-44c) is an analytic approximation, valid for large κ. This pioneering work of Maki was subsequently extended by Eilenberger,[3] whose work showed that the different temperature dependences of the various values of κ_i should depend not only on ℓ/ξ_0, but also on the degree of anisotropy in the impurity scattering. Roughly speaking, these differences among the κ_i reflect the different degrees of sensitivity of the various magnetic properties of a superconductor to the degree of nonlocality of the electrodynamics. Only as $T \to T_c$ does the true electrodynamics become fully local, and then all the κ_i approach a common limiting value, usually denoted simply κ, without subscript. Since the various κ_i in type II materials usually differ from κ by less than 20 percent, our results for a single, constant value retain semiquantitative validity despite all these complications. Hence, we shall not go further into these matters here.

[1] K. Maki, *Physics* **1**, 21, 127 (1964).
[2] J. L. Harden and V. Arp, *Cryogen.* **3**, 105 (1963).
[3] G. Eilenberger, *Phys. Rev.* **153**, 584 (1967).

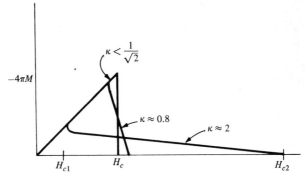

FIGURE 5-2
Comparison of magnetization curves for three superconductors with the same value of thermodynamic critical field H_c, but different values of κ. For $\kappa < 1/\sqrt{2}$, the superconductor is of type I and exhibits a first-order transition at H_c. For $\kappa > 1/\sqrt{2}$, the superconductor is type II and shows second-order transitions at H_{c1} and H_{c2} (for clarity, marked only for the highest κ case). In all cases, the area under the curve is the condensation energy $H_c^2/8\pi$.

5-4 FLUX PINNING, CREEP, AND FLOW

From the practical point of view, the most useful aspect of type II superconductivity to date has been the ability to make superconducting solenoids which can supply steady fields of over 100,000 G (gauss) without dissipation of energy, because of the resistanceless persistent current. A comparable field produced by a water-cooled copper solenoid would require a steady dissipation of 2 MW (megawatts) of power, with attendent cooling problems, and it would not have the essentially infinite stability of the superconducting magnet.

Making such magnets has not come easily. The superconducting material must not only have a critical field substantially higher than the field to be produced, but it must also be able to carry a high current in that field without resistance. The first requirement is well met by many dirty superconductors, since

$$H_{c2} = \frac{\Phi_0}{2\pi\xi^2} \approx \frac{\Phi_0}{2\pi\xi_0\ell} \approx \frac{3ck}{e}\frac{T_c}{v_F\ell} \approx 3 \times 10^4 \frac{T_c}{v_F\ell} \qquad (5\text{-}45)$$

Thus given a high T_c and low Fermi velocity, a low mean free path can lead to a value of H_{c2} of up to $\sim 250,000$ Oe (oersteds) for such brittle materials as V_3Si and Nb_3Sn, and many materials with much more convenient mechanical properties have $H_{c2} \sim 100,000$ Oe.

The real problem lies in finding a material able to carry a usefully high current in the presence of this strong, penetrating field without dissipation of energy. Any appreciable dissipation leads to heating, which degrades the perform-ance further, leading to intolerable catastrophic flux jumps. The origin of the dissipation is the Lorentz force density

$$\mathbf{F} = \mathbf{J} \times \frac{\mathbf{B}}{c} \qquad (5\text{-}46)$$

between the current in the superconductor and the flux threading through it. We derived this earlier (5-23) in the form

$$\mathbf{f} = \mathbf{J} \times \frac{\mathbf{\Phi}_0}{c} \qquad (5\text{-}46a)$$

which is the force on a single vortex. Because of this force, flux lines tend to move transverse to the current. If they do move, say with velocity $\mathbf{v}$, they essentially "induce" an electric field of magnitude

$$\mathbf{E} = \mathbf{B} \times \frac{\mathbf{v}}{c} \qquad (5\text{-}47)$$

which is parallel to $\mathbf{J}$. This acts like a resistive voltage, and power is dissipated.

Looking into this more carefully, the Maxwell equation relating the current density to the curl of the field can be written in two ways. The classic form is

$$\operatorname{curl} \mathbf{H} = 4\pi \frac{\mathbf{J}_{ext}}{c} \qquad (5\text{-}48a)$$

in which $\mathbf{J}_{ext}$ represents only externally imposed currents, not the currents arising from the equilibrium response of the medium. For example, we saw that in the equilibrium intermediate state of a type I superconductor in a magnetic field, H was H_c everywhere, and $\mathbf{J}_{ext}$ was zero. Alternatively, the Maxwell equation can be written microscopically, taking the medium as vacuum, so that it has the form

$$\operatorname{curl} \mathbf{h} = 4\pi \frac{\mathbf{J}}{c} \qquad (5\text{-}48b)$$

where $\mathbf{J}$ is now the *total* current, including the equilibrium response of the medium. Again taking the case of the intermediate state, this equation describes the variation of $\mathbf{h}$ between $\mathbf{h}_n$ and 0 as a result of the microscopic screening currents in the London penetration depth of each superconducting lamina. Finally, $\mathbf{B}$ is the spatial average $\bar{\mathbf{h}}$ over the laminar (or other) structure.

If we pass now to the case of an ideal type II superconductor in the mixed (i.e., vortex) state with no transport current imposed, H is again everywhere equal to the applied value, since $\mathbf{J}_{ext} = 0$, while $\mathbf{B} = \bar{\mathbf{h}}$ drops from H to $B_{eq}(H)$ in a surface layer of depth $\sim \lambda$, where there is a microscopic surface current $\mathbf{J}$. Since this situation is, by definition, the equilibrium one, it is clear that there can be no net force on any vortices, even those which are exposed to this surface current. We conclude from this example that the current density which determines the net

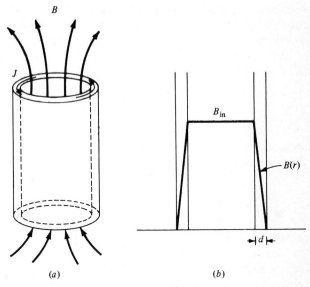

FIGURE 5-3
Flux trapped in hollow cylinder of type II superconductor. (*a*) Sketch of overall geometry. (*b*) Local flux-density profile.

driving force on a vortex is not the total **J**, but only the nonequilibrium part. By convention, we shall denote this by $\mathbf{J}_{ext}$, even though in some cases it may reflect only a nonequilibrium state of an isolated system. Thus, precisely speaking, the force density on the vortices which tends to make them move is

$$\boldsymbol{\alpha} = \mathbf{J}_{ext} \times \frac{\mathbf{B}}{c} = (\text{curl } \mathbf{H}) \times \frac{\mathbf{B}}{4\pi} \qquad (5\text{-}49)$$

This effective force density $\boldsymbol{\alpha}$ differs from the force density (5-46) in leaving out equilibrium forces sustained by the medium, which do not tend to displace flux lines relative to the medium. Although conceptually it is important to make this careful distinction between **B** and **H** and between **J** and $\mathbf{J}_{ext}$, it should be recognized that in most applications of high-κ type II superconductors, $\mathbf{B} \approx \mathbf{H}$, because the microscopic screening currents associated with $4\pi M$ in equilibrium are much less than the useful transport currents. Thus, in the following discussion, we shall for simplicity sometimes ignore the difference between **B** and **H** and between **J** and $\mathbf{J}_{ext}$.

A convenient example is a hollow superconducting cylinder containing trapped flux held in by a current circulating in the wall. (See Fig. 5-3.) This model problem can be considered to be an idealization of a superconducting solenoid operating in the persistent current mode, i.e., with the leads connected by a super-conducting short circuit. If the wall thickness d is small compared to the radius R, we may reduce the problem to a one-dimensional one by neglecting the curvature of the wall. In this case the Maxwell equation given as Eq. (5-48a) can be written as

$$\frac{dH}{dx} = -4\pi \frac{J_{ext}}{c}$$

with H in the z direction and J_{ext} in the y direction. There is an outward force on each vortex of $f = (\Phi_0/4\pi)\,|\,dH/dx\,|$ per unit length. Summed over all the vortices, this gives a force per unit volume of

$$\alpha = \frac{B}{4\pi}\left|\frac{dH}{dx}\right| \qquad (5\text{-}50)$$

To the approximation that $B \approx H$, this is equal to the gradient of the magnetic pressure $B^2/8\pi$, the difference being the part of the magnetic pressure taken up by the ideal medium rather than by the vortex pinning at material defects. Unless this force α is balanced by some other force, we must expect the vortices to move outward through the wall, each carrying one quantum of flux with it, reducing the trapped flux inside and resulting in a corresponding decrease in J. The decreasing flux implies an induced emf around the ring in the direction of the current and given by

$$\mathscr{E} = 2\pi RE = -\frac{1}{c}\frac{d\Phi}{dt} = \frac{1}{c}2\pi RBv \qquad (5\text{-}51)$$

where E is the local electric field, and v is the outward velocity of the flux density B. This result is seen to be equivalent to (5-47). It could also have been arrived at by applying the Lorentz transformation for a change of coordinates from a frame moving with the vortices to one at rest in the laboratory.

Although derived by an induction argument for a transient situation, (5-51) is also correct in steady-state situations in which an external energy source (such as a battery) maintains $\mathbf{B}$ and $\mathbf{J}$ steady on a macroscopic scale while vortices steadily move through the medium.* In either case, the electric field leads to energy dissipation at the rate $\mathbf{E} \cdot \mathbf{J}$. This corresponds to the decrease in magnetic-field energy in the hollow cylinder, or to the energy abstracted from the external source in the steady-state case.

The expression (5-51) for the emf $\mathscr{E}$ can be recast in a useful and significant way if we express the flux Φ in terms of the flux quantum so that $\Phi = n\Phi_0$. Then

$$\mathscr{E} = -\frac{1}{c}\frac{d\Phi}{dt} = -\frac{\Phi_0}{c}\frac{dn}{dt} = \frac{\hbar\omega}{2e} \qquad (5\text{-}52)$$

---

* An interesting analysis of the corresponding case of the driven motion of domains in the intermediate state has been given by P. R. Solomon and R. E. Harris, *Phys. Rev.* **B3,** 2969 (1971).

where $\omega/2\pi$ is the frequency with which vortices leak out, or equivalently, the rate at which the fluxoid quantum number of the circuit is decreasing. Since the integrated phase change of $|\psi| e^{i\varphi}$ around a path, $\Delta\varphi = \oint \mathbf{V}\varphi \cdot d\mathbf{s} = 2\pi n$, where n is the fluxoid quantum number, we can also write (5-52) in the form

$$2e\mathscr{E} = e^*\mathscr{E} = \hbar \frac{d\,\Delta\varphi}{dt} = \hbar\omega \qquad (5\text{-}52a)$$

This is a form of the so-called Josephson frequency relation, which will be discussed at length in the next chapter.

From the above arguments, we conclude that (above H_{c1}) a type II superconductor will show resistance and be unable to sustain a persistent current unless some mechanism exists which prevents the Lorentz force from moving the vortices. Such a mechanism is called a "pinning" force, since it "pins" the vortices to fixed locations in the material. Pinning results from any spatial inhomogeneity of the material, since local variations of ξ, λ, or H_c due to impurities, grain boundaries, voids, etc., will cause local variations of ϵ_1, the free energy per unit length of a flux line, causing some locations of the vortex to be favored over others. To be most effective, these inhomogeneities must be on a scale of the order of λ or ξ, that is, $\sim 10^{-6}$ to 10^{-5} cm, rather than on the atomic scale where inhomogeneity causes electronic scattering which limits the mean free path ℓ. If the pinning is sufficiently strong, vortex motion can be made small enough that the superconductor acts very much like a perfect conductor. However, there will always be thermally activated flux "creep," in which vortices hop from one pinning site to another, and in some cases this will occur at a measurable rate. If the pinning is weak compared to the driving force, the vortices move in a rather steady motion, at a velocity limited by viscous drag. This regime is referred to as flux "flow," and usually gives a "flow resistance" ρ_f which is comparable with ρ_n, the resistance of the material in the normal state. Hence, for practical applications flux flow must be avoided and the creep rate held to a low level.

5-5 FLUX FLOW

Let us first consider the ideal case where there is no pinning and find what properties should be expected in the case of pure flux flow, in which vortex motion is retarded only by viscous damping. One can write down a simple phenomenology to start, simply assuming a viscous drag coefficient η such that the viscous force per unit length of a vortex line moving with velocity $\mathbf{v}_L$ is $-\eta\mathbf{v}_L$. Equating this to the driving force (5-46a), we find the magnitudes related by

$$J \frac{\Phi_0}{c} = \eta v_L$$

Combining this with (5-47), we find the space-averaged fields related by

$$\rho_f = \frac{E}{J} = B \frac{\Phi_0}{\eta c^2} \qquad (5\text{-}53)$$

Thus, to the extent that η is independent of B, ρ_f should be proportional to B.

This analysis reduces the problem to that of finding η, which can be expressed in energetic terms by noting that the rate of energy dissipation per unit length of vortex is

$$W = -\mathbf{F} \cdot \mathbf{v}_L = \eta v_L^2 \qquad (5\text{-}54)$$

However, we still must find how the dissipation actually occurs due to a moving vortex. Two general approaches have been made to this problem: a relatively elementary model developed first by Bardeen and Stephen,[1] and somewhat later a more rigorous analysis by Schmid,[2] by Caroli and Maki,[3] and by Thompson and Hu,[4] using time-dependent Ginzburg-Landau theory. For the present, we shall content ourselves with the former approach, since it gives a clearer picture of the actual dissipation mechanism; the more sophisticated treatment is outlined in Chap. 8.

5-5.1 The Bardeen-Stephen Model

This model makes the approximation that the superconductor is local. Further, it assumes that there is a finite core of radius $\sim \xi$ which is fully normal, and that the dissipation occurs by ordinary resistive processes in this core. Since the idea of a fully normal core is central to this model, while clearly only a simplifying approximation, it is important that we scrutinize the assumption to see how well justified it is.

The basis in microscopic theory for the concept of the normal core is the calculation of Caroli, de Gennes, and Matricon[5] of the quasi-particle excitation spectrum in a pure superconductor containing a vortex. They found that although $\psi(r)$ vanishes only at $r = 0$, rising roughly as $\psi_\infty(r/\xi)$ in the core region, there is a density of low-lying excitations localized in the vortex which is about the same as that of a normal cylinder of radius $\sim \xi$. This illustrates that $\Delta(r) \propto \psi(r)$ need not correspond to an actual local energy gap if ψ is varying in space. We shall return to this point in Chap. 8.

[1] J. Bardeen and M. J. Stephen, *Phys. Rev.* **140**, A1197 (1965).
[2] A. Schmid, *Phys. Kondensierten Materie* **5**, 302 (1966).
[3] C. Caroli and K. Maki, *Phys. Rev.* **159**, 306, 316 (1967); **164**, 591 (1967); **169**, 381 (1968).
[4] C. R. Hu and R. S. Thompson, *Phys. Rev.* **B6**, 110 (1972).
[5] C. Caroli, P. G. de Gennes, and J. Matricon, *Phys. Letters* **9**, 307 (1964). See also, P. G. de Gennes, "Superconductivity of Metals and Alloys," p. 153, W. A. Benjamin, New York (1966).

One can arrive at a similar estimate of the size of the normal core within the framework of the GL theory by finding the radius at which the vortex circulation velocity $v_s = \hbar/m^*r$ is large enough to reduce $|\psi|$ to zero according to the relation (4-34), namely,

$$\frac{|\psi|^2}{\psi_\infty^2} = 1 - \frac{m^{*2}\xi^2 v_s^2}{\hbar^2} = 1 - \left(\frac{\xi}{r}\right)^2$$

Thus, this model gives $r_{\text{core}} = \xi$. It should be emphasized, however, that this use of (4-34) is really unjustified, since it was derived for the case of a velocity that is *uniform* in space. As we have noted, the actual solution of the GL equations for the vortex case shows that ψ is nonzero except for r exactly equal to zero, contrary to the above relation.

An intermediate point of view can be obtained by referring to the shift of the excitation spectrum (by $\mathbf{v}_s \cdot \mathbf{p}_F$) used to obtain (4-41) and discussed further in Sec. 8-1. Again, this is really valid only for a uniform case, but if we apply it locally, we find that there will be gapless excitations inside a radius $\sim \xi$, if $\Delta/\Delta_\infty \approx \tanh r/\xi$ and $v_s = \hbar/m^*r$.

Finally, we note that since $H_{c2} = \Phi_0/2\pi\xi^2$, at H_{c2} normal cores of radius $\sim \xi$ would fill about half the sample volume, again a qualitatively reasonable situation, since at H_{c2} there is a second-order transition to the fully normal state.

From the general agreement of all these approaches, we conclude that the concept of a quasi-normal core of radius $\sim \xi$ should have considerable validity, particularly for excitations, even if it is not strictly justified. For definiteness and simplicity, we take an obviously oversimplified model in which there is a sharp discontinuity at a radius $a \approx \xi$ between a fully normal core and fully superconducting material. We then treat the problem using the London equations outside the core and Ohm's law inside.

We can find the microscopic electric field $\mathbf{e}$ outside the core by using the first London equation

$$\mathbf{e} = \frac{\partial}{\partial t}(\Lambda \mathbf{J}_s) = \frac{\partial}{\partial t}\left(\frac{m^*\mathbf{v}_s}{e^*}\right)$$

$$= -\mathbf{v}_L \cdot \nabla\left(\frac{m^*\mathbf{v}_s}{e^*}\right) = -\mathbf{v}_L \cdot \nabla\left(\frac{\hbar}{2e}\frac{\hat{\boldsymbol{\theta}}}{r}\right) \qquad (5\text{-}55)$$

For example, if $\mathbf{v}_L$ is along the x direction,

$$\mathbf{e} = -\left(\frac{v_{Lx}\Phi_0}{2\pi c}\right)\frac{\partial}{\partial x}\left(\frac{\hat{\boldsymbol{\theta}}}{r}\right) = \left(\frac{v_{Lx}\Phi_0}{2\pi cr^2}\right)(\cos\theta\,\hat{\boldsymbol{\theta}} - \sin\theta\,\hat{\mathbf{r}}) \qquad (5\text{-}55a)$$

where θ is measured from the x direction. This field pattern, sketched in Fig. 5-4, has the form of the field of a line of electric dipoles, and averages to zero over the volume $r > a$ for which it applies. Thus, any overall average electric field will have

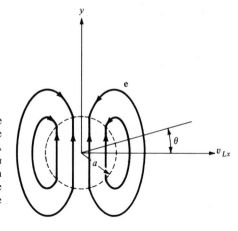

FIGURE 5-4
Schematic diagram of local electric field **e** near a moving vortex line. Dashed circle of radius a marks perimeter of core. A suitable surface charge is required at $r = a$ to be consistent with the discontinuity in the normal component of **e**. In a more exact model, the discontinuity would be smeared out.

to come from the core. Requiring continuity of tangential **e** at $r = a$ with that given by (5-55a), we find a uniform field inside the core, namely,

$$\mathbf{e}_{\text{core}} = \frac{v_{Lx}\Phi_0}{2\pi a^2 c}\hat{\mathbf{y}} \qquad (5\text{-}56)$$

Given $\mathbf{e}_{\text{core}}$, the dissipation of energy per unit length of core is

$$W_{\text{core}} = \pi a^2 \sigma_n e_{\text{core}}^2 = \frac{v_L^2 \Phi_0^2}{4\pi a^2 c^2 \rho_n} \qquad (5\text{-}57)$$

According to Bardeen and Stephen, there is an additional equal amount of dissipation by normal currents in the transition region outside the core. This is readily verified near T_c, since integration using (5-55a) shows that

$$\int_a^\infty \int_0^{2\pi} e^2(r)r\, dr\, d\theta = \frac{v_L^2 \Phi_0^2}{4\pi a^2 c^2}$$

Thus, if $\sigma_{1s} \approx \sigma_n$, as is the case near T_c, there will be exactly as much dissipation outside the core as inside. Away from T_c, the argument is less simple.

Other mechanisms of dissipation have been suggested. For example, even before the advent of the Bardeen-Stephen theory, Tinkham[1] had shown that dissipation comparable to that observed could be explained if the GL wavefunction could adjust to the time-varying field configurations of a moving vortex only

---

[1] M. Tinkham, *Phys. Rev. Letters* **13**, 804 (1964).

in a finite relaxation time τ, such that $\tau^{-1} \approx \Delta(0)/\hbar$ at $T = 0$, decreasing as $(1 - t)$ near T_c. This qualitative conjecture was confirmed by the subsequent work of Schmid and others, cited above, in developing a time-dependent GL theory based on the microscopic theory. In fact, this approach now appears to offer the most rigorous treatment of the problem. Another approach has been given by Clem.[1] He has shown that dissipation results from the irreversible entropy flow from trailing to leading edge of the vortex, corresponding to entropy increase at the leading edge where superconducting material is converted to normal, and the reverse at the trailing edge. It is still not entirely clear to what extent all these various mechanisms are additive, and to what extent they simply provide alternate ways of looking at the same thing.

To maintain the simplicity of our approach, we shall follow the Bardeen-Stephen results, which certainly give the main qualitative features, but it should be borne in mind that the model is rather oversimplified. Thus, we add a factor of 2 to (5-57) to allow for dissipation outside the core and equate the result to (5-54) to obtain

$$\eta = \frac{\Phi_0^2}{2\pi a^2 c^2 \rho_n} \approx \frac{\Phi_0 H_{c2}}{\rho_n c^2} \qquad (5\text{-}58)$$

Substituting this in (5-53), we obtain for the flow resistance ρ_f

$$\frac{\rho_f}{\rho_n} = \frac{2\pi a^2 B}{\Phi_0} = \left(\frac{a}{\xi}\right)^2 \frac{B}{H_{c2}} \approx \frac{B}{H_{c2}} \qquad (5\text{-}59)$$

In writing the final approximate equalities in (5-58) and (5-59), we have used the fact that we expect the core radius a to be approximately equal to ξ. If we simply set $a = \xi$, ρ_f joins smoothly onto ρ_n at H_{c2}, which is reasonable since the transition there is of second order. In fact, the experimental data for temperatures $T \ll T_c$ follow this simple result (5-59) reasonably well over the entire field range.

Note that this simple form would *not* result from a simple, static distribution of normal cores, even if the fraction of normal metal were B/H_{c2}. In that case, the currents would simply avoid the normal cores and flow only through the super-conducting matrix. The *motion* of the vortices is essential. Given this motion [with η given by (5-58)], it turns out that the (normal) current density in the core just equals the applied transport current driving the motion. Thus, the transport current flows right through the moving cores, producing dissipation. If the vortex velocity is not just right, for example because of the other contributions to η mentioned above or because of pinning forces, then these two current densities are not equal. This causes a "backflow" pattern of current to be superimposed on the

[1] J. R. Clem, *Phys. Rev. Letters* **20**, 735 (1968).

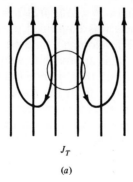
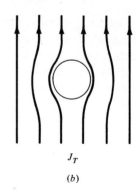

J_T

(a)

J_T

(b)

FIGURE 5-5
Backflow current pattern at pinned vortex. (a) Uniform transport current J_T and backflow current separately. (b) Superimposed current pattern, with zero current in core.

uniform transport current, as indicated in Fig. 5-5. The sense of circulation is such as to give a core current density less than the transport current density J_T. Evidently, this leads to smaller dissipation in the core, i.e., to a lower electrical resistance. In the limit of a stationary core, there would be no resistance.

5-5.2 Onset of Resistance in a Wire

Although not of much practical significance, it is of conceptual interest to apply these ideas to the case of an isolated wire of type II material carrying a current sufficient to cause a field at the surface greater than H_{c1}. Circumferential vortex rings will then form at the surface, shrink, and annihilate at the center, giving resistance. Let us see how the observed behavior should differ from that of type I superconductors, treated in Sec. 3-5 using the London model of a static intermediate state.

First, the critical current for the appearance of resistance will be

$$I_{c1} = \tfrac{1}{2}H_{c1}ca \qquad (5\text{-}60)$$

where a is now the radius of the wire. This I_{c1} will generally be quite small, since $H_{c1} < H_c$. Once a vortex ring is created at the surface, how will it shrink? Equating the rate of decrease of vortex-ring energy to the energy dissipated, we have

$$\frac{d}{dt}(2\pi r \epsilon_1) = 2\pi r \eta \left(\frac{dr}{dt}\right)^2$$

Thus,

$$r\frac{dr}{dt} = \frac{\epsilon_1}{\eta} \qquad (5\text{-}61)$$

and each vortex ring shrinks to annihilation in a time

$$T = \frac{\eta a^2}{2\epsilon_1} = \frac{H_{c2}}{H_{c1}} \frac{a^2}{\rho_n c^2} \qquad (5\text{-}62)$$

where we have used the definitions of H_{c1}, H_{c2}, and η. For typical values, $T \approx 10^{-5}$ sec. A consequence of (5-61) is that the radial velocity $dr/dt \propto 1/r$; this implies that the time spent in any given radial increment δr is proportional to dt/dr or r. Thus, $B(r) \propto r$, just as was the case in the static intermediate-state structure of the London theory.

The analysis above is appropriate only very near I_{c1}, where interactions between vortex rings can be neglected, since $B \approx 0$ in the wire. To derive more general relations, we equate the driving force to the viscous force:

$$\frac{J\Phi_0}{c} = \eta v$$

The inward radial velocity v of the vortex rings can be replaced by cE/B, where E is the induced longitudinal electric field and $\mathbf{B(r)}$ is, as usual, the local value of $\mathbf{h(r)}$ after averaging out the vortex structure. Also, in view of the discussion leading to (5-49), we may replace $\mathbf{J}$ by $(c/4\pi)$ curl $\mathbf{H}$. With these substitutions, we obtain

$$\frac{B}{r} \frac{\partial}{\partial r}(rH) = \frac{4\pi\eta cE}{\Phi_0} \qquad (5\text{-}63)$$

where B and H are circumferential, while E is longitudinal.

Taking first the case of $I \approx I_{c1}$, we may take $H \approx H_{c1}$ throughout the wire, since $dB/dH \to \infty$ at H_{c1}, causing $H \approx H_{c1}$ to be consistent with any small flux density. Then $\partial(rH)/\partial r$ will be simply H_{c1}, and Eq. (5-63) can be written

$$B(r) = \frac{4\pi\eta cEr}{\Phi_0 H_{c1}} \qquad (5\text{-}64)$$

which is proportional to r, as anticipated above. To find E, and hence the resistance, we equate the surface value of B to $B(H_s)$, where $H_s = 2I/ca$ is the surface value of H, and $B(H_s)$ refers to the relationship derived in Sec. 5-3. If we also introduce the expression given in Eq. (5-58) for η, we can finally express E in terms of the corresponding normal-state value, so that

$$\frac{E}{E_n} = \frac{R}{R_n} = \frac{B(H_s)}{2H_{c2}} \qquad I \approx I_{c1} \qquad (5\text{-}65)$$

In view of the form of $B(H)$ at H_{c1}, R/R_n rises with infinite slope at I_{c1}.

The other simple limit occurs at much higher currents (near I_{c2}), where it is

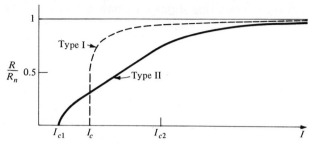

FIGURE 5-6
Onset of resistance in a wire of ideal type II superconductor with no pinning and
$\kappa \approx 1.7$. For comparison, the dashed curve shows the corresponding behavior of a
type I superconductor with the same H_c.

reasonable to make the approximation $B = H$ over most of the radius. In this
case, (5-63) can be integrated (after multiplying through by r^2), with the result

$$H^2 = B^2 = \frac{8\pi\eta cEr}{3\Phi_0} \qquad (5\text{-}66)$$

Note that now B and H are proportional to $r^{1/2}$, whereas near I_{c1}, H was constant
and B varied as r. In both limits, the product BH is proportional to r, with very
nearly the same coefficient. Relating the electric field to that in the normal state at
the same current, as above, we find

$$\frac{E}{E_n} = \frac{R}{R_n} = \frac{3}{4}\frac{H_s}{H_{c2}} = \frac{3}{4}\frac{I}{I_{c2}} \qquad (5\text{-}67)$$

Finally, if $I > I_{c2}$, $H_s > H_{c2}$, and an outer shell of the conductor must be
completely normal, while the inner core has the effective resistivity $\frac{3}{4}\rho_n$, according
to (5-67). Combining these two conductances in parallel, the effective average
conductivity is $(1 + r_1^2/3a^2)\rho_n^{-1}$, where r_1 is the radius of the inner core. Requiring
$H(r_1) = H_{c2}$, we find

$$\frac{r_1}{a} = \frac{2I}{I_{c2}}\left[1 - \left(1 - \frac{3I_{c2}^2}{4I^2}\right)^{1/2}\right]$$

$$\rightarrow \frac{3I_{c2}}{4I} \qquad I \gg I_{c2} \qquad (5\text{-}68)$$

from which the current dependence of the resistance above I_{c2} may be readily
computed. The limiting form at high currents is

$$\frac{R}{R_n} = 1 - \frac{3}{16}\left(\frac{I_{c2}}{I}\right)^2 \qquad I \gg I_{c2} \qquad (5\text{-}69)$$

Thus, just as with the type I superconductor treated in Sec. 3-5, a partially superconducting core persists, which vanishes only as $I \to \infty$, if we neglect heating effects.

These various results are combined graphically in Fig. 5-6, where they are compared with R/R_n for a type I superconductor with the same H_c. For practical purposes, it is important only to note that R is of the same order of magnitude as R_n as soon as I appreciably exceeds I_{c1}, which is typically less than the I_c of a type I superconductor. Hence, a type II superconductor with ideal flux-flow characteristics would be of little value as a lossless current conductor. Pinning is essential for that purpose.

5-5.3 Experimental Verification of Flux Flow

Numerous experiments have been carried out to test the ideas described above. The basic ones were those of Kim and coworkers[1] which measured the resistance of strips of type II superconductors in perpendicular fields. It was these experimental results which stimulated the development of the flux-flow theory in the first place. However, because of the deceptively simple form of the result $\rho/\rho_n \approx B/H_{c2}$, it was thought possible that the reasonable agreement with experiment might be fortuitous, and that another explanation (not involving a time-dependent flux pattern) might hold. With this motivation, other ingenious experiments were performed to test the idea of flux motion more specifically, although the visual studies of domain motion in the intermediate state of type I superconductors made the motion of individual vortices in type II materials seem very plausible. We shall mention only two of these experiments.[2]

The first is the dc transformer of Giaever.[3] This consists of two superconducting films separated by a thin insulating layer. He found that when a current was passed through one film (the "primary"), so as to set up a flux-flow voltage, an equal voltage appeared in the "secondary" film even though no current passed through it. This result is exactly what one would expect from the flux-flow picture, since the moving flux spots in the primary would drag along similar flux spots in the secondary, leading to the same induced voltage. The transformer action fails if slippage occurs[4] between the two vortex arrays because of too large a separation between the films or because of a flux pattern which is too weakly modulated in space. No plausible explanation of this dc transformer action has been given in terms of a stationary resistive structure in the primary.

[1] Y. B. Kim, C. F. Hempstead, and A. R. Strnad, *Phys. Rev. Letters* **12**, 145 (1964); *Phys. Rev.* **139**, A1163 (1965).

[2] For a more complete survey, see Y. B. Kim and M. J. Stephen, in R. D. Parks (ed.), "Superconductivity," chap. 19, Marcel Dekker, New York (1969).

[3] I. Giaever, *Phys. Rev. Letters* **15**, 825 (1965).

[4] For particularly accurate recent work on this point, see J. W. Ekin, B. Serin, and J. R. Clem, *Phys. Rev.* **B9**, 912 (1974).

The other class of experiments we mention is the study of the frequency spectrum of the noise superimposed on the dc resistive voltage in superconductors. Such measurements were initiated by van Ooijen and van Gurp,[1] who reasoned that if vortices move across a conductor of width W at velocity $v_L = cE/B$, then the voltage should really be the sum of a large number of flat-topped pulses of duration $\tau = W/v_L$. If so, there should be a sort of "shot" noise spectrum, which cuts off above $\omega \approx 1/\tau$, and whose amplitude is a measure of the amount of flux moving in each independent, discrete entity. The measurements showed such noise, with a spectrum reasonably much as expected. However, except at the highest currents and fields, they found that the moving entities carried more than a single flux quantum Φ_0; typically as many as 1,000 such quanta appeared to move as a unit. Presumably this result is due to pinning, which is expected to cause flux to move in "bundles" of vortices such that the total force on the bundle is sufficient to overcome a pinning barrier. In fact, the experiments give evidence that some of the vortices stay pinned, while others then must move faster than would be the case if all were moving. The conclusion is that the concept of flux motion is generally correct, but that defects in real material samples considerably complicate the idealized picture.

5-5.4 Concluding Remarks on Flux Flow

In the discussion above, we have tacitly assumed that the vortices would move in a purely transverse direction under the influence of the Lorentz force. This is in contrast with the behavior of vortices in a liquid, which, to a first approximation, drift along with the current. If the flux-bearing vortices move with a component of velocity along $\mathbf{J}_T$, this will lead to a transverse, or "Hall-effect," voltage, given by the same $\mathbf{B} \times \mathbf{v}_L/c$ effect referred to in Eq. (5-47). We have ignored this possibility above because most experimental data indicate that the Hall angle is small. In fact the Bardeen-Stephen model leads to a Hall effect of the same size as in the normal state for fields equal to those present in the core. Since most samples of type II materials studied are alloys or intermetallic compounds with rather short electronic mean free paths, the Hall angle is small and rather hard to measure with confidence. For example, any structural asymmetry left from the fabrication of the samples (such as a rolling direction) will tend to channel the vortices along a particular direction. This "guided motion" determines a spurious Hall angle having no intrinsic significance.

Another problem related to sample quality is caused by the pinning which is present in all real samples. This forces one to use a finite (and perhaps large) current before any flow of vortices occurs. Thus one is not really able to measure the linear magnetoresistive behavior considered in the idealized theory. Efforts have been made to reduce pinning by careful annealing, and it appears that

---

[1] D. J. van Ooijen and G. J. van Gurp, *Phys. Letters* **17,** 230 (1965).

pinning effects can be minimized by superimposing ac currents on the dc one.[1] Still, one is never completely certain that intrinsic properties are being accurately determined. Nonetheless, sufficient progress has been made that a meaningful comparison with the theory is becoming possible. For example, it appears that the Hall angle in alloys[2] increases above the value in the normal state at H_{c2}, a result which is inconsistent with the Bardeen-Stephen model. Maki[3] has shown that an explanation of this effect may be provided by the time-dependent Ginzburg-Landau theory.

Another aspect of flux flow is that the moving vortex transports *entropy* as well as flux. A crude estimate of this transported entropy can be made for isolated vortices by simply taking the excess entropy of the normal core above that of a similar volume of superconducting material. Of course, more rigorous means are needed near H_{c2}, when the vortices overlap strongly. The existence of entropy transport leads to thermomagnetic effects such as the Ettingshausen effect. That is, for steady state in a thermally isolated conductor, a transverse temperature gradient must develop parallel to the vortex flow to give a reverse flow of entropy equal to that carried by the vortices. These effects have been studied particularly extensively by Solomon and Otter,[4] and more recently by Vidal.[5] A reasonably satisfactory theoretical understanding of the quantitative results seems to exist.

As a final remark, we note that rather ideal flux-flow resistance data can be obtained in the presence of pinning by making measurements at microwave frequencies. The point is that pinning force $-k\,\delta x$ are proportional to the amplitude of the displacement of a vortex from its equilibrium position, while the viscous drag force is $\eta v = \eta\omega\,\delta x$. Thus, above a characteristic frequency $\omega_0 = k/\eta$, the pinning force becomes negligible. Gittleman and Rosenblum[6] were able to observe this changeover from a pinned regime showing no resistance to a flow regime with a well-defined ρ_f. It occurred typically near 10^7 Hz in their experiments. Thus measurements of surface resistance of type II superconductors at microwave frequencies ($\sim 10^{10}$ Hz) generally are not complicated by pinning effects.

5-6 THE CRITICAL-STATE MODEL

We now move to the other extreme, in which pinning is strong enough to prevent any substantial vortex motion and associated electrical resistance. Although pinning forces and driving forces presumably act on individual vortices, it is appropriate to adopt a more macroscopic view because the motion of individual

[1] J. A. Cape and I. F. Silvera, *Phys. Rev. Letters* **20**, 326 (1968).

[2] See, for example, C. H. Weijsenfeld, *Phys. Letters* **28A**, 362 (1968).

[3] K. Maki, *Phys. Rev. Letters* **23**, 1223 (1969).

[4] P. R. Solomon and F. A. Otter, Jr., *Phys. Rev.* **164**, 608 (1967); P. R. Solomon, *Phys. Rev.* **179**, 475 (1969).

[5] F. Vidal, *Phys. Rev.* **B8**, 1982 (1973).

[6] J. I. Gittleman and B. Rosenblum, *Phys. Rev. Letters* **16**, 734 (1968).

vortices is largely prevented by their mutual repulsion, which as we have seen tends to impose a crystalline structure on the array. Thus, one must expect flux to move in "bundles" when the driving force per unit volume exceeds the pinning force available in the same volume. This is in agreement with the noise measurements of van Ooijen and van Gurp cited above. Since the force per unit volume is

$$\boldsymbol{\alpha} \equiv \mathbf{J}_{ext} \times \frac{\mathbf{B}}{c} \qquad (5\text{-}70)$$

the condition for zero dissipation is that α never exceed the maximum available pinning force per unit volume, α_c.

The implications of this concept are made clear by returning to consideration of the hollow superconducting cylinder, introduced above in Sec. 5-4. Assume we start with a large field B_0 inside the cylinder (perhaps established by a solenoid in the bore, which is then turned off). So long as $B_0 > H_{c1}$, flux-bearing vortices will immediately start to enter the wall, and the decrease of the flux left in the bore will induce a current in the wall. Since this current density is $J_\theta = (c/4\pi)dH/dr$, there will be a very strong current if $H(r)$ drops abruptly from B_0 to 0 in the wall. When put into (5-70) the combination of high field and high current will lead to $\alpha > \alpha_c$. In this case, vortices will penetrate further into the wall, tending to reduce the gradient term. This process will continue until

$$\alpha \equiv |\boldsymbol{\alpha}| = \frac{J_{ext} B}{c} = \frac{1}{4\pi} B \frac{dH}{dr} \approx \frac{d}{dr}\left(\frac{B^2}{8\pi}\right) \le \alpha_c \qquad (5\text{-}71)$$

everywhere. This situation is called the critical state.

Depending on the relation between B_0 and the wall thickness, the critical state will be reached in one of two ways, as illustrated in Fig. 5-7: (a) α may drop below α_c before the flux penetrates all the way to the outside, so that all the initial flux is still in the bore or in the wall; or (b) if the wall thickness is inadequate to confine the initial amount of flux, flux will leak out through the wall until that which remains can be retained with $\alpha \le \alpha_c$ everywhere. Evidently the maximum B inside that can be retained (with $B = 0$ outside) is given by integrating the approximate form of (5-71), namely,

$$\frac{B_{max}^2}{8\pi} = \int_{R_{in}}^{R_{out}} \alpha_c[B(r)]\, dr \qquad (5\text{-}72)$$

where we have allowed for the possibility that α_c is a function of the local flux density (as well as temperature, etc.).

If we treat α_c as constant, (5-72) shows that the maximum B that can be held in a hollow cylinder should increase as the square root of the wall thickness. This has an important implication for the design of superconducting solenoids; namely, the winding thickness must increase roughly as the square of the field in the bore. [Since (5-72) assumes complete freedom of the local current density to adjust as a function of r, the actual performance of a solenoid will generally be less

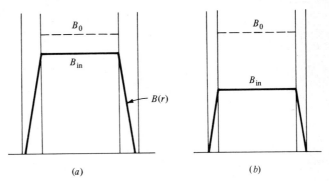

FIGURE 5-7
The critical state in a hollow superconducting cylinder. In (*a*), the wall thickness is sufficient to trap all the initial flux. In (*b*), the walls are too thin to do so. For simplicity, field profiles have been drawn using the Bean model, in which $J_c \propto dB/dr$ is constant. The same value of J_c has been taken in both (*a*) and (*b*).

favorable than (5-72) implies, even if multiple windings are used to give some control over $J(r)$.] Then, since the winding thickness is typically large compared to the bore radius, the mass of superconducting material required rises roughly as the square of the wall thickness, or as B^4 (for constant length). Thus, even before one approaches H_{c2}, where α_c must be expected to go to zero, the size (and hence the cost) of a superconducting solenoid should increase very rapidly with the field capability in the bore. This is, in fact, the case in practice. For example, a series of commercial Nb_3Sn magnets rated at 110, 125, and 130 kG had design weights of 48, 110, and 130 pounds, respectively, not far from the dependence on B^4 derived above.

It is clear that α_c cannot be constant all the way down to $B = 0$, since this would imply an infinite critical current in zero field. This difficulty is avoided in an alternate simple approximation due to Bean,[1] namely that $J_c = $ constant, or $\alpha_c \propto 1/B$. In fact, Bean suggested this model of the critical state before the more extensive work of Kim and coworkers. In the Bean model, the flux-density profiles are simply straight lines of slope $4\pi J_c/c$, which simplifies qualitative discussion. For example, Fig. 5-8*a* illustrates the profiles for flux penetrating into a thick slab of thickness d as the external field H is increased. As is evident from the figure,

$$H_s = \frac{2\pi J_c d}{c} \qquad (5\text{-}73)$$

is the maximum external field that can be completely screened out at the midplane of the superconductor. If the applied field is now reduced and eventually reversed

[1] C. P. Bean, *Phys. Rev. Letters* **8**, 250 (1962).

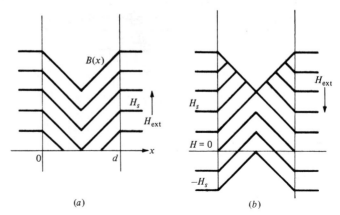

FIGURE 5-8
Internal flux-density profiles in a slab subjected to (*a*) increasing and (*b*) decreasing external field. H_s is the maximum applied field that can be screened at the midplane. Note the occurrence of canceling flux densities in (*b*) when $H_{ext} = -\frac{1}{2}H_s$.

in direction, the successive field profiles are as shown in Fig. 5-8*b*. Note that a very substantial amount of flux may be left trapped in the slab after the external field is reduced to zero. One may even have flux densities which change sign in the interior of the slab. In that case, one would expect some annihilation of opposing vortices to occur, but if the pinning is strong, this would affect only a thin layer at the crossover of B from one sense to the other. As is evident from this figure, there is much hysteresis and associated irreversibility in the cycling of these "hard" superconductors. For example, if the external field is cycled through a maximum field $H_m < H_s$, one can see that the area inside the hysteresis loop $\oint B\, dH$ (and hence the energy Q dissipated as heat per cycle) will increase as H_m^3. On the other hand, if $H_m \gg H_s$, then $Q \propto H_m$. These hysteresis losses limit the value of type II superconductors for ac applications, as will be discussed more completely in Sec. 5-8.

5-7 THERMALLY ACTIVATED FLUX CREEP

At finite temperatures thermal energy may allow flux lines to jump from one pinning point to another in response to the driving force of the flux-density gradient. The resulting flux creep is revealed in two ways: (1) it leads to slow changes in trapped magnetic fields; and (2) it leads to measurable resistive voltages. Before going into details, we briefly compare these two manifestations.

    If flux is trapped in a hollow cylinder, or in a superconducting solenoid in the persistent-current mode, there may be an observable decrease of this trapped field with time. This is in contrast with type I superconductors, in which no such change has ever been detected in macroscopic samples. Actually, even in type II superconductors, this creep is unobservably slow unless the flux-density gradient is very near the critical value. Since any creep that occurs will relieve the gradient, the creep gets slower and slower. In fact, we shall show that the time dependence is logarithmic. This means that as much creep occurs between 1 and 10 seconds as between 1 and 10 years! Under usual conditions for a superconducting solenoid in the persistent-current mode, the flux-density gradient is far enough from critical to assure negligible creep rate.*

If flux is creeping across a current-carrying conductor in response to the driving force caused by the current, there will be a longitudinal resistive voltage proportional to the average creep velocity of the flux-line motion. This is exactly analogous to the flux-flow resistance treated earlier, except that here the velocity of flux motion is much less.

It is of interest to compare the minimum creep velocities which are observable in the two cases. In the magnetic measurement involving persistent currents, one can easily detect motion in which the flux creeps the width of a conductor (~ 0.1 mm) in, say, a day; hence $v_{min} \approx 10^{-7}$ cm/sec. In the resistance measurements with conventional (i.e., nonsuperconducting) electronics, one is typically limited to detecting voltages of $\gtrsim 10^{-7}$ V/cm. With $B \approx 10^4$ G, this corresponds to a minimum detectable velocity of about 10^{-3} cm/sec. Thus, generally speaking, persistent-current measurements are the more sensitive.

5-7.1   Anderson-Kim Flux-Creep Theory

This theory[1] assumes that flux creep occurs by bundles of flux lines jumping between adjacent pinning points. Bundles are assumed to jump as a unit because the range λ of the repulsive interaction between flux lines is typically large compared to the distance between lines; this encourages cooperative motion. The jump rate is presumed to be of the usual form

$$R = \omega_0 e^{-F_0/kT} \qquad (5\text{-}75)$$

where ω_0 is some characteristic frequency of flux-line vibration, unknown in detail, but assumed to lie in the range 10^5 to 10^{11} sec^{-1}, and F_0 is the activation free energy, or barrier energy, i.e., the increase in system free energy when the flux bundle is at the saddle point between two positions where the free energy is at a local minimum. In the absence of any flux-density gradient, jumps are as likely to occur in one direction as the other, and no net creep velocity exists.

[1] P. W. Anderson, *Phys. Rev. Letters* **9**, 309 (1962); P. W. Anderson and Y. B. Kim, *Rev. Mod. Phys.* **36**, 39 (1964).

* The finite decay rate shown by some superconducting solenoids is caused by resistance in the joints between separate superconducting conductors.

Since various lengths, such as λ, ξ_0, ℓ, distance between pinning centers, width of pinning centers, etc., enter the theory in ways which are imperfectly known and depend on the details of the model, we shall introduce a single microscopic length $L \approx 10^{-5}$ cm for all of these.[1] This will keep dimensional arguments straight and will allow adequate estimates of order of magnitude, thus permitting the form of the theory to be developed with a maximum of simplicity.

We now introduce a flux-density gradient, which tilts the spatial energy dependence as indicated schematically in Fig. 5-9, making a jump easier in the direction of decreasing flux density than in the opposite direction. The shift in barrier heights is equal to the work done by the driving force in going over the barrier. Since the force density is α, the force on a flux bundle of cross section L^2 and of length L is αL^3, so the work done in moving it a distance L is $\Delta F = \alpha L^4$. This will lead to a *net* jump rate in the direction of the force α of

$$R = R_+ - R_- = \omega_0 e^{-F_0/kT}\left(e^{\Delta F/kT} - e^{-\Delta F/kT}\right) \qquad (5\text{-}76)$$

This amounts to a net creep velocity of

$$v = 2v_0 e^{-F_0/kT} \sinh \frac{\alpha L^4}{kT} \text{'} 2V_0 \sinh \frac{\alpha L^4}{kT} \qquad (5\text{-}77)$$

where $v_0 = \omega_0 L \approx 10^{3\pm3}$ cm/sec would be the creep velocity if there were no barrier.

To proceed further, we must make at least a rough estimate of the barrier energy F_0. It is useful to write it as

$$F_0 = p\frac{H_c^2}{8\pi}L^3 \qquad (5\text{-}78)$$

with p giving the fractional modulation of the condensation energy in volume L^3 available as a pinning energy. Because any strong pinning centers (such as voids) will typically fill only a small fraction of the volume, while any extended pinning centers (such as strains) will cause only small fractional changes in superconducting properties, p will be small. For an order of magnitude estimate, we take $p \approx 10^{-3}$, and $H_c \approx 2,000$ Oe; then $F_0/k \approx 1200°$K, so F_0 is certainly very large compared to kT. Taking this estimate

$$V_0 = v_0 e^{-F_0/kT} \approx 10^{3\pm3} e^{-1200/T} \approx 10^{3\pm3-500/T} \qquad (5\text{-}79)$$

From this we see that even large uncertainties in $v_0 = \omega_0 L$ are completely overshadowed by small changes in the pinning energy, and we may essentially forget

---

[1] A detailed theoretical analysis of systematic experimental data on flux creep, which allows a more precise characterization of these parameters than we use here, has been given by M. R. Beasley, R. Labusch, and W. W. Webb, *Phys. Rev.* **181**, 682 (1969).

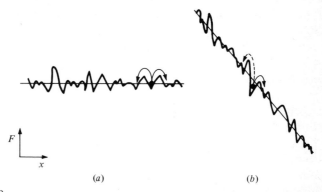

FIGURE 5-9
Schematic representation of flux bundles jumping over barriers to adjacent pinning sites. The ordinate represents the relative value of the total free energy as a function of the position of the center of the flux bundle. (*a*) Zero driving force. (*b*) Driving force due to current (or dB/dx) favoring jumps in "downhill" direction.

about all factors except the exponentials $e^{-F/kT}$. The above estimate shows that in the relevant temperature range of 1 to $10°K$, $V_0 \approx 10^{-500}$ to 10^{-50} cm/sec. Thus, the net creep velocity $v = 2V_0 \sinh(\alpha L^4/kT)$ will be unobservably small unless the driving term $\sinh(\alpha L^4/kT)$ is huge. However, if this is the case, then $2 \sinh(\alpha L^4/kT) \approx e^{\alpha L^4/kT}$, so that in the useful range we have simply

$$v = V_0(T)e^{\alpha L^4/kT} = v_0 e^{-F_0/kT} e^{\alpha L^4/kT} \qquad (5\text{-}80)$$

Consider the situation at $T = 0$, where $v = 0$ unless $\alpha L^4 \geq F_0$. This defines

$$\alpha_c(0) = \frac{F_0(0)}{L^4} \qquad (5\text{-}81)$$

as the critical force-density parameter (at $T = 0$) used in our earlier discussion of the critical state. We can use this relation to eliminate L^4 from (5-80), obtaining

$$v = v_0 \exp\left[-\frac{F_0(T) - F_0(0)\dfrac{\alpha}{\alpha_c(0)}}{kT} \right] \qquad (5\text{-}82)$$

Since some minimum velocity v_{min} is required to give detectable creep effects, we can define an effective $\alpha_c(T)$ as that value of α leading to this minimum detectable creep velocity. From the above expression, we see that this will be

$$\frac{\alpha_c(T)}{\alpha_c(0)} = \frac{F_0(T)}{F_0(0)} - \frac{kT}{F_0(0)} \ln \frac{v_0}{v_{min}} \qquad (5\text{-}83)$$

Now, since H_c and the characteristic lengths approach constant values as $T \to 0$, for $T \ll T_c$ we can write $F_0(T) \approx F_0(0)(1 - \beta t^2)$, where $t = T/T_c$, and β is of the order of unity. Thus, (5-83) can be rewritten as

$$\frac{\alpha_c(T)}{\alpha_c(0)} \approx 1 - \beta t^2 - \gamma t \qquad t \ll 1 \qquad (5\text{-}84)$$

where $\gamma = [kT_c/F_0(0)] \ln (v_0/v_{min}) \approx 0.1$ for our typical orders of magnitude. From this form, we see that at *sufficiently* low temperatures ($t \lesssim \gamma/\beta \approx 0.1$) the linear term dominates. This linear temperature dependence of $\alpha_c(T)$ is that arising from the explicit kT factor in the thermal activation exponential. As was pointed out by Kim and Anderson, this mechanism seems almost unique in being able to account for such a strong variation of α_c at very low temperatures, where H_c, ξ, and λ approach limiting values. Moreover, the order of magnitude of $d\alpha/dT$ obtained in this way was in reasonable agreement with the experimental results of Kim et al.[1] On the other hand, it is clear from (5-84) that for all except the lowest temperatures (typically $\lesssim 1°K$), the temperature dependence of α_c should be dominated by the change in pinning strength (reflected in βt^2), not by the change in creep rate (reflected in γt), if our numerical estimates are approximately correct. This agrees with the conclusion reached by Campbell et al.[2] from comparisons of the temperature dependences of J_c and of the reversible magnetization curves of various materials.

Next we consider the time dependence of the creep phenomenon. Because of the mathematical intricacy of the full treatment, let us first show in a simple way that a logarithmic dependence on the time should be expected. The key idea is that the driving force is roughly proportional to the value of B still trapped in the hollow cylinder of our example. Thus, in view of the exponential dependence of the creep rate on the driving force, we expect

$$\frac{dB}{dt} \approx -Ce^{B/B_0} \qquad (5\text{-}85)$$

where t now denotes time, which has the solution

$$B = \text{const} - B_0 \ln t \qquad (5\text{-}86)$$

Now let us give a more careful treatment. We again neglect the difference between B and H, denoting the local value of both by B. For simplicity we consider a one-dimensional problem: **B** is along $\hat{z}$; ∇B_z and the flux-line velocity **v** are along $\hat{x}$. The conservation of flux lines then requires that

$$\frac{\partial B}{\partial t} = -\frac{\partial (Bv)}{\partial x} \qquad (5\text{-}87)$$

[1] Y. B. Kim, C. F. Hempstead, and A. R. Strnad, *Phys. Rev.* **131**, 2486 (1963).
[2] A. M. Campbell, J. E. Evetts, and D. Dew-Hughes, *Phil. Mag.* **18**, 313 (1968).

The velocity v is related to the force-density parameter α by (5-83), which we write as

$$v = V_0 e^{\alpha/\alpha_1} \qquad (5\text{-}88)$$

where, as in (5-71),

$$\alpha = -\frac{\partial(B^2/8\pi)}{\partial x} \qquad (5\text{-}89)$$

and, following (5-83),

$$\alpha_1 = \frac{kT\alpha_c(0)}{F_0(0)} \approx \frac{\alpha_c(0)}{300} \qquad (5\text{-}90)$$

using our estimated value of $F_0/k = 1200°\text{K}$. Taking the time derivative of (5-89), and interchanging the order of time and space differentiation, we have

$$\frac{\partial \alpha}{\partial t} = -\frac{\partial}{\partial x}\frac{\partial}{\partial t}\frac{B^2}{8\pi} = -\frac{\partial}{\partial x}\frac{B}{4\pi}\frac{\partial B}{\partial t} = \frac{\partial}{\partial x}\frac{B}{4\pi}\frac{\partial(Bv)}{\partial x}$$

$$\approx \frac{B^2}{4\pi}\frac{\partial^2 v}{\partial x^2} = \frac{B^2}{4\pi}V_0 \frac{\partial^2}{\partial x^2}e^{\alpha/\alpha_1} \qquad (5\text{-}91)$$

The justification for dropping spatial derivatives of B compared to those of v is that the exponential dependence of v on derivatives of B is of the order of $\alpha_c(0)/\alpha_1$, or about 300 times, as fast as the direct dependence. In the same spirit, we can locally replace the factor $B^2 V_0/4\pi$ by a constant C. The form of the resulting equation suggests a trial solution of the form

$$Ce^{\alpha/\alpha_1} = (ax^2 + bx + c)g(t) \qquad (5\text{-}92)$$

This satisfies the differential equation (5-91) if $g(t)$ obeys the equation

$$\frac{dg}{dt} = \frac{2ag^2}{\alpha_1}$$

Integrating, we find

$$g(t) = -\frac{\alpha_1}{2at}$$

(To avoid carrying a constant of integration, we have chosen to measure time t from an initial point at which $g = \infty$, corresponding to infinitely rapid initial creep.) Inserting this $g(t)$ in (5-92), and taking the logarithm, we find

$$\alpha = F(x) - \alpha_1 \ln t \qquad (5\text{-}93)$$

where $F(x)$ is a function only of x. In fact, since creep is unobservably slow unless

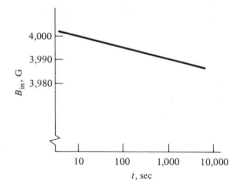

FIGURE 5-10
Decay of "persistent" current in hollow
cylinder of type II superconductor. [*After
Kim, Hempstead, and Strnad, Phys. Rev.
Letters 9, 306 (1962)*.]

α is quite near α_c, it must be that $F(x) \approx \alpha_c$ in the relevant region of space. Thus, (5-93) can be replaced approximately by

$$\alpha = \alpha_c - \alpha_1 \ln t \qquad (5\text{-}93a)$$

Applying this expression to the case of flux trapped in a hollow supercon-ducting cylinder with wall thickness d, where $\alpha \approx B_{in}^2/8\pi d$, we see that

$$B_{in} \approx B_c\left(1 - \frac{\alpha_1}{2\alpha_c}\ln t\right) \qquad (5\text{-}94)$$

where B_c is approximately the value of B_{in} giving the critical-state condition. With the estimate (5-90), $\alpha_1/\alpha_c \approx 1/300$. In using (5-94), it should be borne in mind that the infinity at $t = 0$ is inaccessible because of the choice of origin of time. Also, this simple form will hold only for moderately small fractional changes in B; thus, no significance should be attached to the fact that (5-94) changes sign after extremely long times.

This characteristic and unusual logarithmic time dependence has been well verified experimentally. For example, in the pioneering experiments of Kim, Hempstead, and Strnad[1] on flux trapped in NbZr tubes, this logarithmic decay was followed repeatedly over a period of time from 10 to 5,000 sec after a sudden adjustment of the external field. In the typical example shown in Fig. 5-10, the trapped field B_{in} of some 4,000 G was observed to fall at a rate of about 5 G per decade of time. Comparing this with (5-94), we see that it corresponds to $\alpha_1/\alpha_c = (400 \ln 10)^{-1} \approx 1/900$. Considering the extreme approximations and crude esti-mates that are involved, the order of magnitude agreement with our estimated value of 1/300 for this ratio is satisfactory.

It is interesting to extrapolate the logarithmic decay of the trapped field, to see how long the circulating current will take to die out. Taking (5-93a), with

[1] Y. B. Kim, C. F. Hempstead, and A. R. Strnad, *Phys. Rev. Letters* **9**, 306 (1962).

$\alpha = B_{in}^2/8\pi d$, we see that the crossover would occur when $t = e^{\alpha_c/\alpha_1} \approx e^{900} \approx 10^{400}$ sec $\approx 10^{390}$ years! Of course, the logarithmic dependence is not exact over such large changes in B. For example, when B drops below H_{c1}, flux leakage should effectively stop, since flux lines cannot exist in the volume of the superconductor in equilibrium. Still, this estimate gives an idea of the time scale. The circulating currents are persistent for practical purposes, although in principle they would die out over eons of time. In practice, a superconducting magnet in the persistent-current mode is operated at a low enough value of α to insure that the decay in current over a period of days is unobservable. The state of the superconductor is then roughly equivalent to that of one which started near α_c, but for which a very long decay time has already elapsed, so that the "next decade" will be an extremely long period of time.

5-7.2 Thermal Instability

The dissipation of energy associated with flux creep can lead to disastrous consequences if it leads to a thermal runaway, in which the material rapidly heats up and the entire energy stored in the magnet is suddenly converted into thermal energy. To prevent this, the system must be thermally stable; that is, if some region of the material acquires a temperature increment δT, stability requires that δT decay to zero, not continue to grow. The condition for stability is that for given δT, the increased outflow of heat to the surrounding material be greater than the increased rate of dissipation because of more rapid creep.

We start by considering the effect of δT on P, the power dissipated per unit volume. P can be written simply as

$$P = \alpha v = \alpha v_0 e^{-(F_0 - \alpha L^4)/kT} \qquad (5\text{-}95)$$

since α is the force per unit volume, force times velocity is the rate of doing work, and here all the work goes into heating the system because of dissipative processes in the cores of the flux lines. From this expression we obtain

$$\frac{T}{P}\frac{\partial P}{\partial T} = \frac{F_0}{kT} - \frac{\alpha L^4}{kT} - \frac{\partial F_0}{\partial(kT)} = -\ln\left(\frac{v}{v_0}\right) - \frac{\partial F_0}{\partial(kT)} \approx 100 \qquad (5\text{-}96)$$

for the typical values used above. Thus a small increase in temperature will lead to a very much larger increase in heating power, and instability will result unless cooling is very efficient.

To study this balance, we consider the heat-flow equation

$$C\frac{\partial T}{\partial t} = K\nabla^2 T + P \qquad (5\text{-}97)$$

where C is the specific heat per unit volume, K the thermal conductivity, and P the power input per unit volume. In steady state, $T(\mathbf{r})$ is the solution of $K\nabla^2 T + P = 0$. We now imagine that in a small volume there is a fluctuation $\delta T > 0$. The question is: will $K\nabla^2(\delta T)$ be sufficiently negative to overbalance $(\partial P/\partial T)\delta T$ so that $\partial(\delta T)/\partial t$ is negative, leading to stability. For a fluctuation localized in a volume of radius $\sim r$, the order of magnitude of $K\nabla^2(\delta T)$ will be $-(K/r^2)\delta T$. Combining this estimate with (5-96), we have

$$C\frac{\partial(\delta T)}{\partial t} = \left(-\frac{K}{r^2} + \frac{100P}{T}\right)\delta T \qquad (5\text{-}98)$$

For stability, then,

$$\frac{K}{r^2} > \frac{100P}{T} \qquad (5\text{-}99)$$

From this, we see that the criterion is most demanding when r is taken to be as large as possible, i.e., characteristic of the magnet as a whole if there is no internal cooling. A useful way to estimate the value of (P/T) is from the steady-state temperature rise ΔT above the bath temperature. From $K\nabla^2 T + P = 0$, we have $-K(\Delta T)/r^2 + P = 0$. Using this relation to simplify (5-99), we have

$$\frac{\Delta T}{T} \lesssim \frac{1}{100} \qquad (5\text{-}100)$$

Thus, to avoid thermal runaway, the magnet must always be operated under conditions such that the steady-state dissipation due to flux creep is so small that the resulting temperature rise is less than ~ 1 percent. In fact, magnets are normally operated under conditions in which thermally activated flux creep is negligible, and other considerations actually set the limits on performance. These are discussed further in the next section.

The above analysis indicates that good design requires good thermal conductivity and good contact with the helium bath. Magnet materials with high H_{c2} usually have short electronic mean free paths and hence poor thermal conductivities, making them intrinsically unstable. This tendency is almost invariably reduced in practice by cladding the superconductor with a metal like copper. This copper layer is an electrical insulator compared to a superconductor, but it provides excellent thermal conduction, at the expense of wasting some of the winding volume.

5-8 SUPERCONDUCTING MAGNETS FOR TIME-VARYING FIELDS

In addition to providing thermal stabilization, the addition of a copper coating will also cause eddy-current damping of magnetic-field changes. This helps with stabilizing the magnet against fluctuations, but it also limits the rate at which desired field changes can be made without undue power dissipation. In view of the potential importance of superconducting magnets in pulsed-field applications (in particle accelerators) and in ac applications (at power frequencies), considerable effort has gone into finding ways to make magnets which are both stable and capable of rapid field sweep. At present, the most promising approach involves the use of composite conductors consisting of twisted arrays of fine superconducting filaments—typically ~ 30 μm (micrometers) in diameter—embedded in a matrix of normal metal. An extensive analysis of this scheme has been given by Wilson, Walters, Lewin, and Smith.[1] In this section we review some of the relevant considerations.

A fundamentally important comparison is the rate of movement of heat in relation to that of magnetic flux through the magnet material. From the heat-flow equation (5-97), we may define the thermal diffusivity constant

$$D_T = \frac{K}{C} \qquad (5\text{-}101)$$

The physical significance of this parameter is that the time τ required for heat to diffuse a distance L to relieve a temperature gradient is of the order of

$$\tau_T = \frac{L^2}{D_T} \qquad (5\text{-}101a)$$

By applying Maxwell's equations, a similar magnetic-diffusion equation can be derived, in which the diffusion constant is

$$D_M = c^2 \frac{\rho}{4\pi} = 10^9 \frac{\rho}{4\pi} \qquad (5\text{-}102)$$

where the two forms hold hold if ρ is in cgs units and in ohm-cm, respectively. The time required for magnetic flux to diffuse a distance L is then of the order of

$$\tau_M = \frac{L^2}{D_M} \qquad (5\text{-}102a)$$

Typical numerical values for pure metals at $4°$K are $D_T = 10^3$ cm^2/sec and $D_M = 1$ cm^2/sec, so heat moves much faster than magnetic flux. On the other hand, in alloys (such as the superconducting material in the normal state or normal metal alloys) these numerical values are roughly interchanged, so that magnetic flux moves much more rapidly than heat. Note that in a composite filamentary conductor, $D_M(\text{Cu}) \approx D_T(\text{core}) \approx 1$ cm^2/sec. Taking account of the factor L^2 in the respective times, we see that this implies that heat can escape from the filamentary cores faster than flux can diffuse through the pure copper matrix from one filament to another. This efficient local cooling of the filaments helps make such composite materials highly stable.

[1] M. N. Wilson, C. R. Walters, J. D. Lewin, and P. F. Smith, *J. Physics* **D3**, 1518 (1970).

5-8.1 Flux Jumps

We now examine the control of flux jumps, replacing the flux-creep analysis of the previous section by a more elementary one, in which the material is described by the Bean model of the critical state, with a critical current density J_c below which there is no resistance and above which the material is normal. Since creep is important only in a small range of temperature or field, this simplification is appropriate over the important practical range of conditions.

For analytical simplicity, we take a one-dimensional model, with superconducting layers of thickness d parallel to the field, and we take x normal to the plane, measured from the midplane of a typical conductor. We restrict attention here to the usual case of a conductor sufficiently thin that the field penetrates throughout; that is, we assume $H > H_s = 2\pi J_c d/c$, where H_s is the screening field defined in (5-73). In this case, the internal field is

$$B(x) = H - \frac{4\pi J_c}{c}\left(\frac{1}{2}d - |x|\right) \qquad (5\text{-}103)$$

In writing this symmetrical form, with H the same at both surfaces, we have made the approximation of ignoring the self-field of the transport current carried by the conductor under consideration.

Now assume there is a small upward fluctuation in temperature δT, causing a decrease $\delta J_c = (dJ_c/dT)\delta T$ in the critical current. This changes $B(x)$, inducing an electric field E. As has been shown by various authors, the power dissipation $\dot{Q}$ associated with the nonequilibrium current J_c can be written as the product of J_c and the electric field induced by the changing field. Averaged over unit volume, this heat is

$$\delta Q = \frac{\pi}{3c^2}d^2 J_c \delta J_c \qquad (5\text{-}104)$$

The resulting temperature rise is $\delta T' = \delta Q/C$, where C is the specific heat per unit volume. So long as $\delta T' < \delta T$, the process converges to a finite multiple of the initial δT, and the system is stable against such fluctuations. As can be seen from the above argument, this will be the case if

$$d^2 < \frac{3c^2 C}{\pi J_c}\left(\frac{-dJ_c}{dT}\right)^{-1} \qquad (5\text{-}105)$$

This is the condition for *adiabatic* stability, since we have assumed no heat leaves the volume in question during the duration of the fluctuation. As such, it is a conservative criterion, since removal of heat by the copper matrix will tend to reduce the size of the fluctuation. With typical values for NbTi filaments, this criterion requires filament diameters $d \lesssim 0.01$ cm for stability; experimental results are consistent with this conclusion. In practice, to reduce losses in swept-field applications, d is usually chosen to be even smaller than is required by this stability criterion, so there is a substantial margin of safety.

It is of interest to work out the total heat released if J_c goes all the way down to zero, so that the flux jump goes to completion. Integrating (5-104), we have

$$Q = \frac{\pi}{6c^2} d^2 J_c^2 \qquad (5\text{-}106)$$

Assuming a constant specific heat C, the temperature rise will be $\Delta T = Q/C \approx 3 \times 10^4 d^2 (°\text{K})$, for typical numerical values. For any d large enough to be unstable according to the criterion (5-105), ΔT will be comparable with T itself. For such large temperature rises, C ($\propto T^3$) will increase significantly, making it possible that (5-105) will become satisfied, and the growth of the fluctuation stopped, before J_c is reduced all the way to zero. Such *partial flux jumps* allow some relaxation of the internal shielding currents without necessarily interrupting the flow of transport current.

A *dynamic* stability criterion may be derived in the same general way, but taking account of the time required for the flux to change and of the rate of diffusion of the heat into the copper. The result is similar to (5-105), apart from a factor of the order of the ratio of the diffusivities $D_T(\text{core})/D_M(\text{Cu})$, which is typically near unity, as noted above. In deriving this result, it was assumed that the copper temperature did not rise. This assumption breaks down for an edge-cooled tape-wound magnet, since the heat must diffuse a long way to reach the bath, and other parameters of the system enter into the stability criterion in that case, as is discussed by Wilson, et al.

5-8.2   Twisted Composite Conductors

The arguments above have shown that a composite conductor containing tiny superconducting filaments in a copper matrix should have excellent stability against flux jumps. We now turn to the case of time-varying fields, in which case the copper will carry electric current as well as heat current. There will then be two types of losses: eddy-current losses in the copper, which depend on sweep rate, and hysteresis losses in the superconductor, which are intrinsically independent of sweep rate. As will be shown in detail, normal eddy-current losses can be kept small by using fine wire, with the turns insulated from one another. Thus, hysteresis losses typically dominate. We now compute how large they are.

Returning to our model with thin superconducting sheets, in which the internal field is given by (5-103), we see that $\dot{B}(x) = \dot{H}$, so the induced electric field is $E_y = \dot{H}x/c$, by Faraday's law. Setting P, the local rate of energy dissipation per unit volume, equal to $J_c E$ as before, and averaging over the thickness of the superconducting sheet, we have

$$P = \dot{Q} = \frac{J_c \dot{H} d}{4c} \qquad \text{per unit volume} \qquad (5\text{-}107)$$

Thus, the heat released for any given change ΔH is independent of the rate of change, as expected for a hysteresis loss, and we can write the dissipation for a complete cycle in which the field changes by $\pm \Delta H$ about an operating point as $Q = J_c d\, \Delta H/c$. This result shows that by making the superconducting filaments thin enough, this loss can be made very small.

In arriving at (5-107), we made the assumption that the changing field penetrated all the way through the filament. This assumption will be justified in applications in which the magnetic field is swept over a wide range, such as in a magnet pulsed between $H = 0$ and $H_{\max}$. It will *not* be justified, however, in applications which are basically dc, with an ac ripple of small amplitude $\Delta H < H_s$. For these, it is necessary to take account of the fact that the changing field does not penetrate uniformly. The resulting more accurate expressions for loss per cycle of $\pm \Delta H$ about an operating point are

$$Q = \frac{J_c d\, \Delta H}{c} - \frac{4\pi J_c^2 d^2}{3c^2} \qquad \Delta H > H_s = \frac{2\pi d J_c}{c} \qquad (5\text{-}108a)$$

and

$$Q = \frac{(\Delta H)^3 c}{12\pi^2 d J_c} \qquad \Delta H < H_s \qquad (5\text{-}108b)$$

Our approximation (5-107) is equivalent to keeping only the leading term in (5-108a). The correction term in (5-108a) reduces this leading term by a factor of 3 when $\Delta H = H_s$, the lowest value of ac field for which complete penetration occurs. At this point a continuous transition is made to (5-108b), and the loss thereafter decreases as $(\Delta H)^3$. Thus, the true loss can be much lower than would be estimated by using (5-107).

Going further, (5-108) indicates that $Q \to 0$ in either limit, $d \to 0$ or $d \to \infty$. Dissipation is greatest for intermediate values. This can be made more clear by introducing a dimensionless thickness variable

$$\beta = \frac{2\pi J_c d}{c \Delta H} = \frac{d}{d_s} = \frac{H_s}{\Delta H}$$

which is the ratio of d to the slab thickness d_s penetrated by a field change ΔH. In terms of β, (5-108) becomes

$$Q = \frac{(\Delta H)^2}{6\pi}(3\beta - 2\beta^2) \qquad \beta < 1 \qquad (5\text{-}109a)$$

$$Q = \frac{(\Delta H)^2}{6\pi}\frac{1}{\beta} \qquad \beta > 1 \qquad (5\text{-}109b)$$

These expressions have a maximum at $\beta = 3/4$, where $Q = 3(\Delta H)^2/16\pi$, and they fall to zero as $\beta \to 0$ or $\beta \to \infty$. To minimize Q, there are two strategies: make β either large or small, but not near unity. With typical material parameters, the filament diameter d must be kept below $\sim 200\ \mu$m for thermal stability, and it is

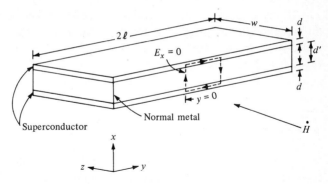

FIGURE 5-11
Laminar model of multifilamentary composite conductor. As discussed in the text, application of Faraday's law to circuits such as the one shown in dashed line enables one to make approximate determinations of current and emf patterns for various cases.

hard to fabricate them smaller than $\sim 10 \ \mu$m. For typical critical current densities and this range of d, it then turns out that the larger filaments give lower losses for $\Delta H < 500$ Oe, whereas the smaller filaments are superior for larger ac amplitudes, such as in pulsed magnets.

An important qualification on the applicability of (5-108) is that each filament must act independently, so that the origin of x can be taken in the center of each one. This would certainly be the case if each filament were electrically insulated from its neighbors, but in practice the copper matrix provides a good electrical contact between all the filaments in one conductor. This profoundly modifies the current patterns at high field sweep rates, because the induced voltages can drive current from one filament to another through the copper. In this case, the filaments are said to be "coupled," and the effective diameter d is no longer that of one filament; rather, it can be as large as that of the entire bundle of 10 to 1,000 filaments in a conductor, and the effectiveness of the multiplicity of fine filaments is lost. As first pointed out clearly by P. F. Smith, this degradation in performance at high sweep rates can be greatly reduced by twisting the conductors with a pitch less than a characteristic length l_c, which depends on sweep rate and other parameters. We now analyze a simplified model which demonstrates these properties and leads to a derivation of the characteristic length.

Consider two thin superconducting sheets of thickness d, width w, and finite length $2l$, separated by a layer of normal metal of resistivity ρ and thickness d', where for algebraic simplicity we take $l \gg d' \gg d$. The geometry is illustrated in Fig. 5-11. Further, we assume there is an applied field parallel to w which is changing at a constant rate $\dot{H}$. We now apply Faraday's law to various circuits to compute the induced emf and the resulting currents.

First, let us compute the eddy-current loss in the normal metal if there were no superconductor present. If we measure x from the midplane of the normal metal layer, $E_y \approx Hx/c$ and $E_x \ll E_y$. The local rate of power dissipation is then $P = JE = E^2/\rho$. Averaging over x, this gives

$$P = \frac{\dot{H}^2 d'^2}{12\rho c^2} \quad (5\text{-}110)$$

as the average dissipation per unit volume. If we take d' equal to the overall diameter of the composite conductor, this will give an estimate of the eddy-current loss with no superconductors present. To get some feeling for the size of this loss, consider an example of a magnet wound with 0.05-cm-diameter wire with $\rho = 0.01\rho_{Cu}(300°K)$. Then $P \approx 10^{-12} \dot{H}^2$ watts/cm^3, where $\dot{H}^2$ refers to an average over the volume of the magnet winding. Even at a sweep rate as high as 10^3 G/sec, this loss would be only 1 mW for a magnet with 10^3 cm^3 of winding. On the other hand, in an ac application, where $\dot{H} = \omega H_{max}$, this loss can become very large at useful power frequencies. For example, at 60 Hz, with $H_{max} = 10^4$, P would be 10^4 W for a magnet winding of 10^3 cm^3, a prohibitive value for cooling at 4°K. To reduce this figure to a more manageable one, one could use somewhat finer wire, and replace the copper matrix by an alloy with much higher low-temperature resistance. However, it would be difficult to reduce the loss by more than a factor of 10^3 in this way. Our general conclusion is that normal-metal eddy-current losses can be made negligible at slow-to-moderate sweep rates, but that above about 10^4 G/sec, they will be increasingly hard to overcome.*

Now, let us take account of the presence of the superconducting layers. We apply Faraday's law to a circuit passing through both superconducting sheets, as indicated in Fig. 5-11. So long as the current density in the superconducting sheets is less than J_c, the entire induced voltage is dropped across the normal metal, leading to a current density between the sheets of $J_x = \dot{H}y/c\rho$. Integrating out from $y = 0$, the total current from one sheet to the other equals the critical current of a sheet when $y = l_c$, where

$$l_c^2 = \frac{2cJ_c d\rho}{\dot{H}} \quad (5\text{-}111)$$

Thus, if $l < l_c$, our assumption that no voltage was dropped in the superconductor is self-consistent, and there is no dissipation in the superconductor due to these circulating currents. However, the eddy-current loss in the normal metal is greatly

---

*One might wonder why these losses are not an equal problem in the windings of ordinary power transformers. The basic difference is that even for $B_{max} \approx 10^4$ G in the iron core, B_{max} in the copper winding may be only 10 G. Thus, eddy-current loss is primarily a problem in the core. This is reduced by lamination to a thickness of about 0.04 cm, and, of course, ρ of silicon steel is about 500 times greater than that of copper at 4°K. Taken together, these factors reduce the dissipation per unit volume by a factor of about 1,000, to a level which is tolerable for cooling at room temperatures.

increased, because the induced emf is dropped over a much shorter distance. As a result, the factor d'^2 in (5-110) is replaced by $(2l)^2$.

On the other hand, if $l > l_c$, the current density in the center of the superconductor would apparently have to exceed J_c, indicating that our assumption of no dissipation in the superconductor is no longer self-consistent. The actual physical situation is that in the central part of the layers, further than l_c from the ends, the supercurrent density is constant at J_c. Consequently, no current enters the normal metal from the superconductor, and accordingly none of the emf is dropped there. Thus, in the central part all the emf is dropped in the superconductor, giving an electric field $\dot{H}d'/2c$. The resulting dissipation is $J_c\dot{H}d'/2c$ per unit volume of superconductor, which is just as high as if the normal-metal layer were replaced by solid superconductor. If one averages the enhanced eddy-current loss in the copper over the length l_c at each end, where J_x in the normal metal rises as $(\dot{H}/c\rho)[y - (l - l_c)]$, one finds that it is 2/3 as great (per unit length) as is the dissipation in the superconductor in the central part. Thus, essentially all the benefit of having the superconductor in thin layers is lost if $l > l_c$.

For sweep rates of 10 to 10^4 G/sec, this length l_c is typically in the range 10 to 100 cm, so it is clear that practical magnet windings are much longer than l_c. However, as pointed out by Smith, it is possible to decouple the filaments by twisting the wire with a pitch which is much shorter than l_c. This effectively reverses the sign of the induced emf each time the filaments are interchanged in position, allowing the induced current in the superconductor to be kept always below J_c. In this case, the dissipation is given by the enhanced eddy-current loss in the normal metal, referred to above.

Specific assumptions about the geometry are required to scale our results for two flat layers up to a wire containing many layers of filaments. Nevertheless, one can estimate a new length l_c' by equating the total induced circulating current with the total critical current of the filaments on one side of the center of the wire. If d' is now taken to be the diameter of the bundle of filaments in the wire, and if a filling factor of $\frac{1}{2}$ is assumed, we find

$$l_c'^2 \approx \frac{cJ_c d'\rho}{2\dot{H}} = \frac{d'}{4d}l_c^2 \qquad (5\text{-}112)$$

as an extension of (5-111). Thus, we expect the excess loss to be reduced by a factor of about $l^2/l_c'^2$ from the value for a solid superconductor of diameter d'. Note that with our definitions of l_c and l_c', the pitch for a twist by 2π is denoted $4l$.

If we now add this extra dissipation due to coupling to the value (5-107) for completely decoupled filaments, the total dissipation per unit volume of superconductor can be conveniently written as

$$P = \frac{J_c\dot{H}d_{\text{eff}}}{4c} \qquad (5\text{-}113)$$

where $$d_{eff} \approx d' \qquad l > l'_c \qquad (5\text{-}113a)$$

and $$d_{eff} \approx d\left(1 + \frac{4l^2}{l_c^2}\right) = d + \frac{2l^2\dot{H}}{J_c\rho c} \qquad l < l'_c \qquad (5\text{-}113b)$$

[Note that in (5-113), we have reverted to using only the leading term of (5-108a), an approximation appropriate to small d and large ΔH.] As expected, P is proportional to $\dot{H}d$ for small $\dot{H}$, where the filaments are decoupled, and to $\dot{H}d'$ for large $\dot{H}$, when they are fully coupled. In the transition region between these regimes, P increases as $\dot{H}^2$, since $l_c^{-2} \propto \dot{H}$. There will be another transition to an $\dot{H}^2$ dependence when the normal eddy-current term (5-110) dominates. This will occur when $\dot{H} \gtrsim \rho c J_c/d' \gtrsim 10^7$ G/sec. Thus, in the presently practical range of $\dot{H}$, losses are dominated by (5-113).

For typical parameter values, this dissipated power will fall in the range 10^{-6} to $10^{-5}\dot{H}$ W/cm^3, depending on the filament size and the degree of decoupling. Noting that only a fraction of the volume is filled with superconductor and that the average of $\dot{H}$ over the winding is typically less than half the value in the bore, we see that the loss can be as small as $10^{-7}\dot{H}$ W/cm^3 when referred to total winding volume and the field in the bore. Such performance has already been demonstrated by Spurway, Lewin, and Smith.[1]

As a final figure of merit, let us compute the ratio of the energy stored in the magnetic field to the energy dissipated in the process of establishing and removing the field. For simplicity consider a solenoid of length L, radius R, and winding thickness D, with $D \ll R \ll L$. Then the stored energy will be $E \approx \pi R^2 L(H^2/8\pi)$. Assuming complete decoupling of the filaments, we can use (5-107) to compute the dissipation. Canceling the factor of 2 for the loss on removing the field as well as setting it up against the factor $\frac{1}{2}$ which relates the average field in the winding to the field in the bore, the total dissipated energy per cycle is $\Delta E = 2\pi RLD(J_c Hd/4c)$. Taking the ratio, and noting that $H = 4\pi JD/c$, we have

$$\frac{E}{\Delta E} = \frac{J}{J_c}\frac{R}{d} \qquad (5\text{-}114)$$

Thus, with $J \approx J_c$, one should be able to achieve energy storage ratios of the order of 1,000 with presently available materials and $R \approx 2$ cm. In fact, ratios of over 100 were obtained in early experiments by Dahl, et al.,[2] without reaching J_c. For comparison, with a normal resistive coil, this ratio would be given by

$$\frac{E}{\Delta E} \approx \frac{\tau}{\Delta t} \qquad (5\text{-}115)$$

where $\tau = L/R$ is the time constant of the magnet and Δt is the duration of the pulse. Noting that $\tau \approx 2\pi DR/c^2\rho$, which is of the order of 10 sec for typical

[1] A. H. Spurway, J. D. Lewin, and P. F. Smith, *J. Physics* **D3**, 1572 (1970).
[2] P. F. Dahl, G. H. Morgan, and W. B. Sampson, *J. Appl. Phys.* **40**, 2083 (1969).

dimensions and resistivities of pure metals at low temperatures, we see that the superconductive coil should be more efficient than even cryogenic normal coils for pulse durations greater than about 0.01 to 0.1 sec. It is for this reason that super-conductive coils are being developed to generate the intense transient fields required in particle accelerators. This development may well open the way to other applications of high-field superconducting magnets outside the range of essentially dc operation.

6

JOSEPHSON EFFECT AND MACROSCOPIC QUANTUM PHENOMENA

In 1962, Josephson[1] startled the world of superconductivity by proposing that a tunnel junction should show a zero-voltage supercurrent due to the tunneling of condensed pairs, as well as the familiar quasi-particle tunneling first observed by Giaever (discussed in Sec. 2-8). He further predicted that, if a voltage difference V were maintained across the junction, the current would be an alternating current of frequency $v = 2eV/h$.

Although these predictions were originally greeted with considerable skepticism,* they have received copious experimental verification. In fact, the dc Josephson effect is now used in very sensitive galvanometers and magnetometers, while the ac effect is used as the basis for very precise measurements of h/e.

---

[1] B. D. Josephson, *Phys. Letters* **1**, 251 (1962); for an excellent later review, see B. D. Josephson, *Advan. Phys.* **14**, 419 (1965); for an informal account of the actual discovery, as given in his Nobel lecture, see *Rev. Mod. Phys.* **46**, 251 (1974).

* It seemed that if the tunneling amplitude for one electron were $e^{-\alpha d} \ll 1$, so that the tunnel probability were $e^{-2\alpha d}$, then the tunnel probability for *two* electrons should be $e^{-4\alpha d} \ll e^{-2\alpha d}$, so that pair tunneling should be negligible. This argument fails to take account of the coherence of the pair-tunneling process, which causes the pair current to be proportional to the *first* power of the *pair* amplitude, i.e., to $e^{-2\alpha d}$, the same factor as for single-particle incoherent tunneling.

Moreover, it is now clear that the fundamental ideas of Josephson are applicable to any sufficiently localized "weak link" in a superconducting circuit, not just for the tunnel junction with an insulating barrier which was originally envisaged. For example, the weak link can be a short constriction in the cross section of a superconductor, a point contact between two superconductors, or two superconductors separated by a thin layer of normal metal rather than by an insulating layer, as in a tunnel junction.

6-1 THE JOSEPHSON CURRENT-PHASE RELATION

Because of the generality of the concepts, we shall not give a rigorous microscopic derivation based on some specific case, such as the tunnel junction. Rather, we shall follow the phenomenological approach we used in setting up the Ginzburg-Landau theory, basically relegating to the microscopic theory only the task of evaluating parameter values. At the phenomenological level, the Josephson junction is replaced by a "black box" which carries a current density determined by the values at the junction of the ψ functions of the strong superconductors on either side; that is, $J = f(\psi_1, \psi_2)$. What is the functional dependence f?

A heuristic approach is to consider the junction as the limit of an ordinary superconducting layer of thickness d as $d \rightarrow 0$ and $|\psi|^2 \rightarrow 0$ in such a way as to give finite results. Thus, following (4-32) and taking x as the direction of current flow through the weak link, we write

$$J_x = \frac{2e\hbar}{m^*}|\psi|^2\left(\frac{d\varphi}{dx} - \frac{2\pi A_x}{\Phi_0}\right) \qquad (6\text{-}1)$$

where φ is the phase of $\psi(x) = |\psi|e^{i\varphi(x)}$. Assuming that $|\psi|^2$ as well as J_x is constant through the thickness d, this can be written as

$$J_x = J_0\gamma \qquad (6\text{-}2)$$

where J_0 is a constant (nominally $2e\hbar|\psi|^2/m^*d$) characterizing the junction and

$$\gamma = (\varphi_2 - \varphi_1) - \frac{2\pi}{\Phi_0}\int_1^2 A_x \, dx \qquad (6\text{-}3)$$

is referred to as the "gauge-invariant phase difference" between points 1 and 2 on opposite sides of the junction. (Note that unless $\mathbf{A}$ is infinite at the junction in the gauge choice used, the vector-potential term drops out as $d \rightarrow 0$.) Although (6-2) would be correct for a long wire, in which the phase $\varphi(x)$ can be followed as a continuous variable along the wire, it is not satisfactory for a localized weak link, in which only the total phase difference γ is defined, since the phase difference between two points is defined only modulo 2π. This suggests replacing (6-2) by

$$J_x = J_0 \sin \gamma \qquad (6\text{-}4)$$

which agrees with (6-2) for small values of γ, but is appropriately periodic to reflect the fact that $\gamma + 2\pi$ is indistinguishable from γ in the present case. The form (6-4) is that actually derived by Josephson for the case of a weakly coupled tunnel junction; more generally, an arbitrary Fourier sine series is permitted, and for some sorts of weak links small deviations from (6-4) have been observed.[1]

For the case of a tunnel junction at $T = 0$, Anderson[2] found $J_0 = \pi\Delta/2eR_n$, where R_n is the tunneling resistance per unit area of the junction when both metals are in the normal state. This result was generalized by Ambegaokar and Baratoff[3] to all temperatures. For two identical superconductors, the result is

$$J_0 = \frac{\pi\Delta(T)}{2eR_n} \tanh \frac{\Delta(T)}{2kT} \qquad (6\text{-}5)$$

which, incidentally, has the same temperature dependence as found in (2-125) for σ_2/σ_n in a homogeneous superconductor. Note that, for $T \ll T_c$, J_0, the maximum zero-voltage supercurrent allowed by (6-4), should be just $\pi/4$ times the normal-state tunneling current for an applied voltage such that $eV = 2\Delta$, the gap energy. This prediction, illustrated in Fig. 6-1, is quite well confirmed in practice, with maximum supercurrents often within a few percent of the theoretical value. For typical junctions, the maximum supercurrent is of the order of 1 mA (milliampere).

If, instead of a tunnel junction, one considers a weak link formed by a short narrow constriction, the coefficient I_0 of $\sin \gamma$ is typically comparable to the critical current $I_c = J_c A$, where J_c is the ordinary critical current density and A is the cross-sectional area of the weak link. However, in very short bridges,[4] the proximity to the strong superconductors at the ends may increase I_c. On the other hand, heating effects at finite voltages can cause the effective value of I_0 to be considerably smaller than the current required to establish the resistive state originally.[5] Thus, the interpretation of the exact value of I_0 is not entirely straightforward except in the case of a tunnel junction.

[1] See, for example, T. A. Fulton and R. C. Dynes, *Phys. Rev. Letters* **25**, 794 (1970).

[2] P. W. Anderson, Lectures at Ravello Spring School, 1963. (Josephson's original paper obtained this result, but with a numerical error in the coefficient.)

[3] V. Ambegaokar and A. Baratoff, *Phys. Rev. Letters* **10**, 486 (1963); erratum, **11**, 104 (1963).

[4] A discussion of the current-phase relationship in short weak links is found in A. Baratoff, J. A. Blackburn, and B. B. Schwartz, *Phys. Rev. Letters* **25**, 1096 (1970).

[5] W. J. Skocpol, M. R. Beasley, and M. Tinkham. *J. Appl. Phys.* **45**, 4054 (1974).

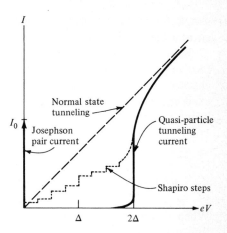

FIGURE 6-1

Comparison of maximum Josephson tunneling current of pairs at zero voltage with quasi-particle tunneling current and with tunneling current when both superconductors are normal. The dotted curve shows the Shapiro steps which are induced by presence of a microwave field. (See Sec. 6-4.) The step interval is $\Delta V = h\nu/2e$.

The discussion above has focused on the supercurrent behavior of the weak link. But with a finite voltage V across the junction, there will also be a dissipative "normal" current, which we may denote $G_0 V$, carried by quasi-particles. Near T_c, this G_0 will be well approximated by simply the normal-state conductance G_n, as measured just above T_c. For a tunnel junction, this conductance falls as T is reduced, as discussed in connection with Giaever tunneling in Sec. 2-8; for other weak links, the behavior may be different.

In addition to this simple quasi-particle conductance, Josephson remarked that there is another *phase-dependent* term, which can be written $(G_{int} \cos \gamma)V$, which reflects an interference term between the pair and quasi-particle currents. Thus, for $V \neq 0$, and for an arbitrary type of weak-link element, we may generalize (6-4) to

$$I = I_0 \sin \gamma + (G_0 + G_{int} \cos \gamma)V \quad (6\text{-}4a)$$

where G_0 and G_{int} themselves may be functions of V. Experiments of Pedersen et al.[1] on tunnel junctions, of Falco et al.[2] on thin-film weak links, and of Vincent and Deaver[3] on point-contact weak links have all demonstrated the existence of this pair–quasi-particle interference term, and all have shown that $G_{int}/G_0 \approx -1$. This result has drawn much recent attention, because it appears that the microscopic theory[4] yields a positive sign for this ratio. On the other hand, one can give a

[1] N. F. Pedersen, T. F. Finnegan, and D. N. Langenberg, *Phys. Rev.* **B6**, 4151 (1972).
[2] C. M. Falco, W. H. Parker, and S. E. Trullinger, *Phys. Rev. Letters* **31**, 933 (1973).
[3] D. A. Vincent and B. S. Deaver, Jr., *Phys. Rev. Letters* **32**, 212 (1974).
[4] D. N. Langenberg, *Revue de Physique Appliquée* **9**, 35 (1974); U. K. Poulsen, ibid., **9**, 41 (1974);
[5] R. E. Harris, *Phys. Rev.* **B10**, 84 (1974).

simple Kramers-Kronig argument[1] which indicates not only the necessary existence of the cos γ term, but also suggests the correct sign and order of magnitude.
Thus the situation is still quite unsettled, and further work will be needed to
resolve the open questions. We shall generally ignore this term in the rest of this
chapter, since its effects can be detected only by rather subtle analysis of selected
experiments. Even the simple term $G_0 V$ plays no role in the zero-voltage equilibrium properties which comprise the dc Josephson effects.

6-2 EFFECT OF MAGNETIC FIELD

When supercurrent tunneling was first observed, it was erroneously attributed to
tiny superconducting short circuits through the tunnel barrier. The error of this
identification was shown by the observation that the maximum supercurrent was
very strongly affected by even weak magnetic fields, contrary to what would be
expected for superconducting short circuits. Let us work out the field dependence
expected on the basis of (6-4).

Consider a tunnel junction normal to the x axis, with face dimensions Y and
Z, in a magnetic field H parallel to the z axis. Assume the thickness of the
insulating barrier is d and that of the two superconducting elements is much
greater than λ. Further assume that the tunneling currents are negligible
compared to the diamagnetic screening currents. We choose a gauge for $\mathbf{A}$ such

---

[1] M. Tinkham and M. R. Beasley, unpublished. In this approach, one restricts attention to the
low-voltage limit, where the problem can be linearized; in particular, one considers a small ac
voltage $V(\omega) \cos \omega t$ and zero dc voltage, so that the phase difference γ makes only small
excursions about its operating point γ_0, as given by (6-27b) below. Then the term $I_0 \sin \gamma$
becomes $I(\omega) = (2eI_0/\hbar\omega) \cos \gamma_0 V(\omega) \sin \omega t$. This supercurrent can be represented by the imaginary admittance term $Y_2^0 = (2eI_0/\hbar\omega) \cos \gamma_0$, reminiscent of $\sigma_2/\sigma_n \propto 1/\omega$ in (2-125).
Moreover, as noted in connection with (6-5), the temperature dependence of the coefficient I_0 is
the same as that of σ_2/σ_n. The Kramers-Kronig conjugate [see (2-96) and associated
discussion] of Y_2^0 is a term $Y_1^0 = (\pi eI_0/\hbar) \cos \gamma_0 \, \delta(\omega)$, which represents the absorption of
energy to provide the kinetic energy of the Josephson supercurrent. If one then postulates that
the superfluid response cuts off for $\hbar\omega > 2\Delta$, just as σ_2/σ_n does, the Kramers-Kronig relations
(2-96) imply that between $\omega = 0$ and $2\Delta/\hbar$ there must be a negative contribution to Y_1 large
enough that its integral over frequency will compensate for Y_1^0. This argument accounts for the
observed negative sign and order of magnitude of the interference term $G_{\text{int}} \cos \gamma$ in (6-4a).

How can this result be reconciled with the microscopic theoretical prediction of the
positive sign? If Y_2 were to rise to a positive singular peak at the gap frequency, as suggested by
the work of E. Riedel [*Z. Naturforschung* **19a**, 1634 (1964)] and of N. R. Werthamer [*Phys. Rev.*
147, 255 (1966)], before falling off to zero at higher frequencies, this same Kramers-Kronig
argument would give the positive sign. But it is far from clear that such a Riedel peak should
occur in Y_2 at an applied frequency $\hbar\omega = 2\Delta$; the singularity actually occurs for an applied
voltage such that $eV = 2\Delta$, and in the experiments the dc voltage is zero and the ac voltage is
very small. This suggests that the theoretical result may be inapplicable because it is computed
for conditions which differ in a critical way from those of the experiments.

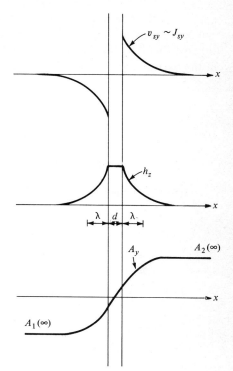

FIGURE 6-2
Spatial dependences of supercurrent density, magnetic field, and vector potential in, and near, the barrier between two superconductors in a Josephson junction.

that $\mathbf{A} = A_y(x)\hat{\mathbf{y}}$, and $h_z = dA_y/dx$. With this gauge choice, φ is a function of y alone, since

$$\nabla\varphi = \frac{m^*\mathbf{v}_s}{\hbar} + \frac{2\pi}{\Phi_0}\mathbf{A} \qquad (6\text{-}6)$$

has only a y component (because $\mathbf{A}$ is along $\hat{\mathbf{y}}$, as is $\mathbf{v}_s$ for the screening currents giving the Meissner effect in the two superconductors). The x-dependence of these quantities is illustrated in Fig. 6-2. Since $\varphi(y)$ is independent of x, we can evaluate it at any x. It is convenient to do so deep in the superconductors, where $\mathbf{v}_s = 0$, $h_z = 0$, and $A_y = A(\infty)$. Then, from (6-6) we see that $d\varphi/dy = (2\pi/\Phi_0)A_y$, so that

$$\varphi(y) = \varphi(0) + \frac{2\pi}{\Phi_0}A(\infty)y \qquad (6\text{-}7)$$

Accordingly, the phase difference* across the junction is

$$\gamma(y) = \varphi_2(y) - \varphi_1(y) = \gamma(0) + \frac{2\pi}{\Phi_0}[A_2(\infty) - A_1(\infty)]y \qquad (6\text{-}8)$$

---

* Note that in the present gauge, the integral of $A_x\,dx$ in (6-3) is zero, since $A_x = 0$. Thus, $\gamma = \varphi_2 - \varphi_1$, as written here.

It is convenient to rewrite this expression in a clearly gauge-invariant form. This can be done by noting that $\oint \mathbf{A} \cdot d\mathbf{s}$ is the flux enclosed within the path of integration, so that $[A_2(\infty) - A_1(\infty)]y = \Phi(y)$ is the flux enclosed between $y = 0$ and y in the barrier and penetration layers. Thus, we can write (6-8) as

$$\gamma(y) - \gamma(0) = \frac{2\pi\Phi(y)}{\Phi_0} = \frac{2\pi H(2\lambda + d)y}{\Phi_0} \tag{6-9}$$

which involves only gauge-invariant quantities.

As an example of the gauge invariance of our results, let us indicate how γ would be computed in the London gauge, in which $A_y(x)$ goes to zero in the interior of the superconductors, and in which ψ is real everywhere, so that $\varphi_1 = \varphi_2 = 0$. For a general change of gauge from $\mathbf{A}$ to

$$\mathbf{A}' = \mathbf{A} + \nabla\chi \tag{6-10a}$$

ψ changes to

$$\psi' = \psi e^{i2\pi\chi/\Phi_0} \tag{6-10b}$$

Thus, if we want ψ' to be real, while ψ has the phase variation (6-7), we must choose

$$\chi_i = -A_i(\infty)y - \frac{\Phi_0 \varphi_i(0)}{2\pi} \tag{6-11a}$$

where $i = 1, 2$, so that in either superconductor

$$\mathbf{A}'_i = \mathbf{A}_i - A_i(\infty)\hat{\mathbf{y}} \tag{6-11b}$$

Note that, unlike A_y, A'_y changes almost discontinuously by $A_2(\infty) - A_1(\infty)$ in the barrier thickness d. But what we need to evaluate (6-3) is A'_x in the barrier. In the original gauge, A_x was zero everywhere. From (6-10a), we then see that $A'_x = d\chi/dx$. According to (6-11a), this is zero in both superconductors, but in the barrier it must be large enough so that

$$\int_1^2 A'_x \, dx = \int_1^2 \frac{d\chi}{dx} \, dx = \chi_2 - \chi_1 = [A_1(\infty) - A_2(\infty)]y + \frac{\Phi_0}{2\pi}[\varphi_1(0) - \varphi_2(0)] \tag{6-12}$$

When this is inserted in (6-3) together with $\varphi'_1 = \varphi'_2 = 0$, we find the same value of γ as found in (6-8). This example illustrates that γ is, indeed, a gauge-invariant quantity, as stated earlier.

Let us now work out the consequences of this spatial variation (6-9) of the relative phase γ. Since γ changes linearly with y, and since the current density (6-4) is a periodic function of γ, it is clear that the current in various parts of the junction area will tend to cancel, giving rise to a sort of diffraction pattern. More precisely, the total current through the junction will be

$$I = Z \int_{-Y/2}^{Y/2} J(y) \, dy = ZJ_0 \int_{-Y/2}^{Y/2} \sin\left[\frac{2\pi H}{\Phi_0}(2\lambda + d)y + \gamma(0)\right] dy$$

$$= YZJ_0 \sin\gamma(0) \frac{\sin(\pi\Phi/\Phi_0)}{\pi\Phi/\Phi_0} \tag{6-13}$$

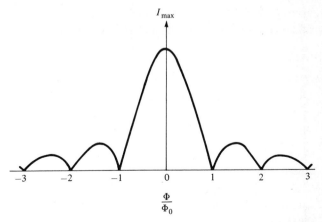

FIGURE 6-3
Dependence of maximum supercurrent through a Josephson junction upon the flux threading the junction. The resemblance to the "single-slit" diffraction pattern of optics is evident.

where YZ is the junction area and $\Phi = HY(2\lambda + d)$ is the total flux threading the junction. For given Φ, the current I still depends on the phase difference $\gamma(0)$ in the center of the junction. This degree of freedom is used by the junction to adjust the current as required by an external current source, just as in a superconducting wire the gradient of the phase adjusts as required by the imposed current. But since the sine function cannot exceed unity in absolute value, (6-13) implies that the maximum supercurrent through the junction as a function of applied field is

$$I_{\max} = YZJ_0 \left| \frac{\sin(\pi\Phi/\Phi_0)}{\pi\Phi/\Phi_0} \right| \qquad (6\text{-}14)$$

This predicted "diffraction pattern," shown in Fig. 6-3, has been very well verified experimentally, providing convincing confirmation of the theory of the effect. At each of the nodal points, Φ is an integral multiple of Φ_0, and the relative phase γ goes through an integral multiple of 2π from one edge of the junction to the other, causing complete cancellation of the current.

If we now relax the restriction that the tunneling current is negligible, a new feature appears. Namely, the Josephson current tends to screen the magnetic field out of the junction region with a weak Meissner effect. In the junction median plane, the screening of the field is governed by

$$\frac{\partial h}{\partial y} = \frac{4\pi}{c} J_x = \frac{4\pi}{c} J_0 \sin \gamma \qquad (6\text{-}15)$$

Noting that $\Phi(y)$ in (6-9) is given by $(2\lambda + d) \int_0^y h(y')\, dy'$ when $h(y)$ cannot be taken equal to the applied field H, we obtain

$$\frac{d\gamma}{dy} = \frac{2\pi}{\Phi_0}(2\lambda + d)h$$

Differentiating this, and using (6-15), we have

$$\frac{d^2\gamma}{dy^2} = \frac{1}{\lambda_J^2}\sin\gamma \qquad (6\text{-}16)$$

where

$$\lambda_J = \left[\frac{c\Phi_0}{8\pi^2 J_0(2\lambda + d)}\right]^{1/2} \qquad (6\text{-}17)$$

The quantity λ_J is of the order of 1 mm for typical parameter values. Clearly, it plays the role of a penetration depth if $\gamma \ll 1$, since then (6-16) reduces to

$$\frac{d^2\gamma}{dy^2} = \frac{\gamma}{\lambda_J^2} \qquad (6\text{-}18)$$

which has exponential solutions of the form $\gamma \sim e^{\pm y/\lambda_J}$. So long as the external field at the edge of the junction is much less than $4\pi J_0 \lambda_J/c$, such exponential solutions are possible, with γ, h, and J_x all going exponentially to zero in the interior.

Insight into the nature of the solutions when γ is not restricted to be small can be gained by noting that (6-16) has the same form as the differential equation for a pendulum, if we make the transcription $y \to t$, $\gamma \to \theta$, and $\lambda_J^{-2} \to \omega_0^2 = g/L$, where θ is the angle of the pendulum from the *top* of its circular orbit, and ω_0 is its natural frequency. In terms of this transcription, the solution for γ found in (6-9), by neglecting the effect of the Josephson current on the field, corresponds to the motion of the pendulum whirling around and around with so much kinetic energy that gravitation acceleration is negligible. In (6-16), this corresponds to $\lambda_J \to \infty$, so $d^2\gamma/dy^2 = 0$, $d\gamma/dy$ is constant, and there is a sinusoidal current pattern along y. The finite junction width Y implies that the pendulum-analog solution is relevant only for a finite time interval T, in which the pendulum makes as many revolutions as there are oscillations of $\gamma(y)$.

If one now considers a pendulum moving with less energy, but still enough to have nonzero kinetic energy at the top of the circle, the motion $\theta(t)$ [and hence the variation $\gamma(y)$] will be periodic, but anharmonic. This leads to a nonsinusoidal, periodically reversing current distribution $J(y)$.

Finally, the Meissner-effect limit corresponds to a pendulum moving with barely enough energy to go over the top, so that starting with an initial angular velocity $(d\theta/dt)_0$ at an initial angle θ_0, it decelerates nearly exponentially as it rises, moves very slowly for a long time going over the top where θ is small, and then exponentially accelerates down the other side, recovering the initial angular velocity at $-\theta_0$. If the angular velocity at the top is negligible compared to the initial value, then θ_0 and $(d\theta/dt)_0$ are connected by the conservation of energy, and are not independent. Referring back to the junction problem, the corresponding initial condition is

$$\left(\frac{2\pi H}{\Phi_0}\right)^2 (2\lambda + d)^2 = \left(\frac{d\gamma}{dy}\right)_0^2 = \frac{2}{\lambda_J^2} (1 - \cos \gamma_0)$$

Solving, we have

$$\cos \gamma_0 = 1 - \frac{1}{2}\left(\frac{cH}{4\pi J_0 \lambda_J}\right)^2 \qquad (6\text{-}19)$$

Thus, for weak fields, γ_0 (the value of $|\gamma|$ at the edges of the junction) is given by

$$\gamma_0 = \frac{cH}{4\pi J_0 \lambda_J}$$

On the other hand, the strongest field which can be screened is that corresponding to $\gamma_0 = \pi$, namely

$$H_1 = \frac{8\pi J_0 \lambda_J}{c} \qquad (6\text{-}20)$$

which is typically of the order of 1 Oe. This field H_1 corresponds roughly* to H_{c1} in a type II superconductor, since for any higher field the Meissner solution is impossible.

Continuing the analogy with a bulk superconductor, the maximum super-current that a wide junction can carry is essentially limited to J_0 over a surface layer of thickness $\sim \lambda_J$ around the periphery of the junction. Any higher current would lead to magnetic fields greater than H_1, and hence to the establishment of a periodic "vortex" state, which would be dissipative in the presence of the transport current. Since typical junction dimensions are less than λ_J, we defer further discussion of this situation until a later section.

6-3 SUPERCONDUCTING QUANTUM INTERFEROMETERS

An interference pattern giving much higher resolution in a magnetic field than is implied by (6-14) can be obtained by increasing the area in which flux is effective in causing a phase change. This can be done readily by using two separate Josephson junctions in a superconducting circuit,[1] as illustrated in Fig. 6-4. [The same

---

* See, for example, R. C. Jaklevic, J. Lambe, J. E. Mercereau, and A. H. Silver, *Phys. Rev.* **140**, A1628 (1965); J. E. Zimmerman and A. H. Silver, *Phys. Rev.* **141**, 367 (1966); A. H. Silver and J. E. Zimmerman, *Phys. Rev.* **157**, 317 (1967).

[1] More precisely, the screening at H_1 is only metastable. The maximum value of H for which screening is thermodynamically stable is $H_{c1} = 2H_1/\pi$, as Josephson showed by computation of the total free-energy balance.

FIGURE 6-4
Schematic diagram of superconducting quantum interference device. A and B refer to two point-contact weak links. The rest of the circuit is strongly superconducting. The maximum super-current I through the two parallel channels is a periodic function of the flux Φ enclosed in the loop.

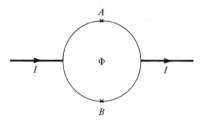

increase in resolution can be obtained using a single junction in a superconducting ring, but then the interference pattern can be sensed only by measuring the impedance of the ring at radio frequencies, since for direct current it is always a short circuit. For simplicity, we deal first with the two-junction dc arrangement, and defer discussion of the RF case to Sec. 6-6.)

Since in practice point-contact weak links are usually used in this technique, we shall simplify our analysis by assuming that the flux threading each separate junction is negligible. (If it is not, it simply superimposes the broad single-slit diffraction pattern of a single junction on the sharp double-slit pattern of the two junctions.) If we further simplify by assuming that the two junctions have the same value of I_0, we can write the total current through the parallel circuit as

$$I = I_0(\sin \gamma_A + \sin \gamma_B) \qquad (6\text{-}21)$$

By the same arguments used to reach (6-9), we can say that γ_A and γ_B are not independent phase differences but are related by

$$\gamma_A - \gamma_B = \frac{2\pi\Phi}{\Phi_0} \qquad (6\text{-}22)$$

where now Φ is the total flux enclosed by the circuit. If we maximize I subject to this constraint, we find

$$I_{\max} = 2I_0 \left| \cos \frac{\pi\Phi}{\Phi_0} \right| \qquad (6\text{-}23)$$

and the phase differences at the two junctions are $\frac{1}{2}\pi \pm \pi\Phi/\Phi_0$.

As is illustrated in Fig. 6-5, $I_{\max}$ is greatest when an integral number of flux quanta are enclosed. It is then simply the sum of the critical currents of the two junctions. On the other hand, $I_{\max}$ goes to zero whenever there are a half-integral number of quanta enclosed. (More generally, if the junctions have different values of I_0, then $I_{\max}$ goes to a nonzero minimum value $|I_{0A} - I_{0B}|$ when Φ/Φ_0 is half-integral.) If the area of the magnetometer loop is ~ 1 cm^2, the "fringe" interval in the interference pattern corresponds to a field change of the order of 10^{-7} G. In fact, with areas of ~ 0.1 cm^2 (which are more typical in practice), it is

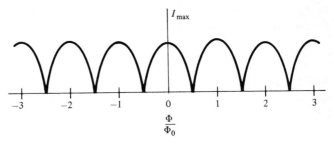

FIGURE 6-5

Dependence of maximum supercurrent through symmetrical two-junction superconducting interferometer (SQUID), shown schematically in Fig. 6-4.

possible to observe the fringes with such high signal-to-noise ratio that the interference pattern can be read to about a thousandth of a fringe. In other words, one can resolve flux increments of $\sim 10^{-10}$ G-cm^2, corresponding to field increments of $\sim 10^{-9}$ G. Because of their obvious potential for applications, these superconducting interferometers have been widely discussed, using the acronymn SQUID (for Superconducting QUantum Interference Device).

An early application of the SQUID magnetometer by Zimmerman and Mercereau[1] was to detect the entrance of individual Abrikosov vortices into a thin wire of niobium. Later, a SQUID magnetometer was used by Beasley et al.[2] to observe the characteristic time dependence of flux creep in hard superconductors, discussed in Sec. 5-7. A recent application to superconductivity, utilizing the full sensitivity of the SQUID, is the measurement by Gollub et al.[3] of the enhancement of the normal state diamagnetism *above* T_c due to thermodynamic fluctuations, a topic we shall take up in the next chapter.

In addition to their practical use for sensitive measurements, the successful operation of these superconducting interferometers provides the most direct and conclusive proof that the quantum phase coherence of ψ extends over macroscopic distances, or in Casimir's picturesque phrase, "over miles of dirty lead wire."

6-4   THE AC JOSEPHSON EFFECT

If the two superconducting elements of a junction are at a different electric potential, the transfer of a pair of electrons from one side to the other involves an energy change of $2eV$. If the process is truly a superfluid process, free of dissipation, this

[1] J. E. Zimmerman and J. E. Mercereau, *Phys. Rev. Letters* **13**, 125 (1964).
[2] M. R. Beasley, R. Labusch, and W. W. Webb, *Phys. Rev.* **181**, 682 (1969).
[3] J. P. Gollub, M. R. Beasley, and M. Tinkham, *Phys. Rev. Letters* **25**, 1646 (1970).

energy must appear elsewhere as a unit (so that the process could in principle be microscopically reversed). In fact, we expect it to appear as a photon of energy

$$h\nu = 2eV \qquad (6\text{-}24)$$

Another way of arriving at this conclusion is to note that the phase of a quantum state evolves in time as $e^{-iEt/\hbar}$. The tunnel current arises from a transition between a state in which a pair is on one side and a state in which it is on the other. Thus, the relative phase of the two states will oscillate at the beat frequency corresponding to the energy difference (6-24), giving a current that oscillates at that frequency just as in an atomic transition.

The ac Josephson effect was first detected by Shapiro,[1] who noted that "steps" could be induced in the dc tunneling characteristic of a Josephson junction subjected to microwave radiation. The steps, shown schematically in Fig. 6-1, occurred very precisely at voltages which were integral multiples of $h\nu/2e$. Two years later the effect was seen in emission[2] as well as absorption. The observation of the very weak emitted signal is made difficult by the poor impedance match between the junction and free space, which can be overcome in absorption by applying a sufficiently intense microwave field.

Before going further into this topic, we should make more precise what is meant by the potential difference V in (6-24), since this formula underlies the very precise measurements[3] of h/e by Josephson tunneling. Roughly speaking, it is, of course, just the difference in electrostatic potential of the two superconductors, $U_1 - U_2$. But this is not gauge-invariant. To make it so, we could modify it to

$$V' \equiv \int_1^2 \mathbf{E} \cdot d\mathbf{s} = U_1 - U_2 - \frac{1}{c}\int_1^2 \frac{\partial \mathbf{A}}{\partial t} \cdot d\mathbf{s}$$

where the integration path follows the tunnel-current direction. Even this is not correct, however, since it considers only the work done on the electron by electric forces. It is well known that equilibrium may be maintained by balancing electric forces against other forces, such as gravity, or against a concentration gradient with concomitant diffusion. These are all included if we replace U by μ/e, where μ is the electrochemical potential. In other words, if $\partial A/\partial t = 0$, it is the equality of μ, not U, at both points which assures no flow of electrons from one point to another, and it is the difference of μ from one point to another which determines the overall energy change when an electron is transferred. Since all ordinary voltmeters (except electrostatic ones) involve electrical contact allowing electron

[1] S. Shapiro, *Phys. Rev. Letters* **11**, 80 (1963). For more recent work, see C. A. Hamilton, *Phys. Rev.* **B5**, 912 (1972), and references cited therein.

[2] I. M. Dmitrenko and I. K. Yanson, *Soviet Phys.—JETP* **22**, 1190 (1965); D. N. Langenberg, D. J. Scalapino, B. N. Taylor, and R. E. Eck, *Phys. Rev. Letters* **15**, 294 (1965).

[3] See, for example, W. H. Parker, D. N. Langenberg, A. Denenstein, and B. N. Taylor, *Phys. Rev.* **177**, 639 (1969).

transfer, they really measure differences in μ, not of U. We conclude, then, that when we write V in (6-24) we mean

$$V \equiv \frac{\mu_1 - \mu_2}{e} - \frac{1}{c} \int_1^2 \frac{\partial \mathbf{A}}{\partial t} \cdot d\mathbf{s} \qquad (6\text{-}25)$$

A final word of caution: In writing (6-25) we have implicitly assumed that μ was a well-defined unique quantity for each superconductor. This is not the case in the immediate vicinity of a region in which normal current is being converted to supercurrent,[1] as for example at the point where current leads of normal metal are attached to the superconductors. In this nonequilibrium situation, the pairs and quasi-particles effectively have slightly different values of μ. Since the Josephson frequency is determined by the difference in μ_p, the electrochemical potential of the paired electrons, while a voltmeter with normal leads measures the difference in μ_{qp}, the quasi-particle potential, this difference is a potential source of error. However, since μ_p is constant in space within each superconductor, and since μ_{qp} relaxes toward it exponentially over a distance of a few microns, this error in chemical potential is easily made utterly negligible in the vicinity of the tunnel junction itself, where the voltage leads are attached.

6-4.1 General Equations

Now let us set up the general differential equations governing the variation of fields and currents in the junction. Unlike the dc case treated earlier, the current now varies in time as well as space. As before, we let the junction lie in the yz plane, with the static magnetic field in the z direction and the phase variation in the y direction. (See Fig. 6-6.) Then, in addition to Maxwell's equations, our basic equation is

$$J_x(y, t) = J_0 \sin \gamma(y, t) \qquad (6\text{-}26)$$

where the phase difference γ satisfies

$$\frac{\partial \gamma}{\partial y} = \frac{2\pi}{\Phi_0} (2\lambda + d)h \qquad (6\text{-}27a)$$

as follows from (6-9), and also

$$\frac{\partial \gamma}{\partial t} = \omega(t) = \frac{2eV}{\hbar} \qquad (6\text{-}27b)$$

which simply generalizes (6-24) in terms of an instantaneous frequency.

---

[1] T. J. Rieger, D. J. Scalapino, and J. E. Mercereau, *Phys. Rev. Letters*, **27**, 1787 (1971); M. L. Yu and J. E. Mercereau, *Phys. Rev. Letters* **28**, 1117 (1972); J. Clarke, *Phys. Rev. Letters* **28**, 1363 (1972); M. Tinkham and J. Clarke, *Phys. Rev. Letters* **28**, 1366 (1972).

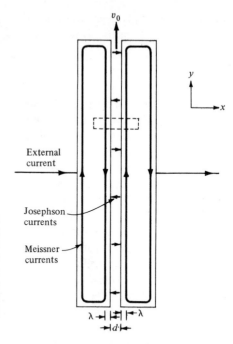

FIGURE 6-6
Diagram of an extended Josephson junction. The periodically reversing Josephson current pattern moves with velocity v_0 in the y direction if a voltage exists across the junction. The Meissner currents screen the applied field from the interior of the superconductors. The dashed contour is used to obtain (6-31a) from Faraday's law.

As a first approximation, we neglect the effect of the Josephson current on the fields, just as we did in the dc case. Then $h = H$ and $V = V_0$ everywhere, and our equations are satisfied by

$$\gamma(y, t) = \gamma_0 + \omega_0 t + k_0 y \qquad (6\text{-}28)$$

where
$$\omega_0 = \frac{2eV_0}{\hbar} \quad \text{and} \quad k_0 = \frac{2\pi(2\lambda + d)H}{\Phi_0} \qquad (6\text{-}29)$$

This γ gives a periodic current distribution of exactly the same form as in the dc case, except that it is moving in the y direction with a phase velocity

$$v_0 = \frac{\omega_0}{k_0} = \frac{cV_0}{(2\lambda + d)H} \qquad (6\text{-}30)$$

Effectively, the vortices set up by the periodic currents in the barrier are flowing through the junction with this velocity. The analogy with the flux-flow regime in a type II superconductor is evident.

Now let us work out how this pattern of Josephson currents couples with the electromagnetic field in the junction. Let $\mathbf{e}^0$ and $\mathbf{h}^0$ refer to the local values of $\mathbf{e}$ and $\mathbf{h}$ in the barrier; $\mathbf{h}$ will extend a distance λ into the superconductors, but $\mathbf{e}$ will be negligible except in the barrier. Then, integrating the Maxwell equation curl $\mathbf{e} = -(1/c)\,\partial\mathbf{h}/\partial t$ over the narrow area indicated by the dashed contour in Fig. 6-6, we obtain

$$\frac{\partial e_x^0}{\partial y} = \frac{1}{c}\frac{2\lambda + d}{d}\frac{\partial h_z^0}{\partial t} \qquad (6\text{-}31a)$$

Also, in the present case the x component of the Maxwell equation for curl $\mathbf{h}$ reduces to

$$\frac{\partial h_z^0}{\partial y} = \frac{4\pi}{c}J_x + \frac{\epsilon}{c}\frac{\partial e_x^0}{\partial t} \qquad (6\text{-}31b)$$

where J_x is the Josephson driving current given by (6-26) and ϵ is the dielectric constant of the barrier. Combining the time derivative of (6-31b) with the space derivative of (6-31a), we obtain

$$\left(\frac{\partial^2}{\partial y^2} - \frac{1}{\bar{c}^2}\frac{\partial^2}{\partial t^2}\right)V = \frac{4\pi}{c^2}(2\lambda + d)\frac{\partial J_x}{\partial t} \qquad (6\text{-}32)$$

where $V = e_x^0 d$ is the voltage across the barrier, and

$$\bar{c}^2 = \frac{c^2}{\epsilon}\left(1 + \frac{2\lambda}{d}\right)^{-1} \ll c^2 \qquad (6\text{-}33)$$

Since $\lambda \approx 500$ Å, while $d \approx 10$ Å, we see that $\bar{c} \approx c/20$ for the typical value $\epsilon = 4$. Thus, the junction region has a very slow electromagnetic wave associated with it. For example, microwaves of $v \sim 10^{10}$ Hz which have a free-space wavelength of 3 cm would have a wavelength in the junction of only ~ 1 mm. It is this disparity which makes it so difficult to couple electromagnetic energy out of the junction region.

It is convenient to transform (6-32) to a form involving only the phase γ as dependent variable. This can be done by using (6-26) and (6-27b) to eliminate J_x and V, and then integrating with respect to time. The result is

$$\left(\frac{\partial^2}{\partial y^2} - \frac{1}{\bar{c}^2}\frac{\partial^2}{\partial t^2}\right)\gamma = \frac{\sin\gamma}{\lambda_J^2} \qquad (6\text{-}34)$$

6-4.2 Examples

First, we note that if the Josephson coupling is very weak, so that $\lambda_J \to \infty$, then the solutions of (6-34) are simply plane waves with phase velocity $\bar{c}$. These electromagnetic waves will couple strongly to the Josephson currents if this velocity matches v_0, the velocity of the vortex pattern given by (6-30). By conservation of energy, the resulting increased dissipation of energy must show up as a peak in the dc current drawn by the junction as a function of applied voltage V_0; this peak occurs when $v_0 = \bar{c}$, or when

$$V_0^2 = \frac{d(2\lambda + d)H^2}{\epsilon} \qquad (6\text{-}35)$$

Just such a peak was observed by Eck et al.,[1] and it provided one of the first demonstrations of the ac Josephson effect without application of external fields. A closely related effect is the observation of steps in the I-V characteristic by Fiske.[2] These occur when the frequency of the Josephson currents match the frequencies of the various electromagnetic cavity modes of the slow wave in the junction.

Another simple special case of (6-34) is obtained by considering solutions in which γ is uniform over the junction. Then the equation again reduces to the form of the equation of motion for a pendulum:

$$\frac{d^2\gamma}{dt^2} + \omega_J^2 \sin \gamma = 0 \qquad (6\text{-}36)$$

where ω_J is defined by

$$\omega_J^2 = \frac{\bar{c}^2}{\lambda_J^2} = \frac{8\pi e d J_0}{\epsilon h} \qquad (6\text{-}37)$$

Note that if we substitute $J_0 = 2eh|\psi|^2/m^*d$, as suggested by our heuristic development of (6-1), we find $\omega_J^2 = 4\pi(2e)^2|\psi|^2/\epsilon m^*$, which is just the plasma frequency of a charged gas of particles with charge $2e$, mass m^*, and number density $|\psi|^2$. Since J_0 is small compared to J_c for a strong superconductor, $|\psi|^2$ in the junction is small compared to the density of electrons in a metal. Hence, this "Josephson plasma frequency" ω_J is much less than the ordinary plasma frequency for metals, which is in the ultraviolet. In fact, ω_J is typically in the microwave frequency range, and this plasma frequency has been observed experimentally by Dahm et al.[3]

In identifying their observed frequency with ω_J, Dahm et al. took advantage of the fact that the resonant frequency defined by (6-36) can be "tuned" away from ω_J if γ is not small. For example, the pendulum analogy shows that if γ executes large oscillations about zero, its time dependence will become anharmonic and the natural frequency will change. A more convenient test of (6-36), however, is the dependence of the resonant frequency on the strength of an

[1] R. E. Eck, D. J. Scalapino, and B. N. Taylor, *Phys. Rev. Letters* **13**, 15 (1964).

[2] M. D. Fiske, *Rev. Mod. Phys.* **36**, 221 (1964); D. D. Coon and M. D. Fiske, *Phys. Rev.* **138**, A744 (1965).

[3] A. J. Dahm, A. Denenstein, T. F. Finnegan, D. N. Langenberg, and D. J. Scalapino, *Phys. Rev. Letters* **20**, 589 (1968).

imposed dc current. If the dc current density is J_a, the value of γ must adjust to $\gamma_0 = \sin^{-1}(J_a/J_0)$. (We consider only junctions small compared to λ_J, so that we can neglect the partial Meissner effect in the junction which makes γ_0 depend on spatial coordinates.) Linearizing about the operating point, (6-36) becomes

$$\frac{\partial^2(\delta\gamma)}{\partial t^2} + \omega_J^2 \cos \gamma_0(\delta\gamma) = 0 \qquad (6\text{-}38)$$

so that the plasma frequency is tuneable by a current according to the relation

$$\omega^2 = \omega_J^2 \cos\left[\sin^{-1}\frac{J_a}{J_0}\right]$$

$$= \omega_J^2\left[1 - \left(\frac{J_a}{J_0}\right)^2\right]^{1/2} \qquad (6\text{-}39)$$

From this relation, we see that ω^4 should decrease linearly with J_a^2. This property was used by Dahm et al. to conclusively identify their observed signals with the Josephson plasma resonance.

Finally, we consider solutions of (6-34), linearized about $\gamma = 0$, but varying in both space and time as $e^{i(\omega t - ky)}$. Such solutions exist provided $(-k^2 + \omega^2/\bar{c}^2) = \lambda_J^{-2}$. This dispersion relation can be rewritten as

$$\left(\frac{\omega}{\omega_J}\right)^2 = 1 + \lambda_J^2 k^2 \qquad (6\text{-}40)$$

Thus, the Josephson plasma frequency ω_J is the lower limit for frequencies which can propagate in the plane of the junction. As a curiosity, we also note that for $\omega = 0$, (6-40) reduces to $k = \pm i/\lambda_J$. This gives rise to the real exponential solutions giving a Meissner effect in the junction, as found earlier in (6-18).

6-5 GENERAL RELATIONS BETWEEN PHASE, NUMBER, ENERGY, AND CURRENT

Now that we have developed a familiarity with the Josephson effect on the phenomenological level, let us briefly consider it in a more general framework. This will make more explicit some of the underlying interrelations between phase and number variables, and between energy and current.

In our development of the BCS wavefunction (2-14), we noted that the phase of $v_{\mathbf{k}}$ relative to $u_{\mathbf{k}}$ was arbitrary, so long as it was the same as that of $\Delta_{\mathbf{k}}(=\Delta)$ to satisfy (2-44). Thus we can write down a BCS wavefunction in which this phase has a specified value φ, as follows:

$$|\psi_\varphi\rangle = \prod_{\mathbf{k}}(|u_{\mathbf{k}}| + |v_{\mathbf{k}}|e^{i\varphi}c_{\mathbf{k}\uparrow}^* c_{-\mathbf{k}\downarrow}^*)|\phi_0\rangle \qquad (6\text{-}41)$$

As discussed in Sec. 2-3, this wavefunction has a large uncertainty in the number of electrons N, although the mean value $\bar{N}$ is specified by a suitable choice of the chemical potential μ. We further remarked that we could set up a state with *precisely* N particles by forming the superposition

$$|\psi_N\rangle = \frac{1}{2\pi}\int_0^{2\pi} d\varphi\, e^{-iN\varphi/2}\,|\psi_\varphi\rangle \qquad (6\text{-}42)$$

The inverse transformation to (6-42) is

$$|\psi_\varphi\rangle = \sum_{N=0}^{\infty} e^{iN\varphi/2}\,|\psi_N\rangle \qquad (6\text{-}43)$$

so that states with sharp values of N and φ are related by the same kind of Fourier-transform relation that relates eigenfunctions of conjugate operators such as momentum and position in elementary quantum mechanics. Consequently, there is an uncertainty relation

$$\Delta N\,\Delta\varphi \gtrsim 1 \qquad (6\text{-}44)$$

which limits the precision with which N and φ can be specified simultaneously. However, since $\bar{N}$ is typically $\sim 10^{20}$, this inequality can be satisfied even if $\Delta N/\bar{N}$ and $\Delta\varphi$ are both of the order of $\bar{N}^{-1/2} \sim 10^{-10}$. To be concrete, this could be accomplished by inserting a gaussian envelope weighting function $e^{-[(\varphi-\bar{\varphi})^2/(\Delta\varphi)^2]}$ into the integral in (6-42). Thus, we are justified in adopting a semiclassical viewpoint, in which N and φ are simultaneously defined with a precision of 1 part in 10^{10}.

The phase φ discussed above is the relative phase of the probability amplitudes for states differing by one pair of electrons. From (2-44), it is then the phase of Δ. Is it also the phase of the GL wavefunction ψ? Since we have not made a formal deduction of the GL theory from the microscopic theory, we cannot give a formal argument. Suffice it to say that in this deduction, $\psi(r)$ turns out to be $\langle\psi_{op\uparrow}(r)\psi_{op\downarrow}(r)\rangle$, that is, the local expectation value of a pair annihilation operator. Hence its phase will be the phase difference of the amplitudes of states differing by two particles, which is exactly the quantity we have called φ.

Since $\hbar\varphi$ and $N^* = \frac{1}{2}N$ (that is, the number of *pairs*) are semiclassical, canonically conjugate variables, we can apply the hamiltonian equations of motion to write

$$\hbar\frac{\partial\varphi}{\partial t} = -\frac{\partial\mathcal{H}}{\partial N^*} = -2\frac{\partial F}{\partial N} = -2\mu \qquad (6\text{-}45a)$$

and

$$\frac{\partial N^*}{\partial t} = \frac{\partial\mathcal{H}}{\partial(\hbar\varphi)} = \frac{1}{\hbar}\frac{\partial F}{\partial\varphi} \qquad (6\text{-}45b)$$

Here we have replaced the hamiltonian $\mathcal{H}$ by the *free* energy F (rather than the expectation value of the energy E) because the work done on a system in an isothermal process is dF, not dE. Hence F is the appropriate "energy function" for the mechanics of a statistical system.* From (6-45a) it follows that the phase of an isolated superconductor evolves in time as $e^{-i2\mu t/\hbar}$, but this has no physical consequence. Again, for an isolated superconductor, (6-45b) has no content, since N is constant and $\partial F/\partial \varphi = 0$.

To obtain observable consequences of the phase, one must bring a second superconductor into sufficiently close contact to have an energetic coupling F_{12}. Since an overall shift in the phase can have no effect on the energy, F_{12} can depend only on the relative phase $\gamma = \varphi_2 - \varphi_1$. Then, (6-45a) implies the Josephson frequency relation, since

$$\frac{\partial \gamma}{\partial t} = \frac{\partial \varphi_2}{\partial t} - \frac{\partial \varphi_1}{\partial t} = \frac{2}{\hbar}(\mu_1 - \mu_2) \qquad (6\text{-}46)$$

Similarly, from (6-45b) the current between the two superconductors is given by

$$\frac{I_{12}}{2e} = -\frac{\partial N_1^*}{\partial t} = \frac{\partial N_2^*}{\partial t} = -\frac{1}{\hbar}\frac{\partial F_{12}}{\partial \varphi_1} = \frac{1}{\hbar}\frac{\partial F_{12}}{\partial \varphi_2} = \frac{1}{\hbar}\frac{\partial F_{12}}{\partial \gamma} \qquad (6\text{-}47)$$

which makes explicit the equivalence of the Josephson current relation to the existence of a phase-dependent coupling energy $F_{12}(\gamma)$. When F_{12} is at a minimum, $\partial F_{12}/\partial \gamma = 0$, and there is no current. This exactly parallels the situation in a strong superconductor, in which the lowest energy state is one without current.

One can use (6-47) to "derive" the Josephson current-phase relation by noting that $F_{12}(\gamma)$ must be a periodic even function of γ—periodic since γ is defined only modulo 2π, and even since we expect the energy to be unchanged by replacement of ψ by ψ^*. Thus, F_{12} can be expanded in a cosine Fourier series in γ. When the derivative with respect to γ required in (6-47) is taken, this becomes a sine series, of which the leading term is (6-4).

It is perhaps more useful to reverse this procedure and use the measured current relation (6-4) to infer the coupling energy, namely,

$$F_{12}(\gamma) = \text{const} - \frac{\hbar I_0}{2e}\cos\gamma \qquad (6\text{-}48)$$

---

* If this concept is unfamiliar, consider the example of the isothermal compression of a mole of ideal gas from V_1 to V_2. Since the internal energy U is a function of T only, $\Delta U = 0$. The work done on the gas is $\Delta W = -\int P\, dV = RT \ln V_1/V_2$, but this energy leaves the gas (to the thermal reservoir at T) as heat, lowering the entropy of the gas. This increases its free energy $F = U - TS$ by an amount $\Delta F = -T\,\Delta S = \Delta W$. Thus, if one wants to make energy arguments for an isothermal system without taking account of the thermal bath, one must use F, not U, as the energy of the system under consideration.

It is useful to express this coupling energy in terms of a temperature T_0, such that $kT_0 = \hbar I_0/2e$. If this is done, one finds

$$\frac{T_0}{I_0} = \frac{\hbar}{2ek} = 2.4 \times 10^7 \text{ °K/A} \qquad (6\text{-}49)$$

Thus, operating in the liquid helium temperature range, one needs $I_0 \gtrsim 10^{-7}$ A for the coupling to be strong enough to maintain phase coherence across the junction in the presence of thermal noise. If I_0 is below this value, the Josephson effect will be washed out by random variations of γ, and the junction will appear to be "normal." Similarly, any external noise signals reaching the junction from circuitry at higher temperatures or from electrical "pickup" can destroy the phase coherence in the junction.

6-6   PRACTICAL SQUID MAGNETOMETERS

In the course of our discussion of the Josephson effects, we have already indicated two of their most celebrated applications: SQUID magnetometers and the precision determination of h/e (or, conversely, the precision voltage standard). These discussions have involved only the supercurrent properties of the Josephson junctions. However, since almost all practical applications of the Josephson effect involve finite voltages across the junction to facilitate instrumentation, dissipative processes will always be present as well. In this section, we present a more realistic discussion of the operation of Josephson magnetometers, taking dissipation into account.

6-6.1   Typical Parameter Values

The dissipative processes may be due to normal electrons in a metallic weak link, or to quasi-particle tunneling in a tunnel junction; they may also arise from direct emission of photons or phonons by the Josephson current. To avoid dealing with these specific cases, we shall simply approximate them all by including a resistance R in parallel with the supercurrent channel. Under almost all circumstances, this R will be of the same order of magnitude as the resistance of the weak link above T_c, typically $\sim 1 \, \Omega$. This shunt resistance corresponds to the term $G_0 V$ in (6-4a), with $G_0 = 1/R$. For simplicity, we drop the term $G_{\text{int}} \cos \gamma$, since it does not alter the major conclusions.

For analyzing the operation of Josephson devices, it is often convenient to characterize also the supercurrent response by a circuit parameter, namely an equivalent inductance L_J. If we define L_J by the differential relation $V = L_J \, dI/dt$, where $I = I_0 \sin \gamma$ and $d\gamma/dt = 2eV/\hbar$, then

$$L_J = \frac{\hbar}{2eI_0 \cos \gamma} \qquad (6\text{-}50)$$

(If instead we had defined L_J by equating the total increase in junction energy to $\frac{1}{2}L_J I^2$, the factor $\cos \gamma$ would have been replaced by $\cos^2 \frac{1}{2}\gamma$.) Although L_J depends on the phase γ, the order of magnitude will be given by $\hbar/2eI_0$, the value for $\gamma \approx 0$. In practical units, $L_J I_0 \cos \gamma = \hbar/2e = \Phi_0/2\pi = 3.3 \times 10^{-16}$ Wb. Thus, for $I_0 \approx 10^{-6}$ A, $L_J \approx 3 \times 10^{-10}$ H.

If we take the relation between I_0 and R_n for a tunnel junction, we find that the supercurrent response (described by L_J) dominates for frequencies or voltages such that $2eV = \hbar\omega \ll \Delta$; for $2eV = \hbar\omega \gg \Delta$, the junction acts resistive,[*] as might have been expected on general grounds. On the other hand, it is found empirically that in point contacts I_0 is typically (although not invariably) lower by one or two orders of magnitude than the theoretical value for a tunnel junction of the same normal resistance.

Now let us compare this Josephson inductance L_J with the self-inductance of the loop completing the circuit. If it is of radius r and made of wire of radius a, the inductance is (again in mks units)

$$L \approx 4\pi \times 10^{-7} \, r \ln \frac{r}{a} \qquad (6\text{-}51)$$

For a 3-mm-diameter loop, this is of the order of 3×10^{-9} H, typically somewhat larger than L_J.

The value of this inductance must be kept low if the quantum behavior of the device is to be maintained in a resistive mode of operation. To see this, we note that when the junction is resistive we can apply the equipartition theorem to the flux contained in the loop. This states that the mean-square fluctuation is governed by

$$\frac{1}{2}\frac{(\delta\Phi)^2}{L} = \frac{1}{2}kT$$

If this fluctuation amplitude exceeds $\sim \frac{1}{2}\Phi_0$, the quantum periodicity in applied flux will be washed out by the random flux of the thermal currents. This leads to the criterion that

$$L \lesssim \frac{\Phi_0^2}{4kT} \qquad (6\text{-}52)$$

so that $L \lesssim 2 \times 10^{-8}$ H for operation at 4°K. Since the effective "noise temperature" may be an order of magnitude higher, L normally is chosen to be as small as 10^{-9} H to assure reliable operation.

A final quantity of interest is the ratio LI_0/Φ_0. If this is greater than unity, the supercurrent through the weak link will be able to screen an entire flux quantum from the loop before breaking down, becoming resistive, and admitting

---

[*] Actually, at high frequencies the impedance of tunnel junctions is dominated by the capacitance between the two superconductors. Since this is not necessarily the case in point contacts, we shall not consider this further here.

flux discontinuously. This can lead to hysteresis effects and complicate the use of the device in some applications. On the other hand, so long as the device is operated in a resistive mode, it is reasonable to assume that the circulating current in the loop will take on that particular one of the set of allowed quantum values which minimizes the energy. Thus, this sort of hysteresis does not lead to ambiguity in suitably designed devices. In fact, I_0 must always be chosen large enough to satisfy the requirement implied by (6-49) for maintaining phase coherence across the junction; that is,*

$$I_0 \gtrsim \frac{2\pi kT}{\Phi_0} \qquad (6\text{-}53)$$

so that I_0 must exceed $\sim 2 \times 10^{-7}$ A for operation at 4°K. Again, allowing for a higher noise temperature, I_0 is normally chosen to be at least 10^{-6} A. Actually, for practical reasons, devices are usually operated with $LI_0/\Phi_0 \approx 1$. In that case, (6-53) reduces essentially to the same condition as (6-52).

6-6.2 The Two-junction DC SQUID

As a first example of the role played by the shunt resistance, let us give a more detailed analysis of the operation of the two-weak-link dc SQUID, described in Sec. 6-3. In that discussion, we showed that the critical current of the device was a periodic function of the flux enclosed in the loop. But it is more convenient to operate the device in a resistive mode at constant current, in which the voltage across the device is a periodic function of Φ/Φ_0. Let us see how this mode can be understood.

The total supercurrent through the device is still governed by (6-21) and (6-22). But in the resistive mode, the constant current I supplied exceeds this supercurrent. Thus, the excess current must be carried by the normal conductance of the two junctions, and we have $I = I_s + 2V/R$, or

$$V = \tfrac{1}{2}R[I - I_0(\sin \gamma_A + \sin \gamma_B)]$$

Using a trigonometric identity, recalling from (6-22) that $\gamma_A - \gamma_B = 2\pi\Phi/\Phi_0$, and defining $\bar{\gamma} = \tfrac{1}{2}(\gamma_A + \gamma_B)$, this can be written as

$$V = R\left[\frac{1}{2}I - I_0 \cos\left(\frac{\pi\Phi}{\Phi_0}\right) \sin \bar{\gamma}\right] \qquad (6\text{-}54)$$

For $I < 2I_0$, this equation determines the value of $\bar{\gamma}$ giving a purely supercurrent solution, with $V = 0$, as described in Sec. 6-3. For higher currents,

---

* This criterion is somewhat conservative when the two sides of the junction are connected by a loop, since then fluctuations in relative phase imply currents, which are opposed not only by the junction coupling energy (reflected in the Josephson inductance), but also by the inductance of the loop. As a result, the denominator of (6-53) should be replaced by something like $(\Phi_0 + 2\pi LI_0)$. For all values of L small enough to satisfy (6-52), the simple formula (6-53) is within a factor of ~ 2 of the more exact result.

however, no such solution is possible, and $V > 0$. In that case, $\bar{\gamma}$ evolves in time according to the Josephson relation $\dot{\bar{\gamma}} = 2eV(t)/\hbar$. At first glance, it then appears that the supercurrent will simply average to zero over a cycle of the alternating current. But if one looks more carefully, one sees that the *time* average of sin $\bar{\gamma}$ is nonzero, since $\bar{\gamma}$ evolves more slowly when V is small, giving greater weight to that part of the cycle. More quantitatively,

$$\bar{V} = \frac{\hbar}{2e} \frac{1}{T} \oint \frac{d\bar{\gamma}}{dt} \, dt = \frac{\hbar}{2e} \frac{2\pi}{T} \qquad (6\text{-}55a)$$

where the period T of the anharmonic cycle is given by

$$T = \int_0^{2\pi} \frac{d\bar{\gamma}}{d\bar{\gamma}/dt} = \frac{\hbar}{2e} \int_0^{2\pi} \frac{d\bar{\gamma}}{V(\bar{\gamma})} \qquad (6\text{-}55b)$$

Carrying out the integration,[1] one finds

$$\bar{V} = R\left[\left(\frac{1}{2}I\right)^2 - I_0^2 \cos^2\left(\frac{\pi\Phi}{\Phi_0}\right)\right]^{1/2} \qquad (6\text{-}56)$$

Since $\cos^2(\pi\Phi/\Phi_0) = \frac{1}{2}[1 + \cos(2\pi\Phi/\Phi_0)]$, we see explicitly that $\bar{V}$ is periodic in Φ with period Φ_0.

From (6-56) it follows that $\bar{V}$ has its maximum value, $\frac{1}{2}IR$, when a half-integral number of flux quanta are present. The minimum value of $\bar{V}$ is zero if $I < 2I_0$; if $I > 2I_0$, the minimum is clearly $\frac{1}{2}IR[1 - (2I_0/I)^2]^{1/2}$, which occurs for integral numbers of flux quanta. Thus, the periodic change in $\bar{V}$ with Φ is of maximum amplitude for $I = 2I_0$, when $\Delta\bar{V} \equiv \bar{V}_{\text{max}} - \bar{V}_{\text{min}} = I_0 R$. For tunnel junctions, this should be of the order of the gap voltage $\Delta/e \approx 10^{-3}$ V. For point contacts, voltages of the order of 10^{-5} V are more typically seen. When $I \gg 2I_0$, $\Delta\bar{V} = I_0^2 R/I$, so $\Delta\bar{V}$ falls off rather slowly even when the bias current is so high that the device is always in a resistive mode.

6-6.3 Effect of Screening

In the above derivation, we have made the simplifying assumption that we could neglect the self-induced flux due to the screening current I_s circulating in the loop, as well as the flux due to the bias current I. For simplicity of analysis, we shall assume that the current leads are attached symmetrically to the loop, so that the mutual inductance between the bias circuit and the loop is zero. In this case, the latter effect is zero.* The neglect of the screening effect of the circulating current,

---

[1] For a systematic treatment of this and many other examples, see A. Th. A. M. De Waele and R. De Bruyn Ouboter, *Physica* **41**, 225 (1969).

* If this is not the case, asymmetries appear in the periodic response. These can be used to increase the sensitivity of the device. [J. Clarke and J. L. Paterson, *Appl. Phys. Letters* **19**, 469 (1971).]

however, is typically justified only marginally, if at all. Nonetheless, the qualitative results remain essentially as found here.

This can be understood as follows: Equations (6-54) and (6-55) are generally valid, so long as Φ is taken to include the screening flux Φ_s as well as the externally supplied flux Φ_x; that is, in addition to (6-54) we have the relations

$$\Phi = \Phi_x + \Phi_s \qquad (6\text{-}57)$$

and

$$\Phi_s = LI_s = \tfrac{1}{2}LI_0(\sin \gamma_B - \sin \gamma_A) \qquad (6\text{-}58)$$

$$= -LI_0 \sin\left(\frac{\pi\Phi}{\Phi_0}\right) \cos \bar{\gamma}$$

Equation (6-58) implies that Φ_s can remain the same* if Φ_x is changed by a multiple of the flux quantum. Then the factor $\cos (\pi\Phi/\Phi_0) \equiv \cos [\pi(\Phi_x + \Phi_s)/\Phi_0]$ in (6-54) is also unchanged for such a change in Φ_x, and V remains periodic in Φ_x with unchanged period despite the inclusion of Φ_s. The effect of including Φ_s is only to change the form of $V(\gamma)$ in a complicated way, so that the integration of (6-55b) is no longer elementary but requires numerical or graphical methods. As a result (6-56) no longer holds, but it is replaced by a qualitatively similar dependence, with the same periodicity.

Although we do not require a detailed solution of the coupled transcendental equations determining the relation between Φ and Φ_x, it is useful to work out some aspects of the zero-voltage case, when all currents are supercurrents. Setting $V = 0$ in (6-54), we have

$$\sin \bar{\gamma} = \frac{I}{2I_0 \cos (\pi\Phi/\Phi_0)}$$

Inserting this in (6-58), and using (6-57) to eliminate Φ_s, we are led to

$$\Phi_x - \Phi = \pm LI_0' \sin\frac{\pi\Phi}{\Phi_0} \qquad (6\text{-}59)$$

where

$$I_0' = I_0\left[1 - \frac{I^2}{4I_0^2 \cos^2 (\pi\Phi/\Phi_0)}\right]^{1/2} \le I_0 \qquad (6\text{-}59a)$$

Given I and Φ_x, these relations determine the set of possible values for Φ. An important special case occurs when Φ is close enough to $n\Phi_0$ that we can approximate $\sin (\pi\Phi/\Phi_0)$ by $\pm(\pi/\Phi_0)(\Phi - n\Phi_0)$. Then (6-59) can be reduced to

$$\Phi - n\Phi_0 = \frac{\Phi_x - n\Phi_0}{1 + \pi LI_0'/\Phi_0} \qquad \Phi - n\Phi_0 \lesssim \tfrac{1}{2}\Phi_0 \qquad (6\text{-}60)$$

---

*If $LI_0/\Phi_0 \gtrsim 1$, Φ_s is multivalued as given by (6-58). However, in the resistive state of the junction contemplated here, we presume that Φ_s always takes on the value giving the lowest energy, i.e., the value for which $|I_s|$ or $|\Phi_s|$ is least.

In the case of weak screening ($\pi L I_0'/\Phi_0 \ll 1$), the only solution to (6-60) is one with $\Phi \approx \Phi_x$, with n being the integer nearest to Φ_x/Φ_0. If there is strong screening, however, then $\sim L I_0'/\Phi_0$ values of n on either side of Φ_x/Φ_0 will still give $\Phi - n\Phi_0 \lesssim \frac{1}{2}\Phi_0$, indicating self-consistent solutions. All of these solutions are metastable, apart from the one with n nearest to Φ_x/Φ_0.

The maximum supercurrent through the device is always determined by this most stable solution, since observation of I_c requires driving the loop normal, at least momentarily, which breaks down any metastable solution, allowing the more stable one to form. Thus, Φ is never farther from a quantum value than $\frac{1}{2}\Phi_0/(1 + \pi L I_0'/\Phi_0)$ when the critical current is measured. This implies that the depth of periodic modulation of I_c for the device is not $2I_0$, as when screening may be neglected [see (6-23)], but a much lower value, if screening is important. Computation of this reduced modulation depth requires numerical solutions for the general case, but it becomes simple in the limit of strong screening, where we can assume that Φ is always indistinguishable from a quantum value. Then the maximum screened flux is $\frac{1}{2}\Phi_0$, for which a circulating current $\Phi_0/2L$ is set up, so that the two junction currents must be $(I/2) \pm (\Phi_0/2L)$ for an impressed current I. Since neither junction current can exceed I_0, the maximum total current is $2I_0 - \Phi_0/L$, compared to $2I_0$ when $\Phi = n\Phi_0$. Thus, in this strong-screening limit, the modulation depth $\Delta I_c \lesssim \Phi_0/L \sim 10^{-6}$ A, regardless of how large I_0 is made. Since a detailed analysis of the resistive mode with screening included is considerably more complex than that leading to (6-56), we content ourselves with noting that the flux-periodic modulation of the voltage developed at the critical current will be of the order of $\frac{1}{2}R\,\Delta I_c$. For the small-screening case this will be $\sim I_0 R$, in agreement with our analysis following (6-56). For the strong-screening case, it will be lower than this by a factor of the order of $\Phi_0/2I_0 L$. Thus, it is customary to work with I_0 not much larger than $\Phi_0/2L$ to optimize the signal strength.

6-6.4 Use of Flux Transformer

In actual applications of the SQUID to physical measurements, it is usually desirable to use a superconducting transformer, or "flux transporter," to couple the flux change of interest into the loop of the SQUID. As discussed by Gollub,[1] the advantages of this arrangement include the following: (1) It allows the SQUID to work at a convenient constant temperature and low ambient magnetic field, regardless of the conditions at the other end of the flux transporter where the sample is located. In this way, SQUIDs have been used with samples exposed to kilogauss magnetic fields and temperatures far above the T_c of the niobium in the SQUID. (2) It allows the SQUID to be physically oriented to facilitate adjustment of the weak links. (3) The sensitivity of the SQUID can be readily decreased by

---

[1] J. P. Gollub, "Diamagnetism Due to Fluctuations in Superconductors," Technical Report No. 3, Div. of Eng. and Appl. Phys., Harvard Univ., 1970.

one or two orders of magnitude by inserting a "dropping inductor" into the superconducting circuit of the flux transporter. This makes it possible to follow phenomena in which the size of the effect of interest varies over many orders of magnitude, as near a phase transition. (4) It facilitates the insertion of negative feedback, by means of which the SQUID magnetometer is converted to a null-balancing device. This has the advantage of making the device independent of the exact form of the periodic $V(\Phi_x)$ dependence, since it "locks on" to a sharp feature in this dependence, canceling any further change in the flux to be measured by an equal and opposite change from the feedback circuit. (5) It allows design flexibility in the choice of the area and number of turns in the input coil in the transporter circuit to optimize overall performance.

The basic principle governing the transformer circuit is that the total flux linking the circuit stays constant, as required by fluxoid quantization. Thus, if the applied field at the primary changes by ΔB_1, the flux linkage change is prevented by the establishment of a current ΔI in the circuit, whose magnitude is set by

$$N_1 A_1 \, \Delta B_1 + L_1 \, \Delta I + L_2 \, \Delta I = 0$$

The field change produced in the secondary is then

$$\Delta B_2 = \frac{L_2 \, \Delta I}{N_2 A_2} = -\frac{L_2 N_1 A_1}{N_2 A_2 (L_1 + L_2)} \Delta B_1$$

For given coil dimensions, $L_i = C_i N_i^2$, where to a good approximation, $C_i \propto r_i^2/(r_i + l_i)$, r_i being the radius and l_i the length of the ith coil. It then follows that ΔB_2 is maximized (for given coil dimensions) if the turns ratio N_1/N_2 is chosen to make $L_1 = L_2$, in which case half the input flux appears in L_2. With this optimized turns ratio, we have

$$\frac{\Delta B_2}{\Delta B_1} = -\frac{1}{2} \frac{N_1 A_1}{N_2 A_2} \approx -\frac{1}{2} \frac{r_1}{r_2} \left(\frac{r_1 + l_1}{r_2 + l_2}\right)^{1/2}$$

using the approximation for C_i mentioned above. Thus, for given secondary dimensions as required to fit the SQUID, the flux change for given ΔB_1 can be increased in proportion to $r_1(r_1 + l_1)^{1/2}$, that is, in proportion to the square root of the volume filled by the field of the primary coil. This result could have been anticipated directly from the equal-inductance principle, since $\frac{1}{2}LI^2 = (1/8\pi) \int B^2 \, d\mathbf{r} \approx B^2 V/8\pi$, and the same current passes through both coils.

6-6.5   The Single-Contact RF SQUID

A variation on the SQUID design which offers a number of practical advantages is the symmetric single-point-contact device described by Zimmerman et al.[1] and shown in Fig. 6-7. The entire structure is machined from a single piece of niobium,

---

[*] J. E. Zimmerman, P. Thiene, and J. T. Harding, *J. Appl. Phys.* **41**, 1572 (1970).

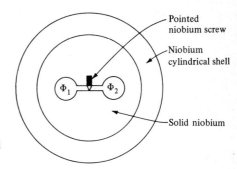

FIGURE 6-7
Schematic diagram of symmetrical single-point-contact RF SQUID.

and the contact is made by a sharply pointed niobium screw. Since only one metal is used in the construction, differential thermal-expansion effects are minimal, and the device can be repeatedly cycled between room temperature and helium temperature without serious degradation. The geometry of the device also provides such a rigid support for the point contact that it is relatively insensitive to damage by mechanical shock. Finally, the presence of the massive niobium shell entirely surrounding the two holes assures that the total flux in the two holes, $\Phi_1 + \Phi_2$, is constant at whatever value was trapped in when the device became superconducting. This makes the device insensitive to uniform external fields, and hence to magnetic "pickup." It is sensitive only to differential fields, which shift flux from one hole to the other. Such a field is normally provided by using a superconducting flux transporter, as described above, with the output coil in one hole of the device.

Apart from the advantages of mechanical stability and insensitivity to uniform ambient fields, this symmetrical two-hole SQUID is equivalent to a loop with a single weak link, embracing the differential flux $\Phi = \Phi_2 - \Phi_1$. In use, the flux state of the ring is interrogated by coupling it to a resonant circuit driven by a constant current source at a convenient RF frequency, such as 30 MHz. (See Fig. 6-8). The properties of such RF-biased SQUID systems were studied earlier by Silver and Zimmerman.[1] Both the Q-factor and the resonant frequency of the resonant circuit are modified by the coupling to the SQUID in a way which depends on the flux through the loop, allowing various schemes to be used to derive from the voltage across the resonant circuit an electrical signal which is periodic in Φ/Φ_0. Such arrangements were described by Zimmerman et al., and more recently detailed efforts at optimization have been described by Giffard, Webb, and Wheatley.[2] An excellent general review has also been given by Webb.[3]

The equations governing the single-junction SQUID are simpler than those derived above for the double-junction SQUID, since there is only one phase

[1] A. H. Silver and J. E. Zimmerman, *Phys. Rev.* **157**, 317 (1967).
[2] R. P. Giffard, R. A. Webb, and J. C. Wheatley, *J. Low Temp. Phys.* **6**, 533 (1972).
[3] W. W. Webb, *IEEE Trans.*, MAG-8, 51 (1972).

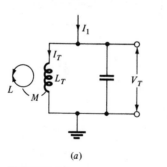

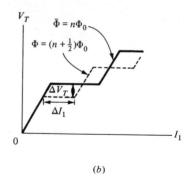

(a)   (b)

FIGURE 6-8

(a) Circuit used for RF-biased SQUID. A constant RF current I_1 is supplied, and the tank voltage V_T is monitored. (b) Relation between V_T and I_1 for the cases of integral and half-integral numbers of flux quanta in the SQUID. For intermediate values of flux, the voltage steps occur at values of V_T intermediate between those shown. The step height ΔV_T and width ΔI_1 are discussed in the text.

variable γ and only a circulating current. The phase difference is given by $\gamma = 2\pi\Phi/\Phi_0$, and the circulating current by

$$I = I_0 \sin \gamma + \frac{V}{R} \qquad (6\text{-}61)$$

Then, since $V = L\dot{\Phi}$, the equation analogous to (6-59), but including the resistive term, is

$$\Phi_x - \Phi = LI_0 \sin\frac{2\pi\Phi}{\Phi_0} + \frac{L\dot{\Phi}}{R} \qquad (6\text{-}62)$$

In this equation, Φ_x includes both the external flux being measured and the RF flux interrogating the device. If we make the same linear approximation to the sine function as was made in reaching (6-60), we obtain

$$\frac{d}{dt}(\Phi - n\Phi_0) + R(L^{-1} + L_J^{-1})(\Phi - n\Phi_0) = \frac{R}{L}(\Phi_x - n\Phi_0) \qquad (6\text{-}63)$$

where $L_J = \Phi_0/2\pi I_0$ in this linearized approximation. Thus, the low-frequency screening relation is

$$\frac{\Phi - n\Phi_0}{\Phi_x - n\Phi_0} = \frac{1}{1 + L/L_J} = \frac{1}{1 + 2\pi LI_0/\Phi_0} \qquad (6\text{-}64)$$

which is rather similar to (6-60). The time constant for relaxation in response to a small change in Φ_x is very short: $\sim L_J/R \sim 10^{-10}$ to 10^{-12} sec. But for large changes associated with a change in the quantum number n, the linear approximation breaks down, and the appropriate time constant can be seen from (6-62) to be essentially the usual one, $L/R \sim 10^{-9}$ to 10^{-10} sec. For typical operating frequencies, $\omega L_J/R \ll 1$, and the device acts essentially like a lossless shorted turn except when it breaks down and changes quantum state.

In the usual mode of operation, the SQUID is loosely coupled to a high-Q tank circuit, fed with a constant current sufficient to induce circulating currents in the loop which exceed the critical current I_0. The energy transfer per cycle from the inductance to the junction can be written

$$\Delta W = \oint IV \, dt = L^{-1} \oint (\Phi_x - \Phi) \, d\Phi = L^{-1} \oint \Phi_x \, d\Phi \qquad (6\text{-}65)$$

since $\oint \Phi \, d\Phi = 0$ for any closed cycle. This is the amount of energy extracted from the tank circuit and dissipated in the junction resistance per cycle. Let us evaluate this energy in the simple approximation of perfect screening, so that Φ is always essentially a quantum value. Then, if $(\Phi_0/2L) < I_0 < (3\Phi_0/2L)$, so that Φ never changes by more than a single quantum when the junction breaks down, we can write (6-65) as

$$\Delta W = \frac{\Phi_0}{L} \sum_n [\Phi_{x\uparrow}(n) - \Phi_{x\downarrow}(n)]$$

where $\Phi_{x\uparrow}(n)$ and $\Phi_{x\downarrow}(n)$ refer, respectively, to the values of Φ_x at which Φ changes from $n\Phi_0$ to $(n+1)\Phi_0$ and the reverse. Since $\Phi_{x\uparrow}(n) = n\Phi_0 + LI_0$, while $\Phi_{x\downarrow}(n) = (n+1)\Phi_0 - LI_0$, each term in the sum contributes $(2LI_0 - \Phi_0)$, and we can write the loss simply as

$$\Delta W = 2I_0\Phi_0 - \frac{\Phi_0^2}{L} \qquad I_0 \geq \frac{\Phi_0}{2L} \qquad (6\text{-}66)$$

for each single-quantum hysteresis loop. This loss will be independent of the operating frequency so long as $\omega\tau = \omega L/R \ll 1$. If $\omega\tau \gtrsim 1$, the flux changes can no longer be treated as occurring instantaneously at appropriate values of Φ_x. Then the simple hysteresis-loss analysis breaks down, and the loss per cycle falls off as $1/\omega\tau$.

Now let us see how the existence of this loss mechanism affects the measured quantity, the voltage across the tank circuit as a function of the RF current supplied I_1 and the dc flux $\bar{\Phi}$ present in the loop. For simplicity, we assume that I_1 is exactly on the resonant frequency of the tank, and that the Q of the tank is high, allowing standard approximations to be made. Let L_T be the inductance in the tank circuit, coupled by a mutual inductance M to the inductance L of the SQUID. Then the current I in the SQUID is related to the current I_T through L_T by the relation $MI_T = LI$, neglecting the impedance of the junction. Consequently, the inductive energy stored in the SQUID is related to the total energy W stored in the tank circuit (including the capacitor) by the relation

$$\tfrac{1}{4}LI^2 = \frac{1}{4}\frac{M^2}{L_T L} L_T I_T^2 = \tfrac{1}{2}k^2 W \ll W \qquad (6\text{-}67)$$

Peak values are used for all ac quantities in this expression and in the following discussion, since they are most physically significant. The quantity k is the coefficient of coupling between the two circuits, normally much less than unity.

The energy stored in the tank circuit can be written in terms of the directly measureable voltage V_T as $W = \frac{1}{2}V_T^2/\omega^2 L_T$. Equating the input power to the loss, the steady-state value of W is determined by

$$\frac{1}{2}V_T I_1 = \frac{\omega W}{Q} + \frac{\Delta W}{\Delta t} \qquad (6\text{-}68)$$

where Δt is the time interval between execution of hysteresis loops. In the absence of hysteresis loss, $V_T = \omega L_T Q I_1$, and $W = W_0 = \frac{1}{2}Q^2 L_T I_1^2$. Hysteresis becomes possible when W exceeds a critical value W_c, at which Φ_x at the SQUID exceeds a critical value. When I_1 is raised above the value giving $W_0 = W_c$, the actual time-average value of W "sticks" at approximately $W_c - \frac{1}{2}\Delta W \approx W_c$, and (6-68) merely determines how quickly W builds up to W_c to give another cycle of hysteresis loss. Thus, as illustrated in Fig. 6-8b, the tank voltage V_T stays at a plateau value $V_{Tc} = \omega(2W_c L_T)^{1/2}$ until I_1 is increased sufficiently that a hysteresis loop is executed in each RF cycle. Putting $\Delta t = 2\pi/\omega$ in (6-68), this gives a plateau width

$$\Delta I_1 = \frac{\omega \Delta W}{\pi V_{T,c}} \qquad (6\text{-}69)$$

At the end of the plateau (or step), V_T again rises with increasing I_1 until a new threshold is reached, at which the SQUID starts to traverse another loop per cycle. [It turns out that if Φ/Φ_0 is exactly integral, the hysteresis loops occur in pairs with the same threshold, so the steps are twice this width. For half-integral Φ/Φ_0, all steps are of double width except the first. But in all other cases, the step width is always given by (6-69).]

Now let us see how the threshold or plateau voltages depend on the average flux Φ, the quantity which is to be measured. For breakdown, LI_0 must be exceeded by the peak value of $\Phi_x = (\Phi - n\Phi) + \Phi_{RF}$. Thus, for the extreme cases $\bar{\Phi} = n\Phi_0$ and $\bar{\Phi} = (n + \frac{1}{2})\Phi_0$, we have $\Phi_{RF} = LI_0$ and $LI_0 - \frac{1}{2}\Phi_0$, respectively. Using the relation $\Phi_{RF} = MI_T = (M/\omega L_T)V_T$, the corresponding tank voltages are $V_T = \omega L_T LI_0/M$ and $(\omega L_T LI_0 M)(1 - \frac{1}{2}\Phi_0/LI_0)$. For the typical case $LI_0 = \Phi_0$, these differ by a factor of 2. In any case, the change in tank voltage between the two plateau levels is

$$\Delta V_T = \frac{1}{2}\frac{\omega \Phi_0 L_T}{M} \qquad (6\text{-}70)$$

On the face of it, it appears that the signal can be made arbitrarily large by making M smaller, by decreasing the coupling between SQUID and tank circuit. But for (6-70) to actually describe an observable flux-periodic modulation of the voltage across the tank circuit, these plateaus must be observable with the same bias current I_1. This is possible only if the plateau width (6-69) is greater than the change δI_1 in threshold current corresponding to the change in V_T given by (6-70), namely, $\delta I_1 = \Delta V_T / \omega L_T Q = \Phi_0 / 2MQ$. From (6-69), we find the width of the bottom step to be $\Delta I_1 = 2M\Phi_0 / \pi L_T L$. Thus the requirement $\Delta I_1 > \delta I_1$ reduces to

$$k^2 Q > \tfrac{1}{4}\pi \approx 1 \qquad (6\text{-}71)$$

If the coupling is reduced below this value, the signal voltage ΔV_T decreases again in proportion to the mutual inductance M. Thus, the optimum coupling condition is $k^2 Q \approx 1$. Although we have derived this result using a very simplified model, it seems to be confirmed empirically and in computer calculations[1] using more realistic models.

If we use the optimum coupling $k^2 Q = 1$, and take $LI_0 \approx \Phi_0$, the signal voltage (6-70) can be rewritten as

$$\Delta V_T = \frac{1}{2} \frac{\omega Q \Phi_0 M}{L} \approx \tfrac{1}{2}\omega Q I_0 M \lesssim \frac{1}{2} \frac{I_0 R Q M}{L} \qquad (6\text{-}72)$$

where in the last form we have used the fact that the operating frequency is limited by the L/R time constant of the SQUID. It is of interest to compare this result with the signal from a dc SQUID, which is limited to $\sim I_0 R \sim \Phi_0 R / 2L$. The ratio is thus

$$\frac{\Delta V_T}{\Delta V_{\text{dc}}} \approx \left(\frac{\omega L_T Q}{R} \right)^{1/2} \left(\frac{\omega L}{R} \right)^{1/2} \qquad (6\text{-}73)$$

The first factor is simply the expected step-up in a resonant transformer with impedance ratio $\omega L_T Q/R$; for typical values it is $\sim 10^2$. The second factor, typically between 0.1 and 1, reflects the loss in performance if the device is operated below the maximum possible frequency. The combination of these two factors typically leads to a voltage gain of a factor of 10 over the dc SQUID. Since even the higher impedance of the tank circuit is only of the order of 10^4 Ω, this voltage step-up is advantageous in practice.

In fact, the voltage plateaus correspond to a dynamic impedance which is much lower than the unloaded tank impedance. If they were truly constant in voltage, as implied above, the incremental impedance would be zero. In fact, typical observed steps have a differential impedance which is lower than the tank impedance by one or two orders of magnitude. (This finite slope occurs because the critical current of the weak link is not absolutely sharp but depends on how rapidly the transition must take place. This reflects the effect of thermally activated fluctuations, a topic to be discussed in the next chapter.) The low differential impedance of the tank circuit makes the device relatively insensitive to noise entering from the drive current source or the amplifier in the signal channel, as well as effectively increasing the available signal power. These features are additional advantages of the RF SQUID.

[1] M. B. Simmons and W. H. Parker, *J. Appl. Phys.* **42**, 38 (1971).

6-6.6 Limit of Sensitivity

In practice, the limit of sensitivity of a SQUID magnetometer is usually set by noise in the room-temperature amplifier or by external noise entering the cryostat. Nonetheless, it is useful to find the intrinsic limit set by Johnson noise in the SQUID.

In the dc SQUID, biased so as to be resistive at all times, it is a good approximation to replace the SQUID loop by a classical circuit of resistance $R' = 2R$ and inductance L. The mean-square Johnson-noise voltage per unit bandwidth $4kTR'$ then gives a mean-square noise current of $4kTR'/(R'^2 + \omega^2 L^2)$. This in turn produces a mean-square noise flux in bandwidth B of

$$(\delta\Phi)^2 = \frac{4kTL^2B}{R'} \qquad (6\text{-}74)$$

where we have specialized to $\omega \ll R'/L$, as is appropriate for measurements of quasi-static flux values. For $L = 10^{-9}$ H, $R' = 2\ \Omega$, and $B = 1$ Hz, this gives

$$\frac{\delta\Phi}{\Phi_0} \approx 0.5 \times 10^{-5} \qquad (6\text{-}75)$$

as the classical limiting sensitivity of a typical device. Actual performance typically falls one or two orders of magnitude short of this value.

The noise analysis of the RF SQUID differs from that of the dc SQUID in that the circuit is effectively resistive only during the short time $(\sim \tau = L/R)$ required for the flux state to change. Only at that time is the SQUID sensitive to the exact value of the external field, and only then are fluctuating "normal" currents important. This reduces the effective averaging time* by a factor of the order of $\omega\tau = \omega L/R$, and causes (6-74) to be replaced approximately by

$$(\delta\Phi)^2 = \frac{4kTLB}{\omega} \qquad (6\text{-}76)$$

Thus, the theoretical sensitivity of an RF SQUID increases with increasing frequency, approaching the value for the dc SQUID as ω approaches R/L, at which point the field is being sensed as frequently as possible with statistical independence. For the typical values $\omega = 2 \times 10^8$ sec^{-1} and $L = 10^{-9}$ H, the expected limiting sensitivity for a 1-Hz bandwidth is then

$$\frac{\delta\Phi}{\Phi_0} \approx 2 \times 10^{-5} \qquad (6\text{-}77)$$

A much more rigorous treatment of the thermal-fluctuation noise in an RF SQUID has been given by Kurkijärvi.[1] In his analysis, he builds in the finite slope

---

[1] J. Kurkijärvi, *Phys. Rev.* **B6**, 832 (1972); *J. Appl. Phys.* **44**, 3729 (1973).

*Except at the high-I_1 end of a step, the sampling rate is less frequent than once per cycle, approaching zero at the beginning of the step. Thus, (6-76) may significantly overestimate the sensitivity.

of the voltage steps in Fig. 6-8*b*, which reflects the statistical nature of the quantum transitions of the ring from one flux state to another. This process is similar to that to be discussed in Sec. 7-1, which gives rise to the finite resistance of a superconducting wire below I_c. In the present context, this more rigorous treatment replaces [in the discussion leading to (6-69)] the sharply defined critical level of stored energy in the tank circuit W_c (or the corresponding critical value of Φ_x, the flux externally applied to the ring) by a statistical distribution of transition points, whose width introduces noise into the measurement process. Despite the much greater complexity of his arguments and the rather different form of his predictions, the numerical conclusion is very similar to (6-77) for typical parameter values. Actual performance within about one order of magnitude of this limit has been reported.

Much of the discrepancy between these theoretical limits and those actually observed in practice can be attributed simply to the noise in the room-temperature amplifier used to observe the SQUID voltages. The quality of an amplifier is often stated[1] in terms of its noise factor

$$F = \frac{V_s^2 + V_a^2}{V_s^2} \qquad (6\text{-}78)$$

where $V_s^2 = 4kT_sR_sB$ is the mean-square Johnson-noise voltage in the source resistance R_s at temperature T_s in bandwidth B, and V_a^2 is the mean-square noise voltage (referred to the source) which is added by the amplifier. If $F \gtrsim 2$, amplifier noise dominates intrinsic source noise, and this will degrade the resolution of the SQUID. Since V_a^2 is independent of T_s, (6-78) can be written

$$F = 1 + \frac{T_a}{T_s}$$

where the noise temperature of the amplifier, defined by $T_a = 300°(F_{300} - 1)$, is the source temperature at which the amplifier noise just doubles the total noise power. The effective noise factor for a source at $4°K$ is then

$$F_4 = 1 + \tfrac{300}{4}(F_{300} - 1) \qquad (6\text{-}79)$$

and the ideal limits for detectable voltages should be increased by $F_4^{1/2}$ above the Johnson-noise limit to take account of the degradation in performance by the room-temperature amplifier.

In the dc SQUID, with $R_s = 1\ \Omega$, F_{300} is typically 4 for the audio frequencies used. Hence $F_4^{1/2} \approx 15$, and instead of (6-75), a resolution of $0.8 \times 10^{-4}\ \Phi_0$ might be expected. Actual performance falls somewhat short of this, perhaps due to excess flux noise introduced into the SQUID by extraneous low-frequency noise currents.

---

[1] An excellent discussion is given by S. Letzter and N. Webster, *IEEE Spec.* **7**, No. 8, 67 (1970).

At the higher frequencies and higher impedance levels ($\sim 10^4\,\Omega$) of the RF-SQUID tank circuit, FET amplifiers with $F_{300} \approx 2$ are available, for which $F_4^{1/2} \approx 9$. With this factor, (6-77) gives a resolution limit of $\sim 2 \times 10^{-4}\,\Phi_0$, very close to actual observed values.

In appraising the significance of these numerical results, it should be noted that in both the dc and RF SQUIDs, the impedance levels are so low that the amplifier noise is close to the value it would have with input shorted, and hence nearly independent of the source impedance. Hence, the performance becomes more ideal as the signal-source impedance, and hence the voltage, is stepped up. For example, at audio frequencies and $R_s = 8 \times 10^5\,\Omega$, a commercial preamplifier is available with room-temperature noise factor $F_{300} = 1.0017$, corresponding to a noise temperature of only 0.5°K. Thus, even for $T_s = 4°K$, the noise factor would be only 1.1, so that the amplifier would contribute little extra noise. But to take full advantage of such an amplifier, it is necessary to raise the impedance of the signal source. This can be done by using an input transformer, but if the transformer is at room temperature, it introduces additional Johnson noise from its winding resistance. This difficulty can be circumvented by cooling the transformer to liquid-helium temperatures. In this way, noise temperatures of about 2°K can be achieved[1] for audio frequencies at the 1-Ω impedance level, allowing a noise factor F_4 of only 1.5 even with this low-impedance source. With such circuitry, it should be possible to measure properties as limited by fluctuations at 4°K while still using a room-temperature amplifier.

6-7 OTHER JOSEPHSON-EFFECT DEVICES

The high sensitivity to magnetic flux of the SQUID devices analyzed in the previous section can be used to permit sensitive measurements of other electrical quantities, such as voltage, resistance, and inductance. Such applications have been discussed by Lukens, Warburton, and Webb,[2] and also by Giffard, Webb, and Wheatley.[3] The latter authors have also discussed[4] applications to noise thermometry at ultralow temperatures, while the use of a SQUID magnetometer to detect nmr signals at low temperatures by observing the change in nuclear magnetization of the sample has been discussed by Day[5] and by Meredith, Pickett, and Symko.[6] We shall not duplicate their discussions here but rather discuss two other types of applications of the Josephson effect in which rather different

[1] D. E. Prober, *Rev. Sci. Inst.* **45**, 849 (1974).
[2] J. E. Lukens, R. J. Warburton, and W. W. Webb, *J. Appl. Phys.* **42**, 27 (1971).
[3] R. P. Giffard, R. A. Webb, and J. C. Wheatley, *J. Low Temp. Phys.* **6**, 533 (1972).
[4] R. A. Webb, R. P. Giffard, and J. C. Wheatley, *J. Low Temp. Phys.* **13**, 383 (1973).
[5] E. P. Day, *Phys. Rev. Letters* **29**, 540 (1970).
[6] D. J. Meredith, G. R. Pickett, and O. G. Symko, *Phys. Letters* **42A**, 13 (1972); also *J. Low Temp. Phys.* **13**, 607 (1973).

geometries are involved. The first is the SLUG (*S*uperconducting *L*ow-inductance *U*ndulating *G*alvanometer) voltmeter, a rather unusual SQUID geometry. The second is the use of Josephson devices as detectors of high-frequency radiation.

6-7.1 The SLUG Voltmeter

This device was first described by Clarke[1] and shortly thereafter its application in a potentiometric voltmeter was described by McWane et al.[2] It consists of a solder blob encompassing a niobium wire (see Fig. 6-9), making some sort of weak-link contact with it in at least two places. As a result, the critical supercurrent between the two superconductors becomes a periodic function of the current I through the niobium wire because of the flux it creates in the penetration depth of the two superconductors between the weak contacts. Since the flux-bearing region is of such small cross section, the inductance L is very small, typically 10^{-11} H, which accounts for the acronym. From the geometry of the junction, one can see that this same area determines the mutual inductance M between the current I and the SQUID loop. Thus, the current to produce a flux quantum is

$$\Delta I = \frac{\Phi_0}{M} \approx 10^{-4} \text{ A} \qquad (6\text{-}80)$$

and for $LI_c \approx \Phi_0$, I_c is typically of this same order.

Because of the low inductance, flux resolution is very good, 10^{-3} to $10^{-4}\ \Phi_0$. Still, this corresponds to current resolution of only $\sim 10^{-7}$ A. Thus, the SLUG is not so sensitive a current-measuring instrument as a conventional galvanometer. Its great virtue is as a voltage-sensing instrument, because a very small voltage from a source with low resistance R_s can set up a large current in the superconducting wire. Because of the near equality of M and L, it turns out that a change ΔI in current through the wire produces an essentially equal change ΔI_c in the critical current of the device. In a suitable circuit, this change in I_c causes an output voltage change $\Delta V \approx R\ \Delta I_c$, where R is the junction resistance. Thus, the device has a voltage (and power) gain of $\sim R/R_s$. For example, if $R \sim 0.1\ \Omega$ while $R_s \sim 10^{-8}\ \Omega$, there is a gain of $\sim 10^7$, which brings a 10^{-15} V signal up to 10^{-8} V, where it can be observed with conventional room-temperature electronics. It should be noted that if the circuit inductance L has a typical value such as 10^{-8} H, the L/R time constant gets awkwardly long for values of R_s below $\sim 10^{-8}\ \Omega$. However, this apparent limitation on useful amplification is made less severe by use of negative feedback to speed up the response.

A particularly imaginative application of the SLUG by Clarke[3] was to verify to high precision that the voltage steps induced in Josephson junctions of different

[1] J. Clarke, *Phil. Mag.* **13**, 155 (1966).
[2] J. W. McWane, J. E. Neighbor, and R. S. Newbower, *Rev. Sci. Inst.* **37**, 1602 (1966).
[3] J. Clarke, *Phys. Rev. Letters* **21**, 1566 (1968).

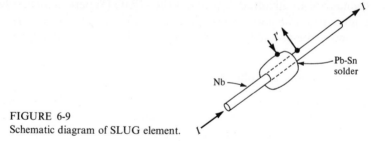

FIGURE 6-9
Schematic diagram of SLUG element.

materials by the same frequency were the same. In this experiment, the circuit is entirely superconducting; so $R_s = 0$. As a result, any difference in potential between the two steps will induce a buildup in circulating current through the SLUG, limited only by the inductance of the circuit. Since no observable current developed in 30 min., given the experimentally known sensitivity of the SLUG (0.3 μA) it was possible to set an upper limit of $\sim 2 \times 10^{-17}$ V on any potential difference. On this basis, it was possible to conclude that the difference in chemical potential induced across both junctions was the same within 1 part in 10^8. This strongly supports the assumption that there are no "solid-state" corrections entering into the value of h/e determined by the experiments of Parker et al., so that e^* is precisely $2e$, where e is the free-electron charge.

6-7.2 Josephson Effect Radiation Detectors

The application of Josephson devices for detection of radiation was pioneered by Grimes et al.[1] They showed that a point-contact Josephson device, biased to a steep part of a step in the I-V characteristic, was a very sensitive detector of radiation in the microwave and submillimeter-wave region. (This fact is inadvertently shown by the fact that SQUID magnetometers do not work well unless completely shielded from stray RF signals.) The device can be used as a broadband detector, as a resonant detector, or as a heterodyne mixer. We shall only touch on these applications here, referring the reader to the literature[2] for details.

    In the broadband mode, the detection occurs through the fact that high-frequency currents reduce the maximum dc supercurrent. This reduction occurs because the high-frequency signal voltage V_s imposes a modulation of amplitude

[1] C. C. Grimes, P. L. Richards, and S. Shapiro, *Phys. Rev. Letters* **17**, 431 (1966).

[2] A useful short review has been given by P. L. Richards, F. Auracher, and T. Van Duzer, *Proc. IEEE* **61**, 36 (1973). A particularly complete and up-to-date survey can be found in the proceedings of the International Conference on the Detection and Emission of Electromagnetic Waves by Josephson Junctions, Perros-Guirec, France, September, 1973, published as tome 9, no. 1, of the *Revue de Physique Appliquée*, January 1974.

$\delta\gamma = 2eV_s/\hbar\omega_s$ on the phase difference γ, which to a first approximation replaces $I_0 \sin\gamma$ by $I_0\langle\cos\delta\gamma\rangle \sin\gamma_0 \approx I_0[1 - (\delta\gamma)^2/2] \sin\gamma_0$. [If the signal is strong enough that the phase modulation $\delta\gamma$ is not small compared to unity, the simple quadratic decrease found here is replaced by a Bessel function $J_0(2eV_s/\hbar\omega)$, as is familiar from FM radio theory.] Thus, a voltage may be developed which is very sensitive to the amount of incident radiation. In this mode, the sensitivity of the device can be comparable with, but not greatly exceed, that of a semiconducting bolometer operating at liquid-helium temperatures.

As a resonant detector, the device is particularly sensitive to frequencies very near those satisfying the Josephson relation $h\nu = 2eV$, where V is the voltage across the junction. In this mode, it appears that the sensitivity may exceed the best far-infrared bolometers by one or two orders of magnitude. The operation of these detectors is complicated by the fact that best performance results only when the Josephson frequency coincides with the resonant frequency of a mode of the electromagnetic field in the cavity surrounding the detector. Moreover, the energy-gap frequency of the superconductor also plays a role in determining sensitivity. A systematic study by Richards and Sterling[1] showed that the bandwidth of the far-infrared response of a point-contact device in a resonant cavity could be much narrower than the cavity mode itself, indicating feedback narrowing of the sort found in a regenerative radio receiver. Because of the critical nature of the adjustments involved, these detectors have not yet been widely used. However, they appear to offer considerable promise as tunable, narrowband, quantum detectors for the far-infrared spectral region, where detector sensitivity is critical because of the weakness of available sources of radiation.

[1] P. L. Richards and S. A. Sterling, *Appl. Phys. Letters* **14**, 394 (1969).

7

FLUCTUATION EFFECTS

In our development and use of the Ginzburg-Landau theory, we have concentrated on finding the properties of that $\psi_0(\mathbf{r})$ which has the minimum free energy. However, thermodynamic fluctuations allow the system to sample other functions $\psi(\mathbf{r})$, and there will be significant statistical weight for any ψ which raises the free energy by only $\sim kT$. We have made use of this concept in our discussion of thermally activated flux creep, for example, where it gave rise to a finite resistance below T_c. In the present chapter, we shall examine more closely the simple case of a thin wire of type I material to see how perfect is the expected absence of resistance as a function of temperature. This is equivalent to the question of how long-lived is the metastable persistent current in a ring against quantum jumps in which the fluxoid quantum number decreases by one or more units. We shall then examine the region *above* T_c, where thermodynamic fluctuations give rise to superconducting effects because $\langle \psi^2 \rangle \neq 0$ although $\langle \psi \rangle = 0$.

7-1 APPEARANCE OF RESISTANCE IN A THIN WIRE

In terms of the Ginzburg-Landau theory, the requirement for a persistent current in a ring is that the line integral $\oint \nabla\varphi \cdot d\mathbf{s}$ around the ring remain an invariant integral multiple of 2π. This corresponds to retaining the same fluxoid quantum number. In a singly connected superconducting wire, fed with current from normal leads, perfect conductivity requires that the potential difference V between the ends be zero. Applying the ideas of Josephson, this implies that the relative phase φ_{12} of the two ends retains a constant value, which will depend on the strength of the supercurrent. More precisely, φ_{12} will fluctuate about a constant mean value, as the supercurrent fluctuates to keep the total current constant by compensating for Johnson noise normal currents. Thus, at any nonzero frequency there will be an ac noise voltage reflecting the real part of the ac impedance of the superconductor, which increases as ω^2. At best a superconductor is really a perfect conductor only for direct current. Consequently, we shall confine our attention to direct current in investigating the appearance of resistance below T_c in superconductors. Even so, there is a subtlety involved, since any real measurement is limited to a finite time span; hence these ac fluctuations will not average out perfectly, and $\bar{V}$ will not be *exactly* zero, even though it may be unobservably small. An appropriate operational definition of zero dc resistance is that V have no measurable average value proportional to a dc applied current. In terms of the phase difference φ_{12}, this means that there should be no measurable secular progression of a short-term average of φ_{12}.

On the other hand, if there *is* resistance, this averaged φ_{12} will increase with time; this would appear to be inconsistent with a steady state. The resolution of this apparent inconsistency is that phase-slip events occur, in which the phase coherence is momentarily broken at some point in the superconductor, allowing a phase slip to occur before phase coherence is reestablished. These events can be spatially localized so long as the phase slip is through an integral multiple of 2π, since a uniform phase change by $2n\pi$ outside the phase-slip region has no physical significance. In fact, we may concentrate on simple phase slips by 2π, since phase slips by multiples of 2π turn out to be most easily accomplished as multiple slips, each of 2π. To maintain a steady state, such events must occur with an average frequency $2e\bar{V}/h$. If V is constant, φ_{12} increases steadily at the rate $2eV/h$, but instantaneously snaps back by 2π when each phase slip occurs. Thus, φ_{12} executes an irregular sawtooth which is equivalent, modulo 2π, to a uniform ramp.

To go beyond this qualitative picture to a quantitative calculation of the frequency of these resistive phase slips, it is convenient to restrict attention to a one-dimensional superconductor. By this, we mean a wire with transverse dimension $d \ll \xi$, so that variations of ψ over the cross section of the wire are energetically prohibited. Then ψ is a function of a single coordinate x, running along the wire. We also assume $d \ll \lambda$, in which case magnetic energies can be neglected compared to kinetic energies.

If we neglect fluctuation effects, $|\psi(x)|$ is constant; this is the problem treated in Sec. 4-4, where we found a nonlinear relation (4-35) between supercurrent density and velocity, and a critical current density (4-36). Above the critical current density, dissipative processes set in strongly, and the resistance rapidly approaches the normal value. What we seek here is the description of the resistive processes when $J < J_c$, so that in the absence of fluctuations there would be perfect conductivity. This regime was first treated in detail by Langer and Ambegaokar.[1]

To visualize the evolution of the complex function $\psi(x)$ during the phase-slip process, it is convenient to depict $|\psi(x)| e^{i\varphi(x)}$ in polar form in a plane perpendicular to the x axis. Then the usual solutions discussed in Sec. 4-4, which have the form $\psi_0 e^{iqx}$, are represented by helices of pitch $2\pi/q$ and radius ψ_0. (See Fig. 7-1a.) These solutions are stationary, equilibrium solutions, representing the flow of supercurrent at zero voltage.

What happens to this picture when a voltage exists between the ends of the wire? The relative phase at the ends of the wire increases at the Josephson rate

$$\frac{d\varphi_{12}}{dt} = \frac{2eV}{\hbar} \qquad (7\text{-}1)$$

We can visualize this as occurring by the phase at one end being steadily "cranked" around the Argand diagram, while the other end is held fixed, thus tightening the helix. So far, this simply describes the accelerative supercurrent of the London equation $\mathbf{E} = \partial(\Lambda \mathbf{J}_s)/\partial t$. That is, the presence of the voltage increases q at such a rate that the total phase difference $\varphi_{12} = qL$ along the wire obeys (7-1). More locally, this is equivalent to

$$\frac{dv_s}{dt} = \frac{eE}{m} \qquad (7\text{-}2)$$

But we know that there is a critical velocity v_c beyond which the simple uniform solution is impossible. Thus, this picture must break down when v_s reaches v_c, if not before. The phase-slip process of Langer and Ambegaokar maintains a steady state with $v_s < v_c$, in the presence of a nonzero voltage V, by annihilating turns of the helix in the interior of the wire at the same rate as new ones are being cranked in at the end. By this means, the energy being supplied at a rate IV is dissipated as heat rather than being converted into kinetic energy of supercurrent, which would otherwise soon exceed the condensation energy.

Although we shall not work through the details of their calculation, we shall review some of the salient points. First, so long as we neglect the normal current

[1] J. S. Langer and V. Ambegaokar, *Phys. Rev.* **164**, 498 (1967).

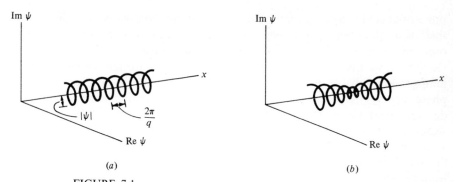

FIGURE 7-1
Graphical representation of complex current-carrying GL wavefunction in one-dimensional superconductors. (a) Uniform solution. (b) Nonuniform solution just before phase-slip event.

(which is nonzero when $E \neq 0$), conservation of current requires that $J_s(x) \propto |\psi(x)|^2 v_s(x)$ be constant. In other words, if $\psi(x) = |\psi(x)| e^{i\varphi(x)}$, then

$$|\psi(x)|^2 \frac{d\varphi}{dx} = \text{const} \propto I \qquad (7\text{-}3)$$

serves as a constraint on possible variations of the complex function $\psi(x)$. In particular, if $|\psi|$ becomes very small in some region, $d\varphi/dx$ must become large there. (See Fig. 7-1b.) As first emphasized by Little,[1] as one approaches the limit $|\psi| \to 0$, it is relatively easy to add or subtract a turn.

What Langer and Ambegaokar did was to find that path through function space, between two uniform solutions with different numbers of turns, which had the lowest intermediate free-energy barrier to overcome. By definition, at this saddle point in the barrier, the GL free energy is again stationary with respect to small changes in ψ, so that ψ should satisfy the usual GL equations which were derived variationally by setting $\delta F = 0$. Using the constraint (7-3), they were able to calculate the saddle-point free-energy increment, namely

$$\Delta F_0 = \frac{8\sqrt{2}}{3} \frac{H_c^2}{8\pi} A\xi \qquad (7\text{-}4)$$

where A is the cross-sectional area of the conductor. This result is very plausible, since it is the condensation energy in a length $\sim \xi$ of the conductor; this is what

[1] W. A. Little, *Phys. Rev.* **156**, 398 (1967).

one would get by arguing that ψ cannot vary more rapidly than in a distance ξ, so that, as a minimum, such a length of wire would have to go almost normal in order to decouple the phase at the two ends of the wire.

Next, we must build in the effect of the current through the wire in making jumps more probable in one direction than the other. In the absence of a current, phase slips by $\pm 2\pi$ are equally likely; this gives a fluctuating noise voltage, but no dc component. Given a driving voltage V, however, the current will build up to a steady-state value at which the $\Delta\varphi_{12} = -2\pi$ jumps outnumber the $\Delta\varphi_{12} = +2\pi$ jumps by an amount $2eV/h$ per sec. The different jump rates arise from a difference δF in the energy barrier for jumps in the two directions, and this difference stems from the electrical work $\int IV \, dt$ done in the process. In view of (7-1), for a phase slip of 2π, the energy difference is

$$\delta F = \Delta F_+ - \Delta F_- = \frac{h}{2e} I \qquad (7\text{-}5)$$

As shown by McCumber,[1] all these arguments carry over exactly to the usual experimental situation, where a constant current rather than a constant voltage source is used.

To complete the theory, it is necessary to introduce an attempt frequency or "prefactor" Ω, so that the mean net phase-slip rate is

$$\frac{d\varphi_{12}}{dt} = \Omega \left[\exp\left(-\frac{\Delta F_0 - \delta F/2}{kT} \right) - \exp\left(-\frac{\Delta F_0 + \delta F/2}{kT} \right) \right]$$

$$= 2\Omega e^{-\Delta F_0/kT} \sinh \frac{\delta F}{2kT} \qquad (7\text{-}6)$$

Substituting (7-5) for δF and equating to the Josephson frequency, this leads to

$$V = \frac{\hbar\Omega}{e} e^{-\Delta F_0/kT} \sinh \frac{hI}{4ekT} \qquad (7\text{-}7)$$

In the limit of very small currents, the hyperbolic sine can be replaced by its argument, and one obtains Ohm's law, with

$$R = \frac{V}{I} = \frac{\pi h^2 \Omega}{2e^2 kT} e^{-\Delta F_0/kT} \qquad (7\text{-}8)$$

However, this approximation is valid only for $I \lesssim I_0$, where

$$I_0 = \frac{4ekT}{h} = 0.013 \ \mu\text{A}/^\circ\text{K} \qquad (7\text{-}9)$$

In this regime, the numbers of jumps with $\Delta\varphi_{12} = \pm 2\pi$ are approximately equal, the current being a small perturbation. At higher currents, a preponderance of

---

[1] D. E. McCumber, *Phys. Rev.* **172**, 427 (1968).

jumps occur in the direction which removes turns from the helix. It is then useful to approximate (7-7) by

$$V = \frac{\hbar\Omega}{2e} e^{-\Delta F_0/kT} e^{I/I_0} \qquad (7\text{-}10)$$

where, in the full theory, ΔF_0 is found to decrease from (7-4) as I^2. In this regime, the superconductor acts like a nonlinear resistor.

These results leave open the value of the prefactor Ω. Evidently it should be proportional to the length of the wire, since one would expect the jump to be able to occur independently at sites all along the wire. This causes the voltage drop to be proportional to the length of the wire for given current, so that the resistance is an extensive variable. In the original work of Langer and Ambegaokar, the attempt frequency was taken rather arbitrarily as nAL/τ, where τ is the electronic relaxation time in the normal state, and n is the electron density. Subsequently, McCumber and Halperin[1] reexamined the problem using the time-dependent GL theory and found a temperature-dependent prefactor of the form

$$\Omega = \frac{L}{\xi} \left(\frac{\Delta F_0}{kT} \right)^{1/2} \frac{1}{\tau_s} \qquad (7\text{-}11)$$

where $1/\tau_s = 8k(T_c - T)/\pi\hbar$ is the characteristic relaxation rate of the superconductor in the time-dependent GL theory, discussed further in Secs. 7-5 and 8-3. This form is plausible, since L/ξ is the number of nonoverlapping locations in which the fluctuations might occur. The factor $(\Delta F_0/kT)^{1/2}$ corrects for the overlap of fluctuations at different places and has little numerical importance. This McCumber-Halperin prefactor is typically smaller than the Langer-Ambegaokar one by a factor of 10^{10}, and it goes to zero as one approaches T_c. Despite the enormous size of this correction factor, its absence was not noticed at first, since it corresponds to a change of only a few millidegrees in the temperature scale because of the exponential dependence of the voltage (7-7) on $\Delta F_0/kT$.

The most reliable and direct tests of these ideas have been the experiments of Lukens, Warburton, and Webb[2] and those of Newbower, Beasley, and Tinkham,[3] both of which were done on tin "whiskers." These are single-crystal, cylindrical specimens, typically $\sim 0.5\ \mu m$ in diameter and a fraction of a millimeter long, grown by applying pressure to a sandwich of tin-plated steel such as is used in tin cans. Even for samples of such small diameter, $\Delta F_0/kT \approx 6 \times 10^6 (1 - t)^{3/2}$, so that the probability of a phase slip becomes astronomically small unless one is within about 1 mK of T_c, where $(1 - t) \sim 0.0003$. The very satisfactory fit between this LAMH (Langer-Ambegaokar-McCumber-Halperin) theory and the experimental data of Newbower et al. is displayed in Fig. 7-2. The LAMH theory is expected to fail, as it does, very near T_c, where its model of isolated phase slips in

[1] D. E. McCumber and B. I. Halperin, *Phys. Rev.* **B1**, 1054 (1970).
[2] J. E. Lukens, R. J. Warburton, and W. W. Webb, *Phys. Rev. Letters* **25**, 1180 (1970).
[3] R. S. Newbower, M. R. Beasley, and M. Tinkham, *Phys. Rev.* **B5**, 864 (1972).

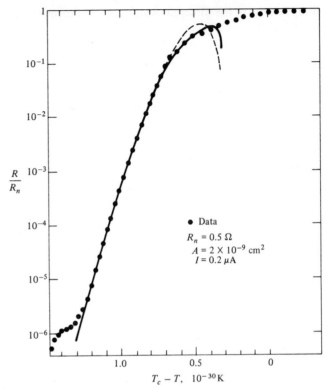

FIGURE 7-2
Decrease of resistance below T_c in a superconducting tin whisker, as measured by Newbower et al. Solid curve is LAMH theory, with only T_c as adjustable parameter. Dashed curve is LAMH theory if parallel normal conduction channel is omitted. "Foot" at $R/R_n \approx 10^{-6}$ is believed to be caused by contact effects.

a superconducting medium is inappropriate, since both the attempt frequency and the free-energy barrier go to zero at T_c. In that case, it may be more appropriate to start with the normal state and consider superconducting fluctuations from it, rather than the reverse.[1] But once the resistance has fallen significantly, the LAMH fit appears to be quantitatively accurate over 6 orders of magnitude. (The "foot" near $R/R_n = 10^{-6}$, where the resistance falls more slowly before resuming its rapid fall, is sample-dependent and believed to be caused by contact effects.)

Given this excellent fit between theory and experiment, it is interesting to use the theory to extrapolate beyond the observable range of resistance. At the lowest

[1] For attempts to work down through T_c from above, see, for example, W. E. Masker, S. Marcelja, and R. D. Parks, *Phys. Rev.* **188**, 745 (1970); J. Tucker and B. I. Halperin, *Phys. Rev.* **B3**, 378 (1971); R. J. Londergan and J. S. Langer, *Phys. Rev.* **B5**, 4376 (1972).

temperature shown in Fig. 7-2, with the measuring current of 0.2×10^{-6} A, the resistive voltage is about 10^{-13} V; this corresponds to about 100 phase slippages by 2π/sec. Extrapolating down another millidegree, the rate is 10^{-11}/sec, or 1 in 1,000 years; in another millidegree, it is 1 in 10^9 years. Thus, in three millidegrees, we have gone from the normal resistance to a regime in which no resistive event would be expected to occur in the age of the universe! Of course, the disappearance of resistance should be even faster in a thicker wire.

In view of these time scales, it is clear that time-average results must be used with care. According to (7-8), there is a finite resistance at all nonzero temperatures, although it becomes astronomically small well below T_c. But this refers to a statistical average over a period long enough for many phase slips to occur. Given an *infinitely* long wire, this would be no problem, and a small resistance should be measurable. With any finite-length wire, however, one rapidly reaches the situation in which *no* phase slip would be expected to occur in any feasible experimental time scale. In that case, the dc resistance would appear to be *strictly* zero, not just small. Thus, the quantized nature of the phase slips provides the key needed to get from a very small to a zero resistance.

7-2 SUPERCONDUCTIVITY ABOVE T_c IN ZERO-DIMENSIONAL SYSTEMS

In the GL theory, T_c is defined as the temperature at which the coefficient $\alpha(T)$ (in the leading term $\alpha |\psi|^2$ in the free-energy expansion) changes sign. Thus, above T_c, F is a minimum when $|\psi| = 0$. However, thermal fluctuations raising the free energy by an amount $\sim kT$ are common, since the probability falls only as $e^{-F/kT}$. This leads to the existence of fluctuation-induced superconducting effects above T_c. These fluctuations are largest in amplitude if confined to small volumes, since the total energy increase must be only $\sim kT$.

We can get a useful orientation on this problem by considering first a particle which is small compared to ξ, so that we can treat ψ as constant over its volume V. This might be called the zero-dimensional limit. Then the GL free energy relative to the normal state (in the absence of any fields) is

$$F = V(\alpha |\psi|^2 + \tfrac{1}{2}\beta |\psi|^4) \qquad (7\text{-}12)$$

where $\alpha \equiv \alpha_0(t - 1)$.

Below T_c, this leads to the usual result that the minimum free energy is

$$F_0 = -\frac{\alpha^2}{2\beta} V = -\frac{\alpha_0^2}{2\beta}(1 - t)^2 V = -\frac{H_c^2}{8\pi} V \qquad (7\text{-}13)$$

and this occurs for

$$|\psi_0|^2 = -\frac{\alpha}{\beta} = \frac{\alpha_0(1 - t)}{\beta} \qquad (7\text{-}14)$$

The fluctuations about this ψ_0 can be estimated by computing

$$\frac{\partial^2 F}{\partial \psi^2}\bigg|_{\psi_0} = -4\alpha V = 4\alpha_0(1 - t)V \qquad (7\text{-}15)$$

and setting

$$\langle F - F_0 \rangle = \frac{1}{2}\frac{\partial^2 F}{\partial \psi^2}\bigg|_{\psi_0} (\delta \psi)^2 \approx kT \qquad (7\text{-}16)$$

This leads to

$$\frac{(\delta \psi)^2}{\psi_0^2} \approx \frac{kT}{4F_0} = \frac{2\pi kT}{H_c^2 V} \approx \frac{10^{-20}}{(1 - t)^2 V} \qquad (7\text{-}17)$$

using numerical values for tin. From this we see that the fluctuations cause a very small fractional change in ψ except *very* near T_c or in a very small sample. Therefore we have generally been well justified in using the "mean field" ψ_0 in our previous work. However, by use of very small particles ($d < 1000$ Å) it has been possible to probe the so-called "critical region," where $(\delta\psi/\psi_0)^2$ is *not* necessarily small, and the mean-field results become inaccurate. In this connection it is important to note that the apparent divergence of (7-17) at T_c is actually cut off by the anharmonic terms in the free-energy expansion, so that even at T_c, $(\delta\psi)^2$ has the finite value

$$(\delta \psi)_{T_c}^2 \approx \left(\frac{2kT_c}{V\beta}\right)^{1/2} \qquad (7\text{-}18)$$

Now let us examine the situation *above* T_c. Here $\alpha > 0$, so that by inspection of (7-12) we see that the minimum free energy is $F_0 = 0$ (relative to the normal state) which occurs for $\psi_0 = 0$. The fluctuations are limited by

$$\frac{\partial^2 F}{\partial \psi^2}\bigg|_{\psi = 0} = 2\alpha V = 2\alpha_0(t - 1)V \qquad (7\text{-}19)$$

which we see differs only by a factor of two from the value given by (7-15) for a temperature an equal distance *below* T_c. The corresponding fluctuation level here is

$$(\delta \psi)^2 \approx \frac{kT}{\alpha V} = \frac{kT}{\alpha_0(t - 1)V} \qquad (7\text{-}20)$$

Again, this is of the same order as the fluctuations below T_c, but since ψ_0 is now zero, all the superconducting effects arise from the fluctuations. Note that $(\delta\psi)^2$ tends to diverge as $(T - T_c)^{-1}$, as in the familiar Curie-Weiss law in the statistical

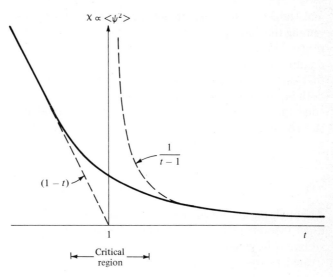

FIGURE 7-3
Temperature dependence of pair density and susceptibility of zero-dimensional
superconductor near T_c.

mechanics of paramagnetism, but this divergence is cut off very near T_c by the
quartic term which leads to (7-18). In fact, by equating (7-12) to kT and solving
exactly, one obtains

$$(\delta\psi)^2 = \frac{\alpha}{\beta}\left[\left(1 + \frac{2\beta kT}{\alpha^2 V}\right)^{1/2} - 1\right] \qquad T > T_c \qquad (7\text{-}21)$$

which reduces to (7-20) and (7-18) in the appropriate limits.

This brings us to the question of how these superconducting fluctuations
above T_c can be observed. The most direct way is a susceptibility measurement on
tiny particles, since χ depends on λ, which is a measure of $\langle\psi^2\rangle$. More explicitly,
for spherical particles of radius $R \ll \lambda$, the London equations lead to a
susceptibility

$$\chi = -\frac{1}{40\pi}\frac{R^2}{\lambda^2} = -\frac{1}{40\pi}\frac{4\pi e^{*2}}{m^*c^2}\langle\psi^2\rangle R^2 \qquad (7\text{-}22)$$

(If $R < \xi_0$, an additional factor of $\sim R/\xi_0$ enters to account explicitly for the
nonlocal electrodynamics.) Thus, if $\langle\psi^2\rangle$ is given by (7-21), χ should rise as
$(t - 1)^{-1}$ as the temperature is reduced, but then rise more slowly once the critical
region is entered; finally, well below T_c, χ and $\langle\psi^2\rangle$ should rise as $(1 - t)$, after
fluctuation effects are swamped by the mean-field superconductivity. These de-
pendences are shown in Fig. 7-3. Exactly this behavior was observed by Buhrman

and Halperin[1] in measurements on fine aluminum powders. In their sample containing the finest particles ($R \approx 250$ Å), the critical region was found to cover the range $0.95 < t < 1.05$, in agreement with theory. Altogether, they obtained such a quantitative fit between their data and exact calculations[2] in the GL framework that one can conclude that the GL free-energy expression (7-12) is satisfactory both inside and outside the critical region, so far as zero-dimensional systems are concerned. However, χ fell below the predicted value for $T \gtrsim 1.5T_c$. This is not surprising, since the GL theory is expected to be reliable only near T_c.

7-3   SPATIAL VARIATION OF FLUCTUATIONS

Although the zero-dimensional case just discussed is simple and permits a rather exact theoretical analysis, it cannot be applied directly to the usual experimental situation in which one or more sample dimensions exceed ξ, since ψ cannot be treated as constant over the sample. Nonetheless, the qualitative ideas can be carried over to some extent by treating a macroscopic sample as if it were composed of tiny, independent particles, whose size is limited by the correlation length of the fluctuations, typically $\sim \xi$. For example, since $|\alpha| = \hbar^2/2m^*\xi^2$, (7-20) and (7-22) lead in this way to a diamagnetic susceptibility of a bulk superconductor above T_c which is proportional to $kT\xi(T)/\Phi_0^2$; this simple result is confirmed by exact calculations, discussed below. Let us now examine the spatial variations of the fluctuations more closely.

We first consider the case of a bulk sample, far enough above T_c that the effects of the quartic term in the free energy can be neglected. The GL free-energy density relative to the normal state can then be written as

$$f = \alpha |\psi|^2 + \frac{\hbar^2}{2m^*}\left|\left(\frac{\nabla}{i} - \frac{2\pi\mathbf{A}}{\Phi_0}\right)\psi\right|^2 \quad (7\text{-}23)$$

Since $\alpha > 0$, both terms are positive, so that the free energy must exceed that of the normal state for any nonzero ψ. The corresponding linearized GL equation is

$$\left(\frac{\nabla}{i} - \frac{2\pi\mathbf{A}}{\Phi_0}\right)^2 \psi = -\frac{2m^*\alpha}{\hbar^2}\psi = -\frac{1}{\xi^2}\psi \quad (7\text{-}24)$$

Since $1/\xi^2 \equiv 2m^*|\alpha|/\hbar^2$, its sign is changed relative to (4-56), reflecting the sign change of α.

[1] R. A. Buhrman and W. P. Halperin, *Phys. Rev. Letters* **30**, 692 (1973).

[2] See, for example, B. Mühlschlegel, D. J. Scalapino, and R. Denton, *Phys. Rev.* **B6**, 1767 (1972). In these calculations a proper thermally weighted average is taken over ψ^2, rather than simply equating the free-energy increase to kT as we have done here.

Let us first consider the case of $\mathbf{A} = 0$, and expand $\psi(\mathbf{r})$ in Fourier series so that

$$\psi(\mathbf{r}) = \sum_{\mathbf{k}} \psi_{\mathbf{k}} e^{i\mathbf{k} \cdot \mathbf{r}} \qquad (7\text{-}25)$$

Inserting this in (7-23), integrating over unit volume, and using the orthogonality of the terms in a Fourier series, we find

$$f = \sum_{\mathbf{k}} \left(\alpha + \frac{\hbar^2 k^2}{2m^*} \right) |\psi_{\mathbf{k}}|^2 \qquad (7\text{-}26)$$

If we assign an energy kT to each orthogonal mode, i.e., to each $\mathbf{k}$ value, then (in unit volume)

$$|\psi_{\mathbf{k}}|^2 = \frac{kT}{\alpha + \hbar^2 k^2 / 2m^*} = \frac{2m^*}{\hbar^2} \frac{kT}{k^2 + 1/\xi^2} \qquad (7\text{-}27)$$

From this we see that Fourier components describing variations in distances less than ξ come in with reduced weight, as expected. However, the density of modes goes as $k^2\, dk$, so that a stronger cutoff is required to give a finite value for the sum over $\mathbf{k}$ in $\langle \psi^2 \rangle = \sum_{\mathbf{k}} |\psi_{\mathbf{k}}|^2$. Presumably such a cutoff must occur when $k \approx 1/\xi(0)$, since the GL theory is not valid for more rapid variations than that.

The spatial implication of these k-dependent amplitudes is obtained if we consider the correlation function

$$g(\mathbf{r}, \mathbf{r}') \equiv \langle \psi^*(\mathbf{r})\psi(\mathbf{r}') \rangle$$

$$= \left\langle \sum_{\mathbf{k}} \psi_{\mathbf{k}}^* e^{-i\mathbf{k} \cdot \mathbf{r}} \sum_{\mathbf{k}'} \psi_{\mathbf{k}'} e^{i\mathbf{k}' \cdot \mathbf{r}'} \right\rangle \qquad (7\text{-}28)$$

where the angular brackets indicate an average. Changing variables to mean and relative coordinates $\bar{\mathbf{r}} = (\mathbf{r} + \mathbf{r}')/2$ and $\mathbf{R} = \mathbf{r}' - \mathbf{r}$, we can write (7-28) as

$$g(\mathbf{r}, \mathbf{r}') = \left\langle \sum_{\mathbf{k}\mathbf{k}'} \psi_{\mathbf{k}}^* \psi_{\mathbf{k}'} \, \exp \left[\frac{i(\mathbf{k} + \mathbf{k}')}{2} \cdot \mathbf{R} \right] \exp \left[-i(\mathbf{k} - \mathbf{k}') \cdot \bar{\mathbf{r}} \right] \right\rangle$$

Carrying out the average over the mean coordinate $\bar{\mathbf{r}}$ gives zero unless $\mathbf{k} = \mathbf{k}'$, when it gives unity. Thus,

$$g(\mathbf{r}, \mathbf{r}') = g(\mathbf{R}) = \sum_{\mathbf{k}} |\psi_{\mathbf{k}}|^2 e^{i\mathbf{k} \cdot \mathbf{R}} \qquad (7\text{-}29)$$

where $|\psi_{\mathbf{k}}|^2$ is given by (7-27). By the symmetry of this formula (and the underlying physics) it is clear that this depends only on the magnitude of $\mathbf{R}$. Replacing the sum by an integral, we have

$$g(R) = \frac{2m^* kT}{\hbar^2} \iint \frac{e^{ikR \cos \theta}}{k^2 + 1/\xi^2} \sin \theta \, d\theta \, k^2 \, dk$$

The integral on θ is elementary, and that on k can be easily done by contour integration, leading to the result

$$g(R) = \frac{m^*kT}{2\pi\hbar^2} \frac{e^{-R/\xi(T)}}{R} \qquad (7\text{-}30)$$

Thus, in the fluctuation regime, the local values of ψ are correlated over a distance $\sim \xi(T)$, as anticipated above. [The divergence of (7-30) as $R \to 0$ is nonphysical, arising from carrying the integration on k to infinity rather than imposing a cutoff at $\sim 1/\xi(0)$.]

Now let us see what effect the presence of a magnetic field has on these results. Before choosing a specific gauge, we note in general that it will be convenient to work with orthonormal eigenfunctions ψ_ν of the pseudohamiltonian operator $\mathcal{H}$ defined by

$$\mathcal{H}\psi_\nu = \frac{\hbar^2}{2m^*}\left[\left(\frac{\mathbf{\nabla}}{i} - \frac{2\pi\mathbf{A}}{\Phi_0}\right)^2 + \frac{1}{\xi^2}\right]\psi_\nu = \epsilon_\nu\psi_\nu \qquad (7\text{-}31)$$

Comparing this with (7-24), we see that eigenfunctions with $\epsilon_\nu = 0$ satisfy the linearized GL equation. However, above T_c, all the ϵ_ν are positive, and ψ_ν are simply used as basis functions. Returning to the argument used in Sec. 4-8 in the derivation of H_{c2} [where the same operator appears as in (7-31), apart from the sign of $1/\xi^2$], we see that

$$\epsilon_\nu = \epsilon_{n, k_z} = \frac{\hbar^2}{2m^*}\left(\frac{1}{\xi^2} + k_z^2\right) + (n + \tfrac{1}{2})\hbar\omega_c \qquad (7\text{-}32)$$

where $\omega_c = 2eH/m^*c$ is the cyclotron frequency of pairs in the applied field.

If we expand a general $\psi(\mathbf{r})$ in this set as

$$\psi(\mathbf{r}) = \sum_\nu c_\nu\psi_\nu(\mathbf{r}) \qquad (7\text{-}33)$$

and calculate the free energy using (7-23), we find by a partial integration using orthogonality that $F = \sum_\nu |c_\nu|^2\epsilon_\nu$. Assigning an energy kT to each normal mode (as we did above in the absence of a field), we have $|c_\nu|^2 = kT/\epsilon_\nu$.

We can now compute the correlation function

$$g(\mathbf{r}, \mathbf{r}') \equiv \langle\psi^*(\mathbf{r})\psi(\mathbf{r}')\rangle = \sum_{\nu\nu'} c_\nu^* c_{\nu'}\langle\psi_\nu^*(\mathbf{r})\psi_{\nu'}(\mathbf{r}')\rangle \qquad (7\text{-}34)$$

to see how (7-30) is modified by the field. At this point we must make a specific gauge choice. Since the physical problem has axial symmetry about the field, it is convenient to choose

$$\mathbf{A} = \tfrac{1}{2}\mathbf{H} \times \mathbf{r} = \tfrac{1}{2}Hr\hat{\mathbf{\varphi}}$$

where $\hat{\mathbf{\varphi}}$ is a unit vector. Since there is also translational invariance along the field in the z direction, the eigenfunctions must have the form

$$\psi_\nu = f_{mn}(\rho)e^{im\varphi}e^{ik_z z} \qquad (7\text{-}35)$$

where $\rho = (x^2 + y^2)^{1/2}$. Putting this in the differential equation (7-31), we find that the asymptotic form of f_{mn} for all m, n is

$$f(\rho) \to f_1(\rho)e^{-a\rho^2} \qquad \rho \to \infty \qquad (7\text{-}36)$$

where $f_1(\rho)$ is a polynomial and $a = \pi H/2\Phi_0$. This exponential cutoff shows that the wavefunctions are confined to cylindrical regions whose area is such that the flux threading them is of order Φ_0. A particularly simple solution is the lowest one, for which $m = 0$ and $f_1(\rho)$ is constant. The eigenvalue is

$$\epsilon_0 = \frac{\hbar^2}{2m^*}\left(\frac{2\pi H}{\Phi_0} + k_z^2 + \frac{1}{\xi^2}\right) \qquad (7\text{-}37)$$

in agreement with the $n = 0$ case of (7-32).

Now let us return to evaluation of the correlation function (7-34). Transforming to relative and center-of-mass coordinates as above, we find that the average over the center-of-mass coordinate vanishes unless $m = m'$ and $k_z = k_z'$ in the general indices v and v'. A further simplification results because $f_1(0) = 0$ unless $m = 0$, so that

$$g(\rho, Z) = \sum_{n,\,n',\,k_z} c_{nk}^* c_{n'k} f_{0n}(0) f_{0n'}(\rho) e^{ik_z Z} \qquad (7\text{-}38)$$

The one-dimensional integration over k_z can be carried out by noting that $|c_{nk_z}|^2 = kT/\epsilon_{nk_z}$, with ϵ_{nk_z} given by (7-32). Thus, for given n, $|c_{nk_z}|^2 \propto (k_{0n}^2 + k_z^2)^{-1}$, where

$$k_{0n}^2 = \xi^{-2} + \frac{(2n + 1)2\pi H}{\Phi_0} \qquad (7\text{-}39)$$

With this dependence on k_z, the various terms in $g(\rho, Z)$ will fall off roughly as $e^{-k_{0n}|Z|}$. The asymptotic behavior is governed by the smallest k_{0n}, namely $k_{00} = (\xi^{-2} + 2\pi H/\Phi_0)^{1/2}$. Combining this dependence with (7-36), we expect that the correlation function will fall off at large distances roughly as

$$g(\rho, Z) \propto e^{-k_{00}|Z|} e^{-\pi H \rho^2/2\Phi_0} \qquad (7\text{-}40)$$

Thus, the "radius" of the correlated fluctuations shrink below ξ as H gets larger than $\sim \Phi_0/\pi\xi^2$. Note that this characteristic field value is essentially $H_{c2}(\tilde{T})$, where $T_c - \tilde{T} = T - T_c$.

In view of (7-22), we might expect that the shrinkage of the size of the fluctuations with increasing H would cause the susceptibility to be less in a finite field than in the limit of zero field. As discussed in more detail below, this is in fact the case.

7-4 FLUCTUATION DIAMAGNETISM ABOVE T_c

Before discussing the more detailed theory, let us start by reviewing the physical essence of the phenomenon along the lines suggested by A. Schmid.[1] We model the superconductor crudely as a collection of independent fluctuating droplets of superconductivity, with χ given by (7-22). We then estimate $|\psi|^2$ for a typical fluctuation of volume V as in (7-20). This leads to

$$\chi \approx -\frac{\pi^2 kT \xi^2 \langle r^2 \rangle}{\Phi_0^2 V} \qquad (7\text{-}41)$$

where $\langle r^2 \rangle$ is a mean-square radius, and the numerical coefficient is approximate. For a three-dimensional bulk sample in weak fields, we have seen that the correlation function for fluctuations dies out in a length $\sim \xi$. Thus, it is reasonable to take $V = 4\pi\xi^3/3$ and $\langle r^2 \rangle = (\xi/2)^2$. With a minor adjustment of the numerical coefficient, this leads to Schmid's exact result based on the GL theory:

$$\chi = -\frac{\pi}{6}\frac{kT}{\Phi_0^2}\xi(T) \approx -10^{-7}\left(\frac{T_c}{T - T_c}\right)^{1/2} \qquad (7\text{-}42)$$

Note that this susceptibility is of the same order of magnitude as the Landau diamagnetism of normal metals, apart from the temperature-dependent enhancement factor.

Although the susceptibility is formally divergent at T_c, in practice the enhancement factor never gets very large before being limited either y the first-order transition in a magnetic field or by the width of the transition in a real sample. Thus, the susceptibility is always extremely small compared to the Meissner regime, where $\chi = -1/4\pi$. Moreover, only small fields can be used without destroying the effect by shrinking and weakening the fluctuations. Nonetheless, the susceptibility is substantial compared with the background, and it can be isolated by measuring the temperature-dependent part of the magnetization in a magnetic field held absolutely constant by a superconducting coil in the persistent-current mode. Such experiments were first carried out by Gollub et al.,[2] using a SQUID magnetometer.

Since the magnetization must be measured in a finite field, the temperature at which the fluctuation diamagnetism should diverge is decreased from T_c to the nucleation temperature $T_{c2}(H)$, the temperature at which $H = H_{c2}(T)$. In a type I superconductor, $T_{c2}(H)$ is the supercooling temperature for an ideal sample. In typical samples, little supercooling is observed, and superconductivity sets in suddenly with a first-order transition near the shifted thermodynamic critical field $T_c(H) > T_{c2}(H)$. Thus, the divergence point at T_{c2} is experimentally inaccessible.

[1] A. Schmid, *Phys. Rev.* **180**, 527 (1969).

[2] J. P. Gollub, M. R. Beasley, R. S. Newbower, and M. Tinkham, *Phys. Rev. Letters* **22**, 1288 (1969); J. P. Gollub, M. R. Beasley, and M. Tinkham, *Phys. Rev. Letters* **25**, 1646 (1970); J. P. Gollub, M. R. Beasley, R. Callarotti, and M. Tinkham, *Phys. Rev.* **B7**, 3039 (1973).

On the other hand, in type II superconductors, the second-order phase transition at $T_{c2}(H)$ can be approached without discontinuity, since $T_{c2}(H) > T_c(H)$. Unfortunately, the breadth of the transitions ($\sim 5 \times 10^{-3}$ K) in most type II materials obscures the detailed behavior at T_{c2}, since the linear increase in magnetization below T_{c2} in a few millidegrees becomes orders of magnitude larger than the fluctuation diamagnetism a few millidegrees above T_{c2}.

Some typical data on indium are shown in Fig. 7-4. The upper part shows results in relatively low fields; M' increases with H, but less than linearly. The lower part shows higher-field data; here M' *decreases* as H increases, because the higher fields are rapidly extinguishing the fluctuations. Note the discontinuous jump indicated at the left end of the curve for $H = 34.9$ Oe. At this point, M jumps by five orders of magnitude to the Meissner-effect value. But since it is a first-order transition, there is no divergence anticipating the jump. As suggested by this figure, a temperature-dependent M' can be observed out to about $2T_c$.

To compare these results with theory, it was obviously necessary to generalize the Schmid result (7-42), which actually had been obtained first by Schmidt,[1] to the case of finite fields. This was done exactly in the framework of the GL theory by Prange.[2] He found that M' should indeed diverge as $(T - T_{c2})^{-1/2}$, and that it should be a universal result if scaled variables were used. That is, he found that

$$\frac{M'}{H^{1/2}T} = f_P(x) \qquad (7\text{-}43)$$

where f_P is a function of the single variable $x = (T - T_c)/H \times (dH_{c2}/dT)_{T_c}$. Unfortunately, when the data for several materials were plotted in terms of these variables, they did not fall on the theoretical universal curve but instead fell systematically well below it, especially for the higher-field values. This was initially disturbing, since it represented serious disagreement between an exact consequence of GL theory and experimental fact.

The explanation of this disagreement was first suggested by Patton, Ambegaokar, and Wilkins.[3] They pointed out that since the GL theory is based on an expansion of the free energy in derivatives of ψ, it is restricted to treating slow variations in space. Since the vector potential as well as the gradient operator enters in the canonical momentum, the GL theory is also limited to reasonably weak fields. Thus, one might expect it to give a poor account of the short-wavelength $[\lesssim \xi(0)]$ fluctuations which dominate far above T_c and in strong magnetic fields. For example, at $2T_c$, $\xi(T) \approx \xi(0)$, while even at T_c the fluctuation size in a field $\sim H_{c2}(0)$, as governed by (7-40), is also of order $\xi(0)$. Considerations of this general sort led them to attempt to correct the Prange calculation by

[1] H. Schmidt, *Z. Phys.* **216**, 336 (1968). (A numerical error of a factor of 4 occurs in this calculation.)
[2] R. E. Prange, *Phys. Rev.* **B1**, 2349 (1970).
[3] B. R. Patton, V. Ambegaokar, and J. W. Wilkins, *Solid State Commun.* **7**, 1287 (1969).

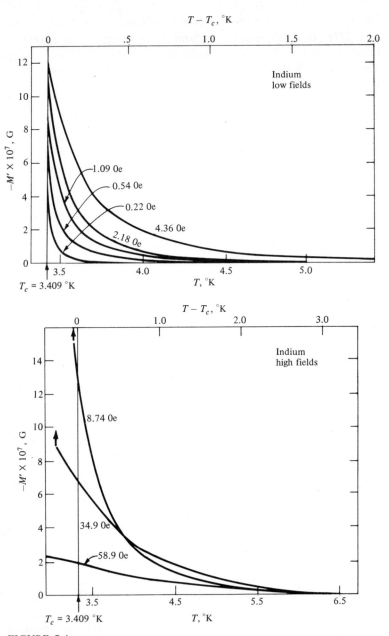

FIGURE 7-4

Fluctuation-enhanced diamagnetism of indium, after Gollub et al. The base line for these curves has been taken as the high-temperature limit, where M becomes independent of T.

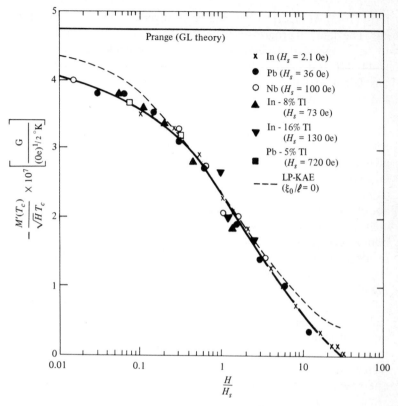

FIGURE 7-5
Universal dependence of $M'(T_c)/H^{1/2}T_c$ on H/H_s. Solid curve is an empirical curve drawn through the data of Gollub et al.; dashed curve is that of LP-KAE in clean limit (see text).

cutting off the short-wavelength fluctuations. In this way, they were led to generalize Prange's result (7-43) to

$$\frac{M'}{H^{1/2}T} = f_{\text{PAW}}(x, H/H_s) \qquad (7\text{-}44)$$

where x is the same reduced temperature variable as before, and H_s is a material-dependent scaling field to be determined by the model.

Gollub et al. tested this idea by plotting data on many materials taken at T_c (where $x = 0$) as functions of H, as shown in Fig. 7-5. According to the Prange form (7-43), $M'(T_c)/H^{1/2}T_c$ would be a universal numerical constant for all materials. The data fall progressively below this value with increasing H in what appears to be a universal dependence on the scaled-field variable H/H_s introduced in (7-44). [H_s is defined as the field for each material for which the observed $M'(T_c)$ has fallen to half the Prange value.] The specific form of falloff predicted by Patton et al. turned out to be qualitatively, but not quantitatively, correct.

Shortly after this universal behavior had been demonstrated experimentally, Lee and Payne[1] (LP) and independently Kurkijärvi, Ambegaokar, and Eilenberger[2] (KAE) produced a theoretical curve (shown dashed in Fig. 7-5) giving good agreement. This theoretical result was based on the microscopic Gor'kov theory in the clean limit. In working out this theory, nonlocal electrodynamic effects played an unexpectedly important role. In effect, they reduce H_s by about an order of magnitude below the value expected from our qualitative arguments, which had suggested that H_s should be of the order of $H_{c2}(0)$. Note that this nonlocality occurs although the field **B** is everywhere uniform (unlike the usual incidence of nonlocal effects only when fields are confined to a thin penetration layer). Although **B** is uniform, **A** is not, and **A** is what matters in the superconducting electrodynamics.

For alloys, it appears experimentally that $H_s/H_{c2}(0)$ approaches a limiting value $\sim 1/2$. This seems intuitively reasonable, since nonlocal effects should drop out with short mean free path. But the calculations of LP-KAE, based on a straightforward application of the Gor'kov theory, appeared to give $H_s/H_{c2}(0)$ increasing without limit as the mean free path was reduced. On the other hand, an alternate calculation by Maki and Takayama,[3] gave a finite limit for $\ell \to 0$ which seems to be in satisfactory agreement with experiment. As Maki[4] has shown, the results of LP-KAE were distorted by inclusion of a zero-point term which reflects only the normal properties of the metal.

We may summarize this discussion by contrasting two regimes: In a type II superconductor, fluctuation diamagnetism can be observed as we approach very close to T_{c2}. In this case, it is dominated by the very lowest energy, long-wavelength modes, which diverge at T_{c2} and are well described by the GL theory. As a result, we expect the Prange (or GL) results to work well near T_{c2}. This is in fact the case, as shown by measurements of Gollub et al. on such type II samples as Pb-5%Tl. But as one goes up in temperature, all fluctuation modes are excited to a comparable extent; statistical weight then favors the short-wavelength ones which are poorly described by the GL theory, and large discrepancies should, and do, appear. It is satisfying that developments in the microscopic theory, spurred by these discrepancies, have now largely succeeded in explaining them.

[1] P. A. Lee and M. G. Payne, *Phys. Rev. Letters* **26**, 1537 (1971); *Phys. Rev.* **B5**, 923 (1972).
[2] J. Kurkijärvi, V. Ambegaokar, and G. Eilenberger, *Phys. Rev.* **B5**, 868 (1972).
[3] K. Maki and H. Takayama, *J. Low Temp. Phys.* **5**, 313 (1971).
[4] K. Maki, *Phys. Rev. Letters* **30**, 648 (1973).

7-4.1 Diamagnetism in Two-dimensional Systems

If a superconductor is in the form of a film of thickness $d \ll \xi$, the fluctuations essentially vary only in the two dimensions of the plane of the film. Such a system is a two-dimensional superconductor. The fluctuation-induced diamagnetism can easily be estimated using (7-41). Assuming a correlation range of $\sim \xi$ in the plane and d in the normal direction, the volume V of such a fluctuation is $\sim \pi \xi^2 d$, so that $\chi \approx -\pi k T \langle r^2 \rangle / \Phi_0^2 d$. Because the fluctuating volume is not spherical, the definition of $\langle r^2 \rangle$ requires care. The following physical argument may be used: The energy density $\chi H^2 / 8\pi$ can also be written as $\mathbf{J} \cdot \mathbf{A}/2c \propto A^2$ in the London gauge. But $\oint \mathbf{A} \cdot d\mathbf{s} \approx B\mathscr{A}$, where $\mathscr{A}$ is the area of the fluctuating region as viewed along the field. Thus, $A \approx B\mathscr{A}/s$, where s is the perimeter of the area. Since $B \approx H$ in the present case of weak-fluctuation diamagnetism, it follows that $\chi \propto (\mathscr{A}/s)^2$. For a sphere, this is simply $\langle r^2 \rangle$. For the disk shape of the fluctuating region, $\langle r^2 \rangle_{\text{eff}} \approx \xi^2$ for $H_\perp$, and $\langle r^2 \rangle_{\text{eff}} \approx (d/2)^2$ for $H_\parallel$. [This general qualitative argument can be confirmed for the parallel-field case by reference to (4-54a), which describes the screening of a parallel field in a thin film. For $d \ll \lambda$, one finds $\bar{h}/H'B/H = 1 + 4\pi\chi$, with $4\pi\chi = -d^2/12\lambda^2 \propto d^2$, as found here.] Thus, in order of magnitude

$$\chi'_\perp \approx \frac{-kT\xi^2}{\Phi_0^2 d} \approx \frac{\xi}{d}\chi'_{3D} \propto \frac{T_c}{T - T_c} \qquad (7\text{-}45a)$$

whereas

$$\chi'_\parallel \approx \frac{-kTd}{\Phi_0^2} \approx \frac{d}{\xi}\chi'_{3D} \propto \text{const} \qquad (7\text{-}45b)$$

In these expressions, χ'_{3D} refers to (7-42).

From these results we conclude that $\chi'_\parallel$ will be essentially unobservable, being extremely small and temperature-independent. On the other hand, $\chi'_\perp$ will be ξ/d times larger than χ'_{3D} per unit volume; but, since the volume falls as d, this gives a thickness-independent total susceptibility equal only to that of a bulk superconductor of thickness $\sim \xi$. For a single film, this small moment would again be hard to observe. However, by constructing a multilayer stack of such films, one could get a larger volume. In this way, one could test the predicted different temperature dependence $\chi'_\perp \propto (T - T_c)^{-1}$ of (7-45a) compared to the dependence of $\chi'_{3D} \propto (T - T_c)^{-1/2}$ of (7-42).

In fact, there exist superconducting layered compounds, such as TaS_2, into which organic compounds such as pyridine can be intercalated to separate each metallic conducting layer ($d \approx 6$ Å) from its neighbors. In an ideal sample, the layers are coupled together only by Josephson-like tunneling through the pyridine intercalate, and one ight imagine that this would give a two-dimensional behavior. Actually, as shown by Lawrence and Doniach,[1] even weak Josephson coupling leads to behavior which is better described near T_c as three-dimensional with anisotropic effective mass than as two-dimensional. In particular, the temperature dependence of the fluctuation susceptibility appears[2] not to differ much from that of a three-dimensional system, at least until well above T_c.

7-5 TIME DEPENDENCE OF FLUCTUATIONS

Since diamagnetism is an equilibrium property, we were able to compute it above using only the time-average quantities $|\psi_k|^2$. When we come to discuss a nonequilibrium property such as electrical conductivity, however, we need a model of the time dependence, since the contribution of a given fluctuation to the conductivity above T_c is proportional to its lifetime, as that time limits the period available for acceleration in an applied field. Such a model is provided by the time-dependent Ginzburg-Landau (TDGL) equations developed by various workers.[3] According to this model, which will be discussed in more detail in the next chapter, the superconducting ψ function relaxes exponentially toward its instantaneous equilibrium value; above T_c, this value is zero. The linearized TDGL equation is then a simple generalization of (7-24), namely

$$\frac{\partial \psi}{\partial t} = -\frac{1}{\tau_0}(1 - \xi^2 \nabla^2)\psi \qquad T > T_c \qquad (7\text{-}46)$$

if electromagnetic potentials are neglected. The parameter

$$\tau_0 = \frac{\pi \hbar}{8k(T - T_c)} \qquad (7\text{-}47)$$

is the temperature-dependent relaxation time of the uniform $(k = 0)$ mode. According to (7-46), the higher-energy modes with $k > 0$ decay more rapidly, with relaxation rate

$$\frac{1}{\tau_k} = \frac{1 + k^2 \xi^2}{\tau_0} \qquad (7\text{-}48)$$

By themselves, these equations imply that (above T_c) any nonzero value of ψ_k will die out exponentially in a time τ_k. To maintain the nonzero thermal average of $|\psi_k|^2$ found in (7-27), one invokes a so-called Langevin force, a completely random (i.e., "white spectrum") driving force which represents the interaction

[1] W. E. Lawrence and S. Doniach, *Proc. 12th Intern. Conf. on Low Temp. Physics*, Kyoto, Japan, 1970 (Keigaku Publ. Co., 1971, p. 361); see also T. Tsuzuki, *J. Low Temp. Phys.* **9**, 525 (1972).
[2] D. E. Prober, M. R. Beasley, and R. E. Schwall, unpublished.
[3] See, for example, A. Schmid, *Phys. Kondens. Mat.* **5**, 302 (1966); C. Caroli and K. Maki, *Phys. Rev.* **159**, 306, 316 (1967); E. Abrahams and T. Tsuneto, *Phys. Rev.* **152**, 416 (1966); J. W. F. Woo and E. Abrahams, *Phys. Rev.* **169**, 407 (1968). A recent critical review has been given by M. Cyrot, *Repts. Prog. of Phys.* **36**, 103 (1973).

between the superconducting electrons and the rest of the thermodynamic system with which it is in equilibrium. The magnitude of this force is fixed by the requirement that it maintain the appropriate value of $\langle |\psi_k|^2 \rangle$, as calculated in (7-27) using only *equilibrium* statistical mechanics. Adding a Langevin force F_k on the right of (7-46), and choosing the value of F_k to give the correct time average,

$$\langle |\psi_k|^2 \rangle = \frac{1}{2\pi} \int_{-\infty}^{\infty} \langle |\psi_{k,\omega}|^2 \rangle \, d\omega$$

we have

$$\langle |\psi_{k,\omega}|^2 \rangle = \langle |\psi_k|^2 \rangle \frac{2\tau_k}{1 + \omega^2 \tau_k^2} \qquad (7\text{-}49)$$

It is easily verified (Wiener-Khintchine theorem)[1] that this power spectrum of ψ_k in frequency space corresponds to exponential decay in time of the correlation function

$$\langle \psi_k^*(0)\psi_k(t) \rangle = \langle |\psi_k|^2 \rangle e^{-t/\tau_k} \qquad (7\text{-}50)$$

Finally, substituting for $\langle |\psi_k|^2 \rangle$ from (7-27) and for τ_k from (7-48), we have, after simplifying,

$$\langle |\psi_{k,\omega}|^2 \rangle = \frac{16 \, k(T - T_c)}{\pi} \frac{kT\tau_k^2}{\hbar\alpha(T)} \frac{1}{1 + \omega^2 \tau_k^2} \qquad (7\text{-}51)$$

Since $\alpha(T) \propto (T - T_c)$, the entire dependence on T, as well as on k and ω, is in the last factor.

7-6   FLUCTUATION-ENHANCED CONDUCTIVITY ABOVE T_c

In the absence of superconducting fluctuations, the normal dc conductivity is given by

$$\sigma_n = \frac{ne^2\tau}{m} \qquad (7\text{-}52)$$

where τ is the mean scattering time of the normal electrons and n is the number of them per unit volume. By analogy, we might expect the fluctuations to contribute an extra term

$$\sigma' \approx \frac{(2e)^2}{m^*} \sum_k \frac{\langle |\psi_k|^2 \rangle \tau_k}{2} \qquad (7\text{-}53)$$

since ordinary scattering processes are ineffective until a given fluctuation relaxes, and $|\psi_k|^2$ will relax twice as fast as ψ_k. Using the values of $\langle |\psi_k|^2 \rangle$ and τ_k from

---

[1] See, for example, C. Kittel, "Elementary Statistical Physics," p. 136, Wiley, New York, 1958.

(7-27) and (7-48) and integrating over k space, this prescription gives results which differ from the exact calculations only by small numerical factors. In particular, the temperature dependence of σ' is correctly found to be $(T - T_c)^{-(4-d)/2}$, where $d\ (= 1, 2, 3)$ is the dimensionality of the system.

Since the calculation is quite tractable and gives some additional insight, let us now compute σ' exactly, within the framework of the GL linearized fluctuation theory. This is done most easily using the Kubo formalism, which relates linear response coefficients to the fluctuations in the unperturbed system, as required by the fluctuation-dissipation theorem. We confine our attention to uniform fields and currents. Then, our general starting point is the Kubo result

$$\sigma_{xx}(\omega) = \frac{1}{kT} \int_0^\infty \langle J_x(0)J_x(t)\rangle \cos \omega t \, dt \qquad (7\text{-}54)$$

[If this approach is unfamiliar, it may be helpful to note that the integral gives the power spectrum of $J_x(t)$; (7-54) is then equivalent to the Nyquist noise current formula: $J_x^2(\omega) = 4kT\sigma_{xx}(\omega)$ per unit bandwidth. Considering a unit cube, this formula is equivalent to the even more familiar expression for the thermal noise voltage in bandwidth B of a resistance R, namely, $V^2 = 4kTRB$.] We assume that the normal quasi-particle current fluctuations are unchanged by the superconducting fluctuations. (This is not strictly correct very near T_c, where the fluctuations are strong.) Thus, to compute σ_{xx}, we include only the fluctuating supercurrent in (7-54). For $\psi = \sum_k \psi_k e^{i\mathbf{k}\cdot\mathbf{r}}$, the current is given by

$$\mathbf{J} = \frac{e\hbar}{m^*i}(\psi^*\nabla\psi - \psi\nabla\psi^*)$$

$$= \frac{e\hbar}{m^*}\sum_{\mathbf{k},\mathbf{q}}(2\mathbf{k} + \mathbf{q})\psi_k^*\psi_{\mathbf{k}+\mathbf{q}}e^{i\mathbf{q}\cdot\mathbf{r}} \qquad (7\text{-}55)$$

Restricting attention to uniform ($\mathbf{q} = 0$) currents in the x direction, this reduces to

$$J_x = \frac{2e\hbar}{m^*}\sum_{\mathbf{k}} k_x |\psi_k|^2 \qquad (7\text{-}56)$$

We now want to compute the current-current correlation function

$$\langle J_x(0)J_x(t)\rangle = \left(\frac{2e\hbar}{m^*}\right)^2 \left\langle \sum_{\mathbf{k},\mathbf{k}'} k_x k_x' |\psi_k(0)|^2 |\psi_{\mathbf{k}'}(t)|^2 \right\rangle \qquad (7\text{-}57)$$

Since ψ_k and $\psi_{\mathbf{k}'}$ are statistically independent, the cross terms average out, and this can be written as

$$\langle J_x(0)J_x(t)\rangle = \left(\frac{2e\hbar}{m^*}\right)^2 \sum_{\mathbf{k}} k_x^2 \langle \psi_k^*(0)\psi_k(t)\rangle^2 \qquad (7\text{-}58)$$

Inserting the exponential time decay (7-50) of the correlation function, and carrying out the cosine Fourier transform (7-54), we obtain

$$\sigma'_{xx}(\omega) = \left(\frac{2eh}{m^*}\right)^2 \frac{1}{kT} \sum_{\mathbf{k}} k_x^2 \langle \,|\psi_{\mathbf{k}}|^2\rangle^2 \frac{\tau_k/2}{1 + (\omega\tau_k/2)^2} \qquad (7\text{-}59)$$

Specializing to the dc case ($\omega = 0$) and inserting $\langle\,|\psi_{\mathbf{k}}|^2\rangle$ and τ_k from (7-27) and (7-48), we obtain

$$\sigma'_{xx}(0) = \frac{\pi e^2}{\hbar}\left(\frac{T}{T - T_c}\right) \sum_{\mathbf{k}} \frac{k_x^2 \xi^4}{(1 + k^2\xi^2)^3} \qquad (7\text{-}60)$$

7-6.1 Three Dimensions

In a three-dimensional bulk sample, k can be taken as a continuous variable, allowing the sum to be performed by integration. Averaged over a sphere, $k_x^2 = k^2/3$; the density of states in unit volume is given by $4\pi k^2\, dk/(2\pi)^3$. After an elementary integration, one obtains

$$\sigma'(0)\bigg|_{3D} = \frac{1}{32}\frac{e^2}{\hbar\xi(0)}\left(\frac{T}{T - T_c}\right)^{1/2} \qquad (7\text{-}61)$$

where as usual $\xi(T) \equiv \xi(0)[T/(T - T_c)]^{1/2}$. Since the conductivity is isotropic, we have dropped the subscripts. Although this result is formally divergent as one approaches T_c, the coefficient is less than the normal conductivity σ_n by a factor of the order of $(kT_c/E_F)(1/k_F\ell) \approx 10^{-7}$. Thus, the fractional change in conductivity at any meaningful temperature interval above T_c will be very small. Note the contrast with the diamagnetic susceptibility (7-42), where the coefficient is comparable to the background normal-state diamagnetism, so that large fractional changes in χ are observable.

7-6.2 Two Dimensions

Now let us consider the case of a film thin enough to justify a two-dimensional approximation, i.e., one in which the thickness $d \ll \xi$. This is the case on which the greatest amount of experimental work has been done. In this case, the variation of ψ perpendicular to the film is limited to a discrete set of standing waves with $k_\perp = v\pi/d$, with $v = 0, 1, 2$, etc. If the film is thin enough to drop all except the $v = 0$ term, the summation in (7-60) becomes a two-dimensional integration in the plane of the film. The average of k_x^2 is then $k^2/2$, and the density-of-states factor becomes $2\pi k\, dk/(2\pi)^2 d$. Carrying out the integration, we obtain the remarkably simple result

$$\sigma'(0)\bigg|_{2D} = \frac{e^2}{16\hbar d}\frac{T}{T - T_c} \qquad (7\text{-}62)$$

Note that this result contains no adjustable parameters (apart from T_c). It is also important to realize that the film thickness d is not critical, since the quantity actually measured experimentally is the conduct*ance* $\sigma'd$ per square, not the conductivity. This result was first derived by Aslamasov and Larkin[1] and is in excellent agreement with the measured results of Glover[2] on thin amorphous films. Such films are particularly favorable, since they have a high normal resistance (low normal conductance) in parallel with the fluctuation term (7-62), which has a universal value, independent of the normal resistance. Thus, the fluctuation conductivity is a larger fractional effect when the background normal conductance is lower.

Before making detailed comparison between (7-62) and experimental data, it is important to assess the size of the error made by dropping all except the $k_\perp = 0$ term in the sum (7-60). The integration over k in the plane can be carried out for arbitrary $k_\perp$, with the value of the integral reduced by a factor $(1 + k_\perp^2 \xi^2)^{-1}$. Thus, for finite d/ξ, (7-62) should be multiplied by a factor

$$\frac{\sigma'}{\sigma'_{2D}} = \sum_{v=0}^{\infty} \frac{1}{1 + (v^2 \pi^2 \xi^2/d^2)} \tag{7-63}$$

Thus, so long as $d \le \xi$, the simple two-dimensional result is accurate to ~ 10 percent. On the other hand, if $d \gg \xi$, this sum can be evaluated by integration; the result is $d/2\xi$, just the factor required to recover the three-dimensional result (7-61) from the two-dimensional result (7-62). Because of the temperature dependence of ξ, films may change from two-dimensional to three-dimensional behavior as one goes farther above T_c. Experimental data appear to follow the transitional behavior predicted by (7-63).

7-6.3   One Dimension

To complete our survey of the important special cases, we now consider the case of a thin wire of cross-sectional area $A \ll \xi^2$, so that a one-dimensional approximation may be made. In this case, only the component of k along the wire (k_x) has a continuous distribution. Retaining only the terms in (7-60) with zero transverse momentum, the density-of-states factor is $dk_x/2\pi A$. Integrating on k_x from $-\infty$ to $+\infty$, we obtain

$$\sigma'(0)\Big|_{1D} = \pi \frac{e^2 \xi(0)}{16\hbar A} \left(\frac{T}{T - T_c}\right)^{3/2} \tag{7-64}$$

with corrections for finite A which are down by a factor of the order of $(A/\pi^2\xi^2)^{3/2}$. As with the two-dimensional case, the measured quantity is the conductance $\sigma'A$, so that the area need not be known accurately to test this theoretical prediction.

[1] L. G. Aslamasov and A. I. Larkin, *Phys. Letters* **26A**, 238 (1968).
[2] R. E. Glover, III, *Phys. Letters* **25A**, 542 (1967).

Nonetheless, the agreement between theory and experiment in the tin whisker crystals which were used in the experiments discussed in Sec. 7-1 is not very good because of the importance of the anomalous contributions (Maki terms) discussed below.

7-6.4 Anomalous Contributions to Fluctuation Conductivity

While the initial comparison between the Aslamasov-Larkin (AL) result (7-62) and the experimental data of Glover and coworkers[1] on thin films showed excellent agreement, later measurements[2] showed σ' values as much as 10 times larger than the AL prediction. These large values were typically found in clean (i.e., low-resistance) aluminum films; lead and bismuth films were close to AL, as found earlier; while tin films showed σ' up to about 4 times the AL value.

At about the same time these anomalously large conductivity enhancements were found experimentally, Maki[3] noted the existence of another term (or Feynman diagram) in the theoretical calculation which had been omitted in the work of AL. Physically, this "Maki term" appears to reflect the increase in the normal-electron conductivity induced by the superconducting fluctuations. A similar increase in σ_1 above σ_n for $\hbar\omega < \Delta < kT$ is familiar just *below* T_c, in the presence of weak but stable superconductivity. In fact, the discussion of (2-93) indicated that, for $\omega = 0$, σ_1 has a logarithmic infinity unless the peaked BCS state density is limited by lifetime effects or gap anisotropy. Although the analogy between this effect and the Maki term is rather superficial, it *is* the case that in one-dimensional or two-dimensional systems the Maki term appears to give an infinite conductivity at all temperatures *above* T_c.

As shown by Thompson,[4] this nonphysical divergence is prevented by the presence of any pair-breaking effect, such as a magnetic field or magnetic impurities, which effectively limits the lifetimes of the evanescent Cooper pairs and cuts off the divergence. A considerable amount of data, especially on films in magnetic fields, could be accounted for by adding this so-called Maki-Thompson term to the AL term treated above.

More recently, Patton[5] and also Keller and Korenman[6] have reexamined this problem and shown that only a finite conductivity is obtained for the Maki term even for the simple BCS model without any extraneous pair breakers, if the so-called vertex corrections are calculated sufficiently carefully. These theories account for an effective Maki-Thompson pair breaker of strength proportional to the resistance per square ($R_\square$) of the film. However, even these improved theories

---

[1] See, for example, D. C. Naugle and R. E. Glover, III, *Phys. Letters* **28A**, 110 (1968).

[2] See, for example, J. E. Crow, R. S. Thompson, M. A. Klenin, and A. K. Bhatnagar, *Phys. Rev. Letters* **24**, 371 (1970).

[3] K. Maki, *Prog. Theor. Phys. (Kyoto)* **39**, 897 (1968); **40**, 193 (1968).

[4] R. S. Thompson, *Phys. Rev.* **B1**, 327 (1970).

[5] B. R. Patton, *Phys. Rev. Letters* **27**, 1273 (1971).

[6] J. Keller and V. Korenman, *Phys. Rev. Letters* **27**, 1270 (1971); *Phys. Rev.* **B5**, 4367 (1972).

require an intrinsic pair breaker to account for the difference in data obtained on different materials with the same $R_\square$, especially in the low resistance films. The strength of this pair breaker is usually specified by a parameter τ_{c0}, which is the fractional depression of T_c due to its action. For aluminum, the data fit with $\tau_{c0} = 2 \times 10^{-4}$, while for tin $\tau_{c0} \approx 0.02$. For the strong-coupling materials lead and bismuth, it appears that τ_{c0} may be as large as 0.1; this, together with the high $R_\square$, probably accounts for the agreement of the data of Glover with the simple AL theory without Maki-Thompson corrections. The trend of the values of τ_{c0} with different materials is consistent with the suggestion of Appel[1] that pair breaking by thermal phonons should give $\tau_{c0} \propto \lambda(T/\theta_D)^2$, where λ is the electron-phonon coupling constant and θ_D is the Debye temperature; but a detailed theory is not yet available. For further details on the status of the fluctuation-enhanced conductivity problem, including many different regimes, the reader is referred to the comprehensive review of Craven, Thomas, and Parks.[2]

7-6.5 High-Frequency Conductivity

Measurements of the frequency dependence of σ' allow a more specific test of the TDGL theory than the simple dc conductivity measurements described above, since from (7-59) it is clear that σ' will fall off for $\omega > \tau_k^{-1}$. More quantitatively, each term in the sum in (7-60) is multiplied by $[1 + (\omega\tau_k/2)^2]^{-1}$ before the integration over k. Specializing to the two-dimensional case, the integration can still be carried out exactly, with the result

$$\sigma'(\omega)\bigg|_{2D} = \frac{e^2}{16\hbar d}\left(\frac{T}{T-T_c}\right)\left[\frac{4}{\omega\tau_0}\tan^{-1}\frac{\omega\tau_0}{2} - \frac{4}{\omega^2\tau_0^2}\ln\left(1 + \frac{\omega^2\tau_0^2}{4}\right)\right] \tag{7-65}$$

The prefactor is recognized as the dc result. If the expression in the square bracket is expanded at low frequencies, one finds

$$\sigma'(\omega) = \sigma'(0)\left(1 - \frac{\omega^2\tau_0^2}{24} + \cdots\right) \tag{7-66}$$

Thus, as expected, the conductivity rolls off when ω exceeds some average τ_k^{-1}, which in turn is somewhat greater than τ_0^{-1} because of the k-dependence of τ_k. Another interesting limit is right at T_c, where $\tau_0 = \infty$. Then for any finite frequency (7-65) reduces to

$$\sigma'(\omega)\bigg|_{T_c} = \frac{e^2}{\hbar d}\frac{kT_c}{\hbar\omega} \qquad \omega > 0 \tag{7-67}$$

Note that this is the same as the dc value $\sigma'(0)$ for $(T - T_c) = \hbar\omega/16k$. Clearly, this fluctuation conductivity is finite even at T_c for all except zero frequency.

[1] J. Appel, *Phys. Rev. Letters* **21**, 1164 (1968).
[2] R. A. Craven, G. A. Thomas, and R. D. Parks, *Phys. Rev.* **B7**, 157 (1973).

These predictions of the theory have been tested by microwave transmission measurements on thin lead films by Lehoczky and Briscoe.[1] They were able to make measurements at frequencies of 24, 37, and 69 GHz, as well as dc measurements. Below T_c, the microwave transmission data agreed well with that computed using the ordinary Mattis-Bardeen complex conductivity function $[\sigma_1(\omega) - i\sigma_2(\omega)]$ of the BCS theory, discussed in Sec. 2-10. Above T_c, the data fit well with the predicted transmission based on (7-65) for σ_1, with $\sigma_2 = 0$. The conductivity increase at T_c was found to be inversely proportional to ω, as expected from (7-67), reaching ~ 11 percent at the lowest frequency used. The good agreement between theory and experimental data as functions of both frequency and temperature indicates that the TDGL model is accurate, at least in these high-resistance lead films. The frequency dependence of the conductivity when the Maki terms are important remains open for further investigation.

[1] S. L. (A.) Lehoczky and C. V. Briscoe, *Phys. Rev. Letters* **23,** 695 (1969).

8

CONCLUDING TOPICS

This final chapter is devoted to brief discussions of several topics which we have skirted earlier to avoid interrupting the more elementary discussions given there. First we discuss the Bogoliubov equations governing the spectrum of excitations for spatially inhomogeneous superconductors. Then we review the effects of magnetic perturbations in modifying the spectrum of excitations, ultimately producing gapless superconductivity. Finally we discuss time-dependent superconductivity: first as described by the time-dependent GL theory, and second in regimes where inelastic phonon processes control the relaxation rate.

8-1 BOGOLIUBOV METHOD: GENERALIZED SELF-CONSISTENT FIELD

In our discussion of the microscopic BCS theory, we considered only pure materials, in which the momentum $\mathbf{k}$ was a good quantum number and in which $\mathbf{k}{\uparrow}$ and $-\mathbf{k}{\downarrow}$ states were occupied in pairs. In 1959 Anderson[1] showed that a more general

---

[1] P. W. Anderson, *J. Phys. Chem. Sol.* **11**, 26 (1959).

prescription, applicable in dirty superconductors as well, was to pair time-reversed states. In a dirty metal, the electronic eigenfunctions are some functions $w_n(\mathbf{r})$ which are certainly far from plane waves, but in the absence of magnetic or other time-reversal noninvariant terms in the hamiltonian, each $w_n(\mathbf{r})$ is degenerate with $w_n^*(\mathbf{r})$ if the spin part of the wavefunction is also reversed. Anderson showed that if this pairing of time-reversed states were followed, one could expect the equilibrium properties like T_c, H_c, and Δ to be essentially independent of the electronic mean free path.

This result pertains to a superconductor which is still essentially homogeneous on the scale of ξ_0 despite the presence of scattering centers. A more general problem arises if the superconducting order parameter varies in space, for example at an interface with another material or in the case of a vortex. To cope with these situations, one may use the Bogoliubov equations, which essentially generalize the ordinary Hartree-Fock equations of many-electron theory to include the effects of the superconducting "pairing potential" $\Delta(\mathbf{r})$ as well as the ordinary scalar potential $U(r)$. Since de Gennes[1] has given a thorough account of this method, we shall content ourselves with sketching some of the results, largely using his notation and conventions to simplify reference to his discussion.

To facilitate treatment of spatial variations, one defines a generalization of the Bogoliubov transformation (2-42) by

$$\Psi(\mathbf{r}\uparrow) = \sum_n [\gamma_{n\uparrow}\, u_n(\mathbf{r}) - \gamma_{n\downarrow}^*\, v_n^*(\mathbf{r})]$$

$$\Psi(\mathbf{r}\downarrow) = \sum_n [\gamma_{n\downarrow}\, u_n(\mathbf{r}) + \gamma_{n\uparrow}^*\, v_n^*(\mathbf{r})] \tag{8-1}$$

where the Ψ's are annihilation operators for position eigenfunctions rather than for momentum eigenfunctions, as were the $c_{\mathbf{k}\sigma}$ used in Chap. 2. The u's and v's are eigenfunctions to be determined so as to diagonalize the hamiltonian

$$\mathscr{H}_{\text{eff}} = \int \left\{ \sum_\sigma \Psi^*(\mathbf{r}, \sigma)\left[\frac{1}{2m}\left(\frac{\hbar}{i}\nabla - \frac{e\mathbf{A}}{c}\right)^2 + U(r) - \mu\right]\Psi(\mathbf{r}, \sigma) \right.$$

$$\left. + \Delta(\mathbf{r})\Psi^*(\mathbf{r}\uparrow)\Psi^*(\mathbf{r}\downarrow) + \Delta^*(\mathbf{r})\Psi(\mathbf{r}\uparrow)\Psi(\mathbf{r}\downarrow) \right\} d\mathbf{r} \tag{8-2}$$

where

$$\Delta(\mathbf{r}) = V\langle \Psi(\mathbf{r}\uparrow)\Psi(\mathbf{r}\downarrow)\rangle = V\sum_n v_n^*(\mathbf{r})u_n(\mathbf{r})(1 - 2f_n) \tag{8-3}$$

again in close analogy to our discussion in connection with (2-38) and (2-40). The diagonalization requires that u and v satisfy the coupled Bogoliubov equations

$$\mathscr{H}_0 u(\mathbf{r}) + \Delta(\mathbf{r})v(\mathbf{r}) = Eu(\mathbf{r})$$

and

$$-\mathscr{H}_0^* v(\mathbf{r}) + \Delta^*(\mathbf{r})u(\mathbf{r}) = Ev(\mathbf{r}) \tag{8-4}$$

[1] P. G. de Gennes, "Superconductivity of Metals and Alloys," chap. 5, W. A. Benjamin, New York, 1966.

where
$$\mathcal{H}_0 = \frac{1}{2m}\left(\frac{\hbar}{i}\nabla - \frac{e\mathbf{A}}{c}\right)^2 + U(\mathbf{r}) - \mu \qquad (8\text{-}5)$$

and $U(\mathbf{r})$ includes the ordinary Hartree-Fock averaged Coulomb interaction between electrons as well as the potential of the ion cores and any overall electrostatic potentials.

We note first that if $\Delta = 0$, the equations (8-4) decouple into the forms

$$\mathcal{H}_0 u = Eu$$

and
$$\mathcal{H}_0^* v = -Ev \qquad (8\text{-}6)$$

so that $u(\mathbf{r})$ and $v(\mathbf{r})$ are the ordinary electron and hole eigenfunctions of the normal state, with energies $\pm E$ relative to the Fermi energy. In general, however, we must seek solutions of the pair of coupled equations (8-4), eventually made self-consistent by computing $\Delta(\mathbf{r})$ from the set of u's and v's by (8-3).

8-1.1   Dirty Superconductors

As a first example, consider the Anderson problem of a dirty, but nonmagnetic, superconductor. Then the normal-state eigenfunctions w_n satisfy

$$\mathcal{H}_0 w_n = \xi_n w_n \qquad (8\text{-}7)$$

where ξ_n is the eigenvalue measured from the chemical potential μ. In a pure metal, the w_n are Bloch functions with well-defined k. In general, they are taken as the *exact*, although unknown, solutions in the presence of whatever (elastic) scattering exists. On the assumption that the metal is still homogeneous on the scale of ξ_0, one may take $\Delta(\mathbf{r})$ to be really a constant. In that case, we can satisfy (8-4) by taking both $u_n(\mathbf{r})$ and $v_n(\mathbf{r})$ proportional to $w_n(\mathbf{r})$: That is, we set $u_n(\mathbf{r}) = u_n w_n(\mathbf{r})$ and $v_n(\mathbf{r}) = v_n w_n(\mathbf{r})$, where u_n and v_n are now simply numbers. Then (8-4) becomes

$$(\xi_n - E_n)u_n + \Delta v_n = 0$$

and
$$(-\xi_n - E_n)v_n + \Delta^* u_n = 0 \qquad (8\text{-}8)$$

whose solution requires

$$E_n = (\xi_n^2 + |\Delta|^2)^{1/2} \qquad (8\text{-}9)$$

as in usual BCS theory. Moreover, when one goes back to find the self-consistent value of Δ, one finds the familiar result that it is determined by

$$\frac{1}{V} = \frac{1}{2}\sum_n \frac{|w_n(\mathbf{r})|^2}{E_n}\tanh\frac{\beta E_n}{2} \qquad (8\text{-}10)$$

Since we assume the scattering does not change the density of states, this will lead to the same results as (2-50). (Some attention to the normalization of w_n and definition of V is needed to show this in detail.) Thus, we do not expect much change in T_c or Δ on going from a clean to a dirty specimen; this agrees with experimental fact.

8-1.2 Uniform Current in Pure Superconductors

In our discussion of the critical current in a thin wire we mentioned, in connection with (4-41), that quasi-particle energies were shifted by an amount $\mathbf{v}_s \cdot \mathbf{p}$. This can be seen using our present methods. If the pairs have center-of-mass momentum* $2\mathbf{q}$, we expect

$$\Delta = |\Delta|^{i2\mathbf{q}\cdot\mathbf{r}} \qquad (8\text{-}11)$$

From (8-3) we see that this will result if

$$v_{\mathbf{k}}(\mathbf{r}) = V_{\mathbf{k}}\,e^{i(\mathbf{k}-\mathbf{q})\cdot\mathbf{r}}$$

and

$$u_{\mathbf{k}}(\mathbf{r}) = U_{\mathbf{k}}\,e^{i(\mathbf{k}+\mathbf{q})\cdot\mathbf{r}} \qquad (8\text{-}12)$$

Note that, with $q \neq 0$, we are no longer pairing time-reversed states. When (8-11) and (8-12) are substituted in the Bogoliubov equations (8-4), they become

$$(\xi_{\mathbf{k}+\mathbf{q}} - E_{\mathbf{k}})U_{\mathbf{k}} + |\Delta|\,V_{\mathbf{k}} = 0$$

and

$$(-\xi_{\mathbf{k}-\mathbf{q}} - E_{\mathbf{k}})V_{\mathbf{k}} + |\Delta|\,U_{\mathbf{k}} = 0 \qquad (8\text{-}13)$$

Solving for the excitation energies $E_{\mathbf{k}}$, we find

$$E_{\mathbf{k}} = \frac{\xi_{\mathbf{k}+\mathbf{q}} - \xi_{\mathbf{k}-\mathbf{q}}}{2} + \left[\left(\frac{\xi_{\mathbf{k}+\mathbf{q}} + \xi_{\mathbf{k}-\mathbf{q}}}{2}\right)^2 + |\Delta|^2\right]^{1/2} \qquad (8\text{-}14)$$

Since $\xi_{\mathbf{k}} = (\hbar^2 k^2/2m) - \mu$, we have

$$\frac{1}{2}(\xi_{\mathbf{k}+\mathbf{q}} - \xi_{\mathbf{k}-\mathbf{q}}) = \frac{\hbar^2}{m}\mathbf{k}\cdot\mathbf{q} = \frac{\hbar}{m}\mathbf{p}_{\mathbf{k}}\cdot\mathbf{q} \qquad (8\text{-}15)$$

Also, so long as $q \ll k_F$, $\xi_{\mathbf{k}+\mathbf{q}} + \xi_{\mathbf{k}-\mathbf{q}} \approx 2\xi_{\mathbf{k}}$. Thus (8-14) can be simplified to

$$E_{\mathbf{k}} = E_{\mathbf{k}}^0 + \mathbf{p}_{\mathbf{k}}\cdot\mathbf{v}_s \qquad (8\text{-}16)$$

where $\mathbf{v}_s = \hbar\mathbf{q}/m$ is the velocity of the supercurrent and $E_{\mathbf{k}}^0 = (\xi_{\mathbf{k}}^2 + |\Delta|^2)^{1/2}$ is the excitation energy in the absence of a current.

To assure self-consistency, these new shifted energies should be used in the Fermi functions in (8-3). However, at low temperatures, the f_n are nearly zero until

---

* Note that here we follow de Gennes' convention of assigning the pairs momentum $2\mathbf{q}$, rather than $\mathbf{q}$, as we have done elsewhere. Also, here we take $m^* = 2m$.

E_n approaches zero, so that $|\Delta|$ will not decrease much with current although the minimum excitation energy

$$E_{min} = \Delta - p_F v_s \qquad (8\text{-}17)$$

does. This provides a simple example of the fact that the pair potential Δ is not necessarily the same as the energy gap in the excitation spectrum. In fact we can obtain a simple concrete realization of *gapless* superconductivity by considering the situation when v_s very slightly exceeds Δ/p_F. Then, for a small number of $\mathbf{k}$ values in the direction opposite to $\mathbf{v}_s$, excitations can be made at zero energy just as in the normal state. But, just as in the normal state, the Fermi statistics limit the occupancy of these zero-energy states to $f_k \sim \frac{1}{2}$. In view of (8-3) that means that these few states contribute nothing to sustaining the pair potential, but all the other regions of the Fermi surface contribute more or less normally. When the self-consistent solution is worked out in detail,[1] it is found that there is a small region of v_s above Δ/p_F for which $E_{min} = 0$ while $\Delta \neq 0$. Thus, there is still a coherent condensed state, with macroscopic quantum properties described by the pair wave function $\psi \propto \Delta \propto e^{2i\mathbf{q}\cdot\mathbf{r}}$, and hence we still expect the perfect-conductivity property to remain. However, the depairing that sets in for $v_s > \Delta/p_F$ limits the maximum supercurrent to only about 1 percent more than the current when $v_s = \Delta/p_F$. For greater values of v_s, $dJ_s/dv_s < 0$, so the regime is unstable and not easily observed experimentally.

8-1.3   Excitations in Vortex

From the GL theory (Sec. 5-1) we know that in the vortex state of a type II superconductor $\Delta(z, r, \theta)$ has the form $|\Delta(r)|\, e^{i\theta}$, where $|\Delta(r)|$ rises from zero at the center of the vortex to Δ_∞ at a distance, the major rise occurring in a distance $\sim \xi(T)$. The GL solution for $\Delta(r)$ should be exact at T_c, where $\xi(T) \gg \xi(0)$, so that the slow-variation requirement of the GL theory is satisfied. However, it is only a qualitative guide at low temperatures where Δ varies on the scale of $\xi(0)$. Given this rapidly varying Δ, what is the nature of the quasi-particle excitations? This problem was first solved by Caroli, de Gennes, and Matricon[2]; the solution is reviewed in Sec. 5-2 of de Gennes' book. Their solution uses the Bogoliubov equations but makes the approximation of $\kappa \gg 1$, so that the magnetic field is negligible over the core region. Since they assume pure material (so that momentum is conserved), as well as $\kappa \gg 1$, their solution is not strictly applicable to any real material. The calculation was extended to all values of κ in an important paper by Bardeen, Kümmel, Jacobs, and Tewordt,[3] who attempted to obtain self-consistent solutions for $\Delta(r)$ and $h(r)$ using a variational expression for the free energy. Their qualitative conclusions are similar to those of Caróli et al.

[1] J. Bardeen, *Rev. Mod. Phys.* **34**, 667 (1962); also, K. T. Rogers, unpublished Ph.D. thesis, University of Illinois, 1960.

[2] C. Caroli, P. G. de Gennes, and J. Matricon, *Phys. Letters* **9**, 307 (1964).

[3] J. Bardeen, R. Kümmel, A. E. Jacobs, and L. Tewordt, *Phys. Rev.* **187**, 556 (1969).

At large distances from the center of the vortex $(r \gg \xi)$, it is quite a good approximation to use our previous result (8-16) for the shifted energy spectrum due to a uniform velocity field, with $v_s = \hbar/2mr = \hbar/m^*r$. As indicated in Sec. 5-5, if this argument is used near the core, it leads to gapless superconductivity inside $r \approx \xi$. Actually, in this region one really must solve the Bogoliubov equations to build in the effects of the rapid spatial variation of $\Delta(r)$. The solutions of Caroli et al. and of Bardeen et al. show that there is indeed a group of low-lying excitations with wavefunctions $u(r)$, $v(r)$ localized near the vortex core. The lowest one lies at only $\sim \Delta_\infty^2/E_F \sim 10^{-4} \Delta_\infty \ll kT_c$, which is effectively gapless. Since the level density is found to correspond roughly to that of a cylinder of normal material $\sim \xi$ in radius, this result forms the most microscopic rationale for the concept of the "normal" core of a vortex.

8-2 MAGNETIC PERTURBATIONS AND GAPLESS SUPERCONDUCTIVITY

In the previous section, we have seen that the excitation spectrum of a superconductor is modified if it carries a current; if the current is sufficiently strong, the spectrum becomes gapless for a finite current range before superconductivity is destroyed. The origin of this change is that adding the common drift momentum $\mathbf{q}$ to the paired electrons with initial momenta $\mathbf{k}$ and $-\mathbf{k}$ gives them different kinetic energies $\xi_{\mathbf{k}+\mathbf{q}}$ and $\xi_{-\mathbf{k}+\mathbf{q}}$, thus lifting the degeneracy of $\xi_{\mathbf{k}}$ and $\xi_{-\mathbf{k}}$, which had been exact because of time-reversal symmetry. In fact, tracing through the deduction of (8-17) from (8-14), we see that the gap reduction $(\Delta - E_{\min})$ is just equal to half the maximum splitting of the time-reversal degeneracy by the drift momentum. As it stands, this result is applicable only to pure superconductors, in which $\mathbf{k}$ is a good quantum number. However, as we shall see, the basic idea is very general.

When we considered (Sec. 8-1.1) Anderson's theory of "dirty" superconductors, i.e., nonmagnetic alloys with mean free path $\ell < \xi_0$, we noted that pairing of time-reversed degenerate states led to the same T_c and BCS density of states as for a pure superconductor. On the other hand, Abrikosov and Gor'kov[1] (AG) showed that *magnetic* impurities (which break the time-reversal symmetry) lead to a strong depression of T_c and a modification of the BCS density of states, so that it becomes gapless for a finite range of concentration below the critical value which destroys superconductivity entirely. Subsequent work by Maki,[2] de Gennes,[3] and others showed that the results of Abrikosov and Gor'kov for the density of states

[1] A. A. Abrikosov and L. P. Gor'kov, *Zh. Eksperim. i Teor. Fiz.* **39**, 1781 (1960); *Soviet Phys.— JETP* **12**, 1243 (1961).

[2] K. Maki, *Prog. Theor. Phys.* (*Kyoto*) **29**, 333 (1963); **31**, 731 (1964); **32**, 29 (1964); K. Maki and P. Fulde, *Phys. Rev.* **140**, A1586 (1965).

[3] P. G. de Gennes, *Phys. Kondens. Materie* **3**, 79 (1964); P. G. de Gennes and G. Sarma, *J. Appl. Phys.* **34**, 1380 (1963); P. G. de Gennes and M. Tinkham, *Physics* **1**, 107 (1964).

and the depression of T_c could be transcribed to describe the effects of many other pair-breaking perturbations, i.e., those which destroy the time-reversal degeneracy of the paired states. For this transcription to work, there must also be rapid scattering, as in a dirty superconductor, to assure "ergodic" behavior of the electrons. Examples of such perturbations include external magnetic fields, currents, rotations, spin exchange and hyperfine fields, in addition to magnetic impurities. It was also shown that spatial gradients in the order parameter, such as those induced by proximity to a boundary with a normal metal or in the surface sheath or vortex state of a type II superconductor, have a pair-breaking effect which can induce gapless superconductivity. In fact, gapless superconductivity turns out to be the rule rather than the exception if the transition to the normal state due to the perturbation is of second, rather than first, order.

Since the general theory of these effects is most naturally couched in the Green's function formalism of Gor'kov, which is beyond the scope of this book, we shall confine our treatment to an outline of some of the major results and their experimental confirmation. For further details, the reader is referred to the reviews by Maki[1] and de Gennes.[2]

8-2.1 Depression of T_c by Magnetic Perturbations

In the AG theory and its extensions to other magnetic perturbations, the pair-breaking strength is characterized by the typical energy difference 2α it causes between time-reversed electrons. Although this pair-breaking energy was simply the constant energy splitting between $\mathbf{k}$ and $-\mathbf{k}$ electrons in the presence of drift momentum $\mathbf{q}$ for the pure metal, in the present ergodic case the scattering is so rapid that the depairing energy is constantly changing, tending to average out. The relevant time scale for averaging is the time required for the relative phase of the two time-reversed electrons to be randomized by the perturbation. This is brought out most clearly in de Gennes' version of the theory, in which the effective pair-breaking energy 2α is called $\hbar/\tau_K$, where τ_K is essentially this time. Evidently, an energy difference $\hbar/\tau_K$ operating over a time τ_K produces a phase shift of order unity.

This idea may be clarified by an example. Consider the effect of a magnetic field on a particle of dirty superconductor small enough that no vortex structure is allowed. The leading term in the magnetic perturbation is $(e/mc)\mathbf{p_k} \cdot \mathbf{A}$, so that the difference in its energetic effect on electrons $\mathbf{k}$ and $-\mathbf{k}$ is $(2e/mc)\mathbf{p_k} \cdot \mathbf{A}$. Since the phase evolves as $e^{-iEt/\hbar}$, the differential change in phase is

$$\frac{d\varphi}{dt} = \left(\frac{2e}{\hbar c}\right)\mathbf{v_k} \cdot \mathbf{A} \qquad (8\text{-}18)$$

[1] K. Maki, in R. D. Parks (ed.), "Superconductivity," vol. II, chap. 18, Marcel Dekker, New York, 1969.

[2] P. G. de Gennes, "Superconductivity of Metals and Alloys," chap. 8, W. A. Benjamin, New York, 1966.

If the scattering time τ is short compared to $(d\varphi/dt)^{-1}$, then the phase change $(d\varphi/dt)\tau$ between collisions is small, and this $d\varphi/dt$ must be integrated over many fragments of trajectory before a phase difference ~ 1 is reached. Since $\mathbf{v_k}$ will change arbitrarily at each scattering event, the phase difference grows as the square root of the number of free paths, by a random-walk process. In time t, the spread in phase will be of the order of $(d\varphi/dt)(\tau t)^{1/2}$. For this to be of order unity, the time τ_K required is given by

$$\frac{1}{\tau_K} = \tau \left\langle \left(\frac{d\varphi}{dt}\right)^2 \right\rangle = \tfrac{1}{3}v_F^2\tau\left(\frac{2e}{hc}\right)^2 \langle A^2 \rangle = D\left(\frac{2\pi}{\Phi_0}\right)^2 \langle A^2 \rangle \qquad (8\text{-}19)$$

where $D = \tfrac{1}{3}v_F\ell = \tfrac{1}{3}v_F^2\tau$ is the electronic diffusion constant. We also recall that $2\alpha = h/\tau_K$ gives the connection between the two notational conventions.

The reduced T_c in the presence of such a pair breaker is found to be given by the implicit relation

$$\ln\frac{T_c}{T_{c0}} = \psi\left(\frac{1}{2}\right) - \psi\left(\frac{1}{2} + \frac{\alpha}{2\pi kT_c}\right) \qquad (8\text{-}20)$$

where our notation is that $T_c = T_c(\alpha)$, $T_{c0} = T_c(0)$, and $\psi(z) = \Gamma'(z)/\Gamma(z)$ is the digamma function. Expanding the digamma function about $\tfrac{1}{2}$ yields the result

$$k(T_{c0} - T_c) = \frac{\pi\alpha}{4} = \frac{(\pi/8)h}{\tau_K} \qquad (8\text{-}20a)$$

so that for weak pair breaking, the depression of T_c is linear in α. On the other hand, one finds that superconductivity is completely destroyed (that is, $T_c = 0$) for

$$2\alpha = \frac{h}{\tau_K} = 1.76kT_c = \Delta_{\mathrm{BCS}}(0) \equiv \Delta_{00} \qquad (8\text{-}20b)$$

The entire dependence of T_c on the pair-breaker strength α, or conversely, the temperature dependence of the critical pair-breaker strength $\alpha_c(T)$, is depicted in Fig. 8-1. The shaded region indicates the range of parameter values for which gapless behavior is predicted.

As emphasized by Maki and de Gennes, to the extent that all pair-breaking ergodic perturbations are equivalent to magnetic impurities in their effect on T_c, this function $\alpha_c(T)$ should be a *universal function* which can be applied to any of them. For the cases treated in Chap. 4 using the linearized GL equation (4-56) to find a critical field, it turns out that 2α is the lowest eigenvalue of that equation, written in the form

$$hD\left(\frac{\nabla}{i} - \frac{2e}{hc}\mathbf{A}\right)^2\Delta = 2\alpha\Delta \qquad (8\text{-}21)$$

Note that the result (8-19) obtained above is a special case of this relation, since Δ

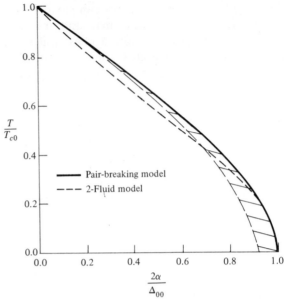

FIGURE 8-1
Universal functional relation between the pair-breaking parameter α and the reduced T_c is shown by solid curve. The shading depicts the gapless region. Values of 2α for various magnetic perturbations are given in (8-22a) to (8-22g). The dashed curve labelled two-fluid is a plot of $2\alpha/\Delta_{00} = (1 - t^2)/(1 + t^2)$, where $t = T/T_{c0}$. This relation reproduces the results of the GL critical field calculations building in the two-fluid temperature dependences $\lambda(t) \propto (1 - t^4)^{-1/2}$, $H_c(t) \propto (1 - t^2)$, and hence $\alpha_c \propto \xi^{-2}(t) \propto \lambda^2 H_c^2 \propto (1 - t^2)/(1 + t^2)$. [See (8-23).]

is constant over a small particle. Using our solutions from Chap. 4, we can write down the following cases:

$$\alpha = \frac{DeH}{c} \qquad \text{bulk type II in vortex state} \qquad (8\text{-}22a)$$

$$\alpha = 0.59\frac{DeH}{c} \qquad \text{surface sheath} \qquad (8\text{-}22b)$$

$$\alpha = \frac{1}{6}\frac{De^2H^2d^2}{\hbar c^2} \qquad \text{thin film, parallel field} \qquad (8\text{-}22c)$$

$$\alpha = \frac{DeH}{c} \qquad \text{thin film, perpendicular field} \qquad (8\text{-}22d)$$

$$\alpha = \frac{2De^2\langle A^2\rangle}{\hbar c^2} \qquad \text{small particle} \qquad (8\text{-}22e)$$

It is interesting to note that the angular dependence of the critical field of a thin film (4-70) can be obtained by adding (8-22c) and (8-22d), using the appropriate component of the total field in each case, and setting the sum equal to $\alpha_c(T)$. Note also that the critical fields found from (4-56) will agree with those found from (8-20) if the temperature-dependent coherence length is defined by

$$\xi^2(T) = \frac{D\hbar}{2\alpha_c(T)} = D\tau_{Kc}(T) \qquad (8\text{-}23)$$

In this way, the linearized GL equation can be used all the way down to $T = 0$ (for dirty materials for which the present theory is valid) even though the GL theory is normally valid only near T_c. It is for this reason that results of the GL theory often are found to have validity over a wider temperature range than might be expected.

The above pair-breaking effects have all acted on the orbital motion of the electrons. The original AG calculation dealt with the effect of a magnetic impurity coupled to the electron spin by an exchange interaction of the form $J(r)\mathbf{S} \cdot \mathbf{s}$, where $\mathbf{S}$ is the impurity spin and $\mathbf{s}$ is the spin of the conduction electron. Apart from numerical factors, the pair-breaking energy is given by

$$2\alpha \approx \frac{xJ^2}{E_F} \qquad (8\text{-}22f)$$

where x is the fractional impurity concentration and J is an average over the atomic volume.

There will also be a pair-breaking effect from the effect of an external magnetic field on the electronic spins. The appropriate $d\varphi/dt$ here is $2\mu_B H/\hbar = eH/mc$, and the appropriate scattering time is τ_{s0}, the time required for spin flip to occur by spin-orbit coupling associated with the scattering process. If these factors are used in (8-19), one finds

$$2\alpha \approx \frac{\tau_{s0} e^2 \hbar H^2}{m^2 c^2} \qquad (8\text{-}22g)$$

This result applies only when spin-orbit scattering is rapid enough that $\tau_{s0} eH/mc \ll 1$. In the other limit, one may neglect spin-orbit scattering. There is then no random-walk averaging, the pair-breaking energy 2α is simply $2\mu_B H$, and zero-energy excitations can occur when $\mu_B H = \Delta$. But before this field is reached, a first-order transition to the normal state occurs when $\mu_B H = \Delta_{00}/\sqrt{2}$ (at $T = 0$), as pointed out by Clogston and by Chandrasekhar.[1] This limits the critical field of a material with negligible spin-orbit scattering to a value $H_p = \Delta_0/\sqrt{2}\mu_B$. With $\Delta_0 = 1.76kT_c$, this works out to

$$\frac{H_p}{T_c} = 18,400 \text{ G}/^\circ\text{K} \qquad (8\text{-}24)$$

[1] A. M. Clogston, *Phys. Rev. Letters* **9**, 266 (1962); B. S. Chandrasekhar, *Appl. Phys. Letters* **1**, 7 (1962).

Many useful type II superconductors have $H_{c2} \gtrsim H_p$, indicating the importance of the randomizing effect of the spin-orbit scattering in reducing the pair-breaking effect of the Zeeman energy of the spins.

When several pair-breaking mechanisms are present, it is usually adequate to simply sum their contributions (8-22) to α in finding the critical condition for the destruction of superconductivity.

8-2.2 Density of States

In interpreting early experimental work on various properties (such as thermal conductivity,[1] microwave absorption,[2] and electron tunneling)[3] of superconducting films in magnetic fields, it was assumed that the density of states could be adequately described by retaining the BCS spectrum with a field-dependent gap $\Delta(H)$. However, the values of $\Delta(H)$ inferred in this way from different sorts of measurements were not really consistent, particularly at low temperatures. The discrepancies were of the sort that could be accounted for if the gap were "fuzzing out," with low-lying excitations coming in before the peak in the spectrum at the gap edge had entirely disappeared.

As noted above, just such a behavior is in fact predicted over a finite range of magnetic perturbation strength. The predicted state density for several representative cases is shown in Fig. 8-2, following the computations of Skalski et al.,[4] which appeared at about the same time. In this diagram, energies are normalized to Δ, the so-called gap parameter, which is a measure of the strength of the pairing potential at temperature T in the presence of the pair-breaking strength α. It is *not* the same as the minimum excitation energy, or spectral gap, which we denote Ω_G. This difference is illustrated in Fig. 8-3, which contrasts the dependence of Ω_G and of Δ (at $T = 0$) on α. Also shown is T_c/T_{c0}, the same curve as presented in Fig. 8-1. Note the sizable departures from the simple BCS theory, in which $\Omega_G(0) = \Delta(0) = 1.76kT_c$, so that all three quantities would scale in the same way. Finally, in Fig. 8-4 we show the temperature dependence of Δ/Δ_{00} for various values of $2\alpha/\Delta_{00}$. In this figure, the shaded area represents the gapless region, which occurs when $\alpha \geq \Delta$.

The detailed calculations leading to the above results are rather formidable. This enhances the interest in the approach of de Gennes,[5] which deals with the gapless regime where $\Omega_G = 0$ and $\Delta/\alpha \ll 1$. In this regime he showed that the pair potential Δ can be introduced by a second-order perturbation calculation

[1] D. E. Morris and M. Tinkham, *Phys. Rev.* **134**, A1154 (1964).

[2] R. H. White and M. Tinkham, *Phys. Rev.* **136**, A203 (1964).

[3] R. Meservey and D. H. Douglass, Jr., *Phys. Rev.* **135**, A24 (1964).

[4] S. Skalski, O. Betbeder-Matibet, and P. R. Weiss, *Phys. Rev.* **136**, A1500 (1964).

[5] P. G. de Gennes, "Superconductivity in Metals and Alloys," p. 265, W. A. Benjamin, New York, 1966.

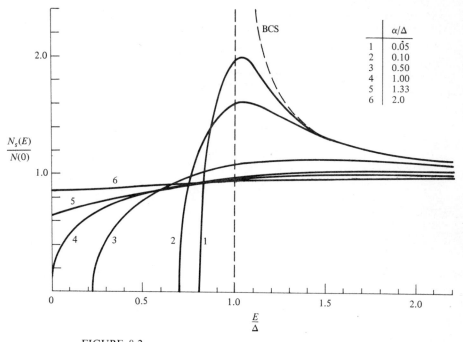

FIGURE 8-2

The density of states as a function of the reduced energy as computed by Skalski et al. for several values of the reduced pair-breaking strength α/Δ. In this diagram, Δ is to be understood as $\Delta(T, \alpha)$.

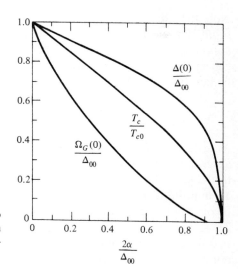

FIGURE 8-3

Decrease of spectral gap Ω_G and gap parameter Δ at $T = 0$, and of transition temperature T_c, with increasing pair-breaking strength α. (*After Skalski et al.*)

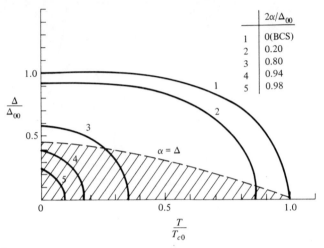

	$2\alpha/\Delta_{00}$
1	0(BCS)
2	0.20
3	0.80
4	0.94
5	0.98

FIGURE 8-4
Temperature dependence of pair potential or gap parameter Δ for various pair-breaker strengths, as computed by Skalski et al. The spectral gap Ω_G is zero in the shaded region defined by $\alpha > \Delta$.

(although this would have failed in the absence of strong pair breaking). The result is that the excitation energies are changed from $|\xi|$ in the absence of Δ to

$$E = |\xi|\left(1 + \frac{1}{2}\frac{\Delta^2}{\xi^2 + \alpha^2}\right) \qquad \Delta \ll \alpha \qquad (8\text{-}25)$$

Note that for $|\xi| \gg \alpha$, $E \approx |\xi| + \Delta^2/2|\xi| \approx (\Delta^2 + \xi^2)^{1/2}$, which is the BCS result for ordinary superconductors. On the other hand, for $|\xi| \ll \alpha$, E approaches $|\xi|(1 + \Delta^2/2\alpha^2)$, which is proportional to $|\xi|$ and hence shows no gap in the spectrum. We can readily compute the density of states from (8-25), finding (to order Δ^2) that

$$\frac{N_s(E)}{N(0)} = \frac{d\xi}{dE} = 1 + \frac{\Delta^2}{2}\frac{E^2 - \alpha^2}{(E^2 + \alpha^2)^2} \qquad \Delta \ll \alpha \qquad (8\text{-}26)$$

Note that $N_s(E) < N(0)$ for $E < \alpha$, but it remains finite, corresponding to the absence of a gap. On the other hand, for $E > \alpha$, $N_s(E) > N(0)$. This is analogous to the peaked density of states above the gap in a BCS superconductor, but with an important difference: the energy scale is set by α, not by Δ. Moreover, this theory applies only when Δ is small, i.e., near the transition to the normal state. Thus, the value of α must be nearly $\alpha_c(T)$. This leads to the remarkable conclusion that the energy scale for the deviations of $N_s(E)$ from $N(0)$ is a function only of T, not of the value of Δ, which goes to zero as the transition is approached. In other words, Δ controls only the amplitude of the deviations of $N_s(E)$ from $N(0)$, not their spectral distribution.

The most detailed comparison between theory and experiment is offered by the electron-tunneling measurements of the density of states. These were first carried out at sufficiently low temperatures by Woolf and Reif[1] on lead and indium films containing magnetic impurities. They found good agreement with the theoretical predictions in the case of a rare-earth impurity (Gd), but with Fe or Mn impurities, the gap seemed to be filled in even more rapidly than predicted. They attributed this difference to the fact that the $4f$ electrons of Gd should closely resemble the localized moments assumed in the AG theory, while the $3d$ electrons of Fe and Mn interact more strongly with the conduction electrons, and thus are less localized.

Tunneling experiments using a thin film in a parallel magnetic field offer a cleaner test of the theory, since they avoid this question of the degree of localization of the magnetic moment. Moreover, the field can be varied at will, whereas a uniform concentration of impurities is notoriously difficult to achieve, and each concentration requires a different specimen. Careful tunneling experiments in a magnetic field were carried out by Levine[2] and by Millstein and Tinkham[3]. In the latter experiments, for example, the films were of tin or of tin-indium alloys, of thickness $d \sim 1000$ Å and with mean free paths $\ell \sim 300$ to 1200 Å. Thus ℓ was less than, but still comparable with, the coherence length $\xi_0 = 2300$ Å. The other film in the tunneling sandwich was aluminum with sufficient Mn impurity to keep it in the normal state. Measurements of the density of states were carried out using ^{3}He cooling at $0.36°$K, about $T_c/10$. Even so, the thermal smearing due to the width of the cutoff of the Fermi function leads to quite a sizable difference between the differential conductance (dI/dV) and the true density of states. [For example, comparing (2-79) and (2-80), we see that the normalized differential conductivity $\sigma(V) = G_{ns}(V)/G_{nn}$ measures an average of $N_s(E)/N(0)$ over an energy range of several kT about eV, weighted with the bell-shaped function df/dE.] Thus, in comparing experimental curves with theoretical ones, it is essential first to fold the thermal smearing into the theoretical curve if any real accuracy is to be obtained. When this was done, the $\sigma(V)$ curves were found to be in semiquantitative agreement with the transcribed Abrikosov-Gor'kov theory, whereas any attempt to fit the data with a BCS density of states with modified gap value was completely hopeless.

Still, there was a residual discrepancy, apparently outside experimental error, in that the observed $\sigma(V)$ curves were even more flattened out than the thermally smeared AG curves. Also, it appeared that the films became gapless for fields substantially less than the $(0.91)^{1/2} H_{c\|}$ expected from the theory. A possible

[1] M. A. Woolf and F. Reif, *Phys. Rev.* **137**, A557 (1965).

[2] J. L. Levine, *Phys. Rev.* **155**, 373 (1967).

[3] J. Millstein and M. Tinkham, *Phys. Rev.* **158**, 325 (1967).

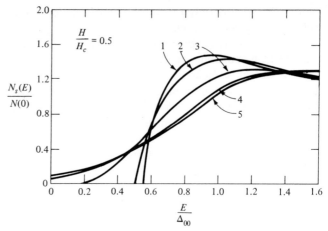

FIGURE 8-5
The density of states as a function of energy E as computed by Strässler and Wyder for several values of the mean free path and a magnetic field of half the critical field of the small spheres. [1: $\ell = 0$; 2: $\ell = (\pi/10)\xi_0$; 3: $\ell = \pi\xi_0$; 4: $\ell = 10\pi\xi_0$; 5: $\ell = \infty$.] Curve 1 corresponds to the calculations of Skalski et al. (Fig. 8-2), based on Abrikosov-Gor'kov theory. Curve 5 corresponds to the limit treated by Larkin.

explanation for these discrepancies was suggested by the work of Larkin,[1] who had treated the very different idealized case of small superconducting spheres in a magnetic field, under the assumption of specular surface scattering and no volume scattering. He found that the spectrum became gapless at about $0.4H_c$, much *lower* than found experimentally. Strässler and Wyder[2] then carried out an extension of Larkin's approach, putting in volume scattering. By varying ℓ/ξ_0 from 0 to ∞, they could compute density-of-states curves, such as those shown in Fig. 8-5 for the particular case where the applied field is half the critical field. The limiting case of $\ell = 0$ corresponds to the curve computed by Skalski et al. (Fig. 8-2), whereas the limit $\ell = \infty$ corresponds to Larkin's result. The differences among these curves can be understood qualitatively by observing that without scattering, some electron states are perturbed very strongly by the field and become gapless in low field; with rapid scattering, each electron feels an averaged strength of the perturbation, effectively eliminating the strong effect on any one. Returning to the experiments, essentially perfect agreement with the data was obtained if ℓ/ξ_0 for the various films was set equal to values in the range 0.3 to 1, which were within a

[1] A. Larkin, *Zh. Eksperim. Teor. Fiz.* **48**, 232 (1965) [*Soviet Phys.—JETP* **21**, 153 (1965)].
[2] S. Strässler and P. Wyder, *Phys. Rev.* **158**, 319 (1967).

factor of 2 of independent estimates of this ratio. Considering the uncertainties in these estimates and the idealizations of the theoretical model, such a degree of agreement is quite satisfactory.

In conclusion, it is interesting to note that the temperature dependence of H_c [or, equivalently, $T_c(H)$] found by Millstein and Tinkham was in quantitative agreement with the AG result. One might ask why the finite mean-free-path effects treated by Strässler and Wyder were not evident here. The answer seems to be simply that $T_c(H)$, being an integral property, is less sensitive to details of the model than is the more microscopic $N_s(E)$.

8-3 TIME-DEPENDENT GINZBURG-LANDAU THEORY

In view of the enormous success of the Ginzburg-Landau theory of thermodynamic equilibrium properties of superconductors near T_c, it is natural to seek a time-dependent generalization leading to a differential equation for the space and time dependence of the order parameter Δ. A number of authors [1, 2] have attacked this problem, and a critical review of this work has been given recently by Cyrot.[3] From this work, it has emerged that it is very difficult to obtain a nonlinear time-dependent Ginzburg-Landau (TDGL) equation of any generality. As noted by Gor'kov and Eliashberg,[2] this difficulty stems essentially from the singularity in the density of states at the gap edge, which leads to slowly decaying oscillatory responses in the time domain. As we have seen in Sec. 8-2, the presence of magnetic impurities or other pair breakers rounds off the singularity in the BCS density of states, and with sufficient pair-breaking strength, the spectrum becomes gapless. In this gapless regime, one can make an expansion in powers of Δ/α and ω/α, where $\alpha = \hbar/2\tau_K$ is the usual pair-breaking parameter discussed in the previous section. In this way, Gor'kov and Eliashberg obtained the only rigorous version of a nonlinear TDGL equation available at this time which is valid for a realistic range of fields and frequencies; but it is of course restricted to a gapless superconductor. Schmid had previously obtained equations of similar form without imposing this restriction, but apparently[4] they lack rigorous justification except in a gapless regime.

[1] See, for example, A. Schmid, *Phys. Kondens. Mat.* **5**, 302 (1966); C. Caroli and K. Maki, *Phys. Rev.* **159**, 306, 316 (1967); **164**, 591 (1967); E. Abrahams and T. Tsuneto, *Phys. Rev.* **152**, 416 (1966); J. W. F. Woo and E. Abrahams, *Phys. Rev.* **169**, 407 (1968); R. S. Thompson, *Phys. Rev.* **B1**, 327 (1970).

[2] L. P. Gor'kov and G. M. Eliashberg, *Zh. Eksperim. i Teor. Fiz.* **54**, 612 (1968) [*Soviet Phys.—JETP* **27**, 328 (1968)].

[3] M. Cyrot, *Repts. Prog. Phys.* **36**, 103 (1973).

[4] G. M. Eliashberg, *Zh. Eksperim. i Teor. Fiz.* **55**, 2443 (1968) [*Soviet Phys.—JETP* **29**, 1298 (1969)].

The result of Gor'kov and Eliashberg is conveniently written in the normalized form adopted by Hu and Thompson,[1] namely, the coupled set of equations

$$D^{-1}\left(\frac{\partial}{\partial t} + i\frac{2e\psi}{\hbar}\right)\Delta + \xi^{-2}(|\Delta|^2 - 1)\Delta + \left(\frac{\nabla}{i} - \frac{2e}{\hbar c}\mathbf{A}\right)^2\Delta = 0 \qquad (8\text{-}27)$$

$$\mathbf{J} = \sigma\left(-\nabla\psi - \frac{1}{c}\frac{\partial\mathbf{A}}{\partial t}\right) + \mathrm{Re}\left[\Delta^*\left(\frac{\nabla}{i} - \frac{2e}{\hbar c}\mathbf{A}\right)\Delta\right]\frac{1}{8\pi e\lambda^2} \qquad (8\text{-}28)$$

$$\rho = \frac{\psi - \varphi}{4\pi\lambda_{\mathrm{TF}}^2} \qquad (8\text{-}29)$$

plus the Maxwell equations coupling the scalar and vector potentials φ and $\mathbf{A}$ to the charge and current densities ρ and $\mathbf{J}$. Here D is the normal-state diffusion constant and ψ is the electrochemical potential divided by the electronic charge. Δ is the gap parameter divided by its equilibrium value in the absence of fields $\Delta_0 = \pi k[2(T_c^2 - T^2)]^{1/2}$, where T is the temperature and T_c is the critical temperature (in the presence of the magnetic impurities); thus $\Delta = 1$ in the absence of fields. The temperature-dependent coherence length is $\xi = \hbar(6D/\tau_s)^{1/2}/\Delta_0$, where τ_s is the spin-flip scattering time, and the temperature-dependent magnetic penetration depth is $\lambda = \hbar c(8\pi\sigma\tau_s)^{-1/2}/\Delta_0$. σ is the normal-state conductivity, and λ_{TF} is the Thomas-Fermi static-charge screening length. It is perhaps appropriate to mention at this point that these equations necessarily assume that heating effects due to the dissipation of energy by the time-varying fields and currents can be neglected. As Gor'kov and Eliashberg remarked, this restricts the applicability of the theory to very near T_c unless there is a very high density of paramagnetic impurities.

Because of the complexity of working with these coupled nonlinear partial differential equations, we shall content ourselves with mentioning a few of the applications which have been made. In their original paper, Gor'kov and Eliashberg first considered the response of a paramagnetic alloy superconductor to a strong variable magnetic field. They worked out an approximate solution for the case of the skin effect with a bulk sample, finding, for example, the amount of third harmonic radiation generated by the nonlinear surface currents. They then treated the case of a thin film sample of thickness d, subjected to an oscillating field which is the same on both sides of the film and parallel to it. For this simple case, they were able to work out rather exact results, which reduce to intuitively plausible limits for frequencies well above and well below a characteristic relaxation rate τ^{-1}. In fact, for $\omega\tau \ll 1$, they found that the order parameter follows the alternating field adiabatically, with $\Delta(t)/\Delta_0 = [1 - h^2(t)]^{1/2}$, where $h = H/H_{c\parallel}$, as expected from (4-52) since Δ is proportional to the GL ψ. In the high-frequency limit, one instead finds $h^2(t)$ replaced by the time average value $\overline{h^2(t)}$. The charac-

[1] C. R. Hu and R. S. Thompson, *Phys. Rev.* **B6,**, 110 (1972).

teristic relaxation frequency separating these regimes is found to be $1/\tau = e^2 d^2 D H_{c\parallel}^2 / 3c^2 \hbar$. Comparing this expression with (8-22c), and recalling that $2\alpha/\hbar = 1/\tau_K$, we see that this τ is exactly the same as τ_{Kc}. Thus, near T_c, we have

$$\tau = \frac{\pi \hbar}{8k(T_c - T)} \qquad (8\text{-}30)$$

using (8-20a). This is also the relaxation-time expression anticipated in (7-47), in our discussion of fluctuation-enhanced conductivity.

Following the pioneering work of Schmid, and of Caroli and Maki, Thompson and collaborators have applied these equations to an extensive analysis of the dynamic structure of vortices moving in type II superconductors in the resistive mixed state or flux-flow regime, discussed more qualitatively in Sec. 5-5. The Schmid-Caroli-Maki solution was obtained by assuming that there was a uniform electric field associated with a uniformly translating Abrikosov vortex lattice; (8-27) was then solved for Δ, (8-28) used to compute $\mathbf{J}(\mathbf{r})$, and the ratio of the averaged current density $\langle \mathbf{J} \rangle$ to the assumed uniform $\mathbf{E}$ used to define an effective flux-flow conductivity. Thompson and Hu showed that for this solution, $\nabla \cdot \mathbf{J} \neq 0$, so that charge would build up to generate a nonuniform $\mathbf{E}$ until a new steady-state solution was established. A central feature of this Thompson-Hu solution is a backflow current $\mathbf{J}_b$, which goes through the vortex cores and returns around their sides. This current is lossless, although it flows directly through the core, which illustrates the degree of oversimplification implicit in the "normal" core of Bardeen and Stephen. The order parameter vanishes only along a line at the axis of the vortex, and the presence of low-lying excitations and the absence of an energy gap do not preclude the existence of supercurrents in the core region. One of the quantitative results of the calculation is that the normalized flux-flow resistance $R = \rho_f/\rho_n$ for a high-κ superconductor (containing a large concentration of paramagnetic impurities) should follow a concave-upward curve between $B = 0$ and $B = H_{c2}$, rather than the simple linear relation (5-59). The computed initial slope $dR/d(B/H_{c2})$ is 0.33 and the slope at H_{c2} is computed to be 5.2. Experimental data have this general shape, but there are experimental difficulties in isolating ideal behavior, free from pinning effects.

Probably the most convincing experimental test of the TDGL theory is actually in the measurements of fluctuation conductivity *above* T_c, as outlined in Sec. 7-6. Above T_c, one automatically has a gapless superconductor, but even so there are complications as discussed there unless the so-called Maki terms are suppressed by residual pair-breaking effects of some sort.

8-3.1 Electron-Phonon Relaxation

In many cases, one is interested in understanding the relaxation processes of superconductors which are *not* doped with magnetic impurities, so the Gor'kov-Eliashberg theory is not directly applicable. In the absence of magnetic pairbreakers, one must rely on inelastic phonon-electron interactions to achieve equilibrium between quasi-particles and condensate (by creation and recombination processes) and within the quasi-particle population (by inelastic scattering). Near T_c the characteristic time for both types of processes is the inelastic phonon-scattering time τ_2 for electrons in the normal state at T_c, although there will be differences depending on the nature of the disequilibrium being relaxed. The most direct way of estimating this time is from the low-temperature phonon-limited electronic thermal conductivity. (One cannot use the scattering time from low-temperature *electrical* conductivity, since small-angle scattering by low-energy phonons is ineffective in producing electrical resistance, but it *is* effective in relaxing a thermal energy distribution.) A simple, but less reliable, estimate can be made from the room-temperature conductivity (where the phonons are equally effective in causing both electrical and thermal resistivity). Assuming a simple Debye phonon spectrum and free-electron Fermi surface, the high-temperature scattering time τ_{300} scales to

$$\tau_2(T_c) = C\tau_{300}\,\frac{300\Theta_D^2}{T_c^3} \qquad (8\text{-}31)$$

where the coefficient C is estimated[1] to be 1/93 for an average thermal quasi-particle but 1/8.4 for a quasi-particle[2] right at the Fermi surface, where the typical energy changes are less. In any case, $\tau_2(T_c)$ varies roughly as $(\Theta_D/T_c)^3$. For tin, it is of the order of 10^{-10} to 10^{-9} sec; it is much shorter in lead, much longer in aluminum.

Because the detailed results depend on exactly what sort of relaxation process is required, and because understanding of this subject is still being developed, we shall content ourselves with a rather qualitative discussion of some of the results that have been obtained. An important early step was made by Schmid,[3] who treated the case of a small deviation from a spatially uniform equilibrium pair density. For the simplest case, in which the deviation as well as the background is spatially uniform, he found a relaxation time

$$\tau \approx \tau_2(T_c)(1-t)^{-1/2} \qquad t = \frac{T}{T_c} \qquad (8\text{-}32)$$

so long as $\Delta > 1/\tau_2$. If this inequality is reversed, the collision broadening of the energy levels exceeds the gap, the spectrum is effectively gapless, and one recovers essentially the time (8-30) appropriate to a gapless superconductor. This latter case typically holds only for $(1-t) \lesssim 10^{-3}$, so (8-32) is the usual temperature dependence. Schmid also noted that if the nonequilibrium density is spatially modulated, the relaxation rate rises to approach a saturation value $\sim 1/\tau_2$. The prediction (8-32) has recently received rather direct confirmation by Peters and

[1] J. M. Ziman, "Electrons and Phonons," p. 391, Oxford University Press, London, 1960.
[2] M. Tinkham, *Phys. Rev.* **B6**, 1747 (1972).
[3] A. Schmid, *Phys. Kondens. Mat.* **8**, 129 (1968).

Meissner,[1] who measured the frequency dependence between 30 and 1,000 MHz of the nonlinearity of the kinetic inductance of a superconducting tin microbridge.

A rather similar temperature-dependent relaxation time τ_Q was inferred by Clarke[2] for the imbalance of the number of excitations on the holelike $(k < k_F)$ and electronlike $(k > k_F)$ branches of the excitation spectrum. This imbalance (denoted Q) was created by tunneling injection of quasi-particles predominantly onto one branch, and the steady-state imbalance was inferred from a measurement of the difference between the chemical potential of the superfluid pairs and the quasi-particle potential, using a superconducting voltmeter. The detailed theory[3] of this relaxation process gives results in quite satisfactory agreement with the experiments. For this process, the temperature dependence of τ_Q can be understood in terms of a simple argument: Scattering from the electronlike to holelike branch is essentially forbidden by a selection rule except for those quasi-particle states with energy between Δ and $\sim 2\Delta$, which have a mixed character; the fraction of scattering processes involving such states is of order $\Delta/kT_c \propto (1 - t)^{1/2}$, leading to the temperature dependence $\tau_Q \approx \tau_2(1 - t)^{-1/2}$.

Very recently, Skocpol, Beasley, and Tinkham[4] have shown that the curiously regular steps in the I-V curves of superconducting microbridges can be understood in terms of the establishment of localized quantum phase-slip centers, in each of which the order parameter executes a relaxation oscillation[5] at the Josephson frequency corresponding to the voltage across the center. This voltage, in turn, is determined by the quasi-particle current which must flow to carry a total current above the critical current I_c of the bridge. According to their analysis, the effectively normal length associated with each center is 2Λ, where $\Lambda = (\frac{1}{3}v_F \tau_2 \ell)^{1/3}$ is the length[6] a quasi-particle diffuses in a random-walk motion before suffering an *in*elastic collision, allowing relaxation in energy. In this case, the experiments show that the temperature-independent $\tau_2(T_c)$ is the relevant time, rather than something like (8-32). It may be that this difference results because the nonequilibrium quasi-particles are mostly created in the energy range near Δ, where the branch-crossing process is uninhibited.

An attempt at formulating rather general kinetic equations for nonequilibrium conductors has been made by Galaiko.[7] He finds that the relaxation of a system to an equilibrium state takes place in two stages. First, a fast quantal rearrangement of the system occurs, a self-consistent field of the condensate is

[1] R. Peters and H. Meissner, *Phys. Rev. Letters* **30**, 965 (1973).

[2] J. Clarke, *Phys. Rev. Letters* **28**, 1363 (1972).

[3] M. Tinkham and J. Clarke, *Phys. Rev. Letters* **28**, 1366 (1972); M. Tinkham, *Phys. Rev.* **B6**, 1747 (1972).

[4] W. J. Skocpol, M. R. Beasley, and M. Tinkham, *J. Low Temp. Phys.* **16**, 145 (1974).

[5] H. A. Notarys and J. E. Mercereau, *Physica* **55**, 424 (1971); T. J. Rieger, D. J. Scalapino, and J. E. Mercereau, *Phys. Rev. Letters* **27**, 1787 (1971); *Phys. Rev.* **B6**, 1734 (1972).

[6] This quasi-particle diffusion length was introduced in connection with S/N interfaces by A. B. Pippard, J. G. Shepherd, and D. A. Tindall, *Proc. Roy. Soc. (London)* **A324**, 17 (1971).

[7] V. P. Galaiko, *Zh. Eksperim. i Teor. Fiz.* **61**, 382 (1971) [*Soviet Phys.—JETP* **34**, 203 (1972)]; *Zh. Eksperim. i Teor. Fiz.* **64**, 1824 (1973) [*Soviet Phys.—JETP* **37**, 922 (1973).]

formed, and a quasi-particle description of the excitations becomes valid. After that, the quasi-particle distribution and the order parameter slowly evolve due to collisions between the excitations. He also finds that the normal-state inelastic scattering time τ_2 sets the time scale for the latter process, but it is not clear whether any simple phenomenological formulation of general validity will emerge from his approach. Thus, the question of general time-dependent generalizations of the Ginzburg-Landau equilibrium theory remains open and of considerable interest at this time.

APPENDIX

UNITS AND NOTATION

This book has been written using gaussian cgs units throughout (except for a few formulas quoted in practical units for convenience of application, in which case the units are explicitly mentioned). This choice conforms to that used in almost all of the literature on which the book is based. Moreover, it has the convenience that $\mathbf{B} = \mathbf{H}$ in vacuum, and that the electric field associated with a moving flux density has the natural form $\mathbf{E} = (\mathbf{v}/c) \times \mathbf{B}$. Readers who are uncomfortable with these units will find an excellent guide to their relation to other systems, such as the mks or practical system, in the appendix of J. D. Jackson, "Electromagnetic Theory," Wiley, New York, 1962, or the appendix of W. K. H. Panofsky and M. Phillips, "Classical Electricity and Magnetism," Addison-Wesley, Reading, Mass., 1962.

Because of the ubiquity of magnetic fields in this subject, special notational conventions have become common to simplify the discussion. Basically we follow the convention of de Gennes (and many other authors) and use $\mathbf{h}(\mathbf{r})$ to denote the local value of magnetic induction or flux density, which may vary on the scale of λ. We reserve the use of $\mathbf{B}$ to denote the value of $\mathbf{h}$ averaged over such microscopic lengths, but still capable of varying smoothly over the macroscopic dimensions of the sample.

In normal metal or vacuum, of course, there is no microscopic variation of $\mathbf{h}$ (we neglect the Landau diamagnetism and Pauli paramagnetism), so $\mathbf{B} = \mathbf{h}$. In these cases, also $\mathbf{B} = \mathbf{H}$, so all three symbols denote equal quantities and may be used interchangeably.

In the Meissner state of a massive superconductor, **h** is reduced to zero in a penetration depth λ by supercurrents in the skin layer, as described by the Maxwell equation

$$\text{curl } \mathbf{h} = \frac{4\pi \mathbf{J}_{tot}}{c} \qquad \text{(A-1)}$$

Hence, $\mathbf{B} = 0$ inside. On the other hand, **H** is governed by the Maxwell equation

$$\text{curl } \mathbf{H} = \frac{4\pi \mathbf{J}_{ext}}{c} \qquad \text{(A-2)}$$

where $\mathbf{J}_{ext}$ excludes currents arising from the equilibrium response of the medium, such as those in the penetration depth described by the London equation. Hence curl $\mathbf{H} = 0$, and the tangential component H_t is constant through the skin depth, retaining the value of H_t $(= B_t = h_t)$ found outside the sample. If the sample is ellipsoidal, H inside is uniform and everywhere equals the equatorial value of H_t. Of course, the actual value of this H_t will in general exceed the uniform applied field H_a by a factor $(1 - \eta)^{-1}$, where η is the shape-dependent demagnetizing factor of the sample. [See discussion associated with (3-47).]

In the intermediate state of a type I superconductor, which is reached when $H_t = H_c$, the magnitude of **h** varies continuously between H_c in the normal laminae and zero in the superconducting ones. **B** is the average of **h** over this laminar structure, and is constant within an ellipsoidal sample. The magnitude of **H** must be H_c for coexistence of superconducting and normal regimes to be possible. These interrelations are illustrated in Fig. 3-7.

In the mixed state of a type II superconductor in a magnetic field, **h** varies on the microscopic scale of the vortex structure, while **B** is the average of **h** over this structure. In the ideal equilibrium case, $|\mathbf{H}|$ is again everywhere equal to H_t at the equatorial surface. In the presence of transport currents and of disequilibrium due to flux pinning, **H** will vary because curl **H** is no longer zero. This situation is discussed more fully in connection with (5-48) in the text.

For notational symmetry, we also can define a microscopically varying electric field **e(r)**, whose macroscopic average value is **E**. But because curl $\mathbf{e} = -\dot{\mathbf{h}}/c$, **e** is uniform in static situations, and it is zero in equilibrium ones. Thus, the distinction between **e** and **E** arises less frequently than the distinction between **h** and **B** (which can result from *equilibrium* supercurrents); consequently, we have normally used **E** for both to avoid confusion with the electronic charge e. The notation **e** is introduced only to describe the electric field distribution about a moving vortex, where the macroscopic average **E** is quite different from **e** and gives the physically important resistive voltage.

The following list of references is not exhaustive, but it includes many of the standard references with some brief indication of their individual features. Because the literature on applications is particularly diffuse and hard to locate, a separate list of such references is also included.

Fundamental physics: books

BLATT, J. M.: "Theory of Superconductivity," Academic Press Inc., New York, 1964. A treatment emphasizing the quasi-chemical equilibrium view of the superfluid state.

DE GENNES, P. G.: "Superconductivity in Metals and Alloys," W. A. Benjamin, Inc., New York, 1966. An excellent, physically motivated treatment.

KUPER, C. G.: "Introduction to the Theory of Superconductivity," Clarendon Press, Oxford, 1968. A somewhat more elementary treatment.

LONDON, F.: "Superfluids," vol. 1, John Wiley & Sons, Inc., New York, 1950. Discusses the London theory and its philosophical background; still valuable for its thoughtful discussions; contains famous footnote on p. 152 predicting fluxoid quantization over a decade before it was discovered experimentally.

LYNTON, E. A.: "Superconductivity," 3d ed., Methuen & Co., Ltd., London, 1969. A good concise survey.

RICKAYZEN, G.: "Theory of Superconductivity," John Wiley & Sons, Inc., New York, 1965. A good general survey.

ROSE-INNES, A. C., and F. H. RHODERICK: "Introduction to Superconductivity," Pergamon Press, New York, 1969. An introductory treatment on roughly the same level as this book.

SAINT-JAMES, D., G. SARMA, and E. J. THOMAS: "Type II Superconductivity," Pergamon Press, New York, 1969. A richly detailed review of this subject, emphasizing theory at the Ginzburg-Landau level.

SCHRIEFFER, J. R.: "Theory of Superconductivity," W. A. Benjamin, Inc., New York, 1964. A good account of the theory by one of its founders, including Green function topics which are not treated in this book.

SHOENBERG, D.: "Superconductivity," Cambridge University Press, New York, 1952. A classic survey, with emphasis on experimental results, pre-BCS.

Fundamental physics: reviews and collections

ANDERSON, P. W.: The Josephson Effect and Quantum Coherence Measurements in Superconductors and Superfluids, in C. J. Gorter (ed.), "Progress in Low Temperature Physics," vol. V, North-Holland Publishing Company, Amsterdam, 1967. Notable for its insight into this topic.

BARDEEN, J.: Theory of Superconductivity, in S. Flügge (ed.), "Handbuch der Physik," vol. 15, Springer-Verlag OHG, Berlin, 1956. A masterful survey of the state of the theory immediately *before* BCS.

BARDEEN, J., and J. R. SCHRIEFFER: Recent Developments in Superconductivity, in C. J. Gorter (ed.), "Progress in Low Temperature Physics," vol. III, North-Holland Publishing Company, Amsterdam, 1961. Reviews initial post-BCS period.

DEWITT, C., B. DREYFUS, and P. G. DE GENNES (eds): "Low Temperature Physics," Gordon and Breach, Science Publishers, Inc., New York, 1962. Contains lectures at the Les Houches summer school of 1961, including a chapter by M. Tinkham which has been reprinted as a separate volume under the title "Superconductivity." Although dated, this contains little that is not still correct, and contains a more detailed treatment of the sum rule and Kramers-Kronig arguments than is included in the present volume.

DOUGLASS, D. H., Jr., and L. M. FALICOV: The Superconducting Energy Gap, in C. J. Gorter (ed.), "Progress in Low Temperature Physics," vol. IV, North-Holland Publishing Company, Amsterdam, 1964. An extensive survey of experimental results and associated theory.

GLOVER, R. E., III: Superconductivity Above the Transition Temperature, in C. J. Gorter (ed.), "Progress in Low Temperature Physics," vol. VI, North-Holland Publishing Company, Amsterdam, 1970. A survey of fluctuation-enhanced conductivity above T_c.

PARKS, R. D. (ed.): "Superconductivity," vols. I and II, Marcel Dekker, Inc., New York, 1969. This two-volume treatise, with chapters written by two dozen distinguished authors on their special areas of interest, is by far the most comprehensive available treatment of the subject as it stood in 1968.

SERIN, B.: Superconductivity, Experimental Part, in S. Flügge (ed.), "Handbuch der Physik," vol. 15, Springer-Verlag OHG, Berlin, 1956. Gives survey of experimental position in 1956.

WALLACE, P. R. (ed.): "Superconductivity," vols. I and II, Gordon and Breach, Science Publishers, Inc., New York, 1969. This is another two-volume treatment, containing the lectures at a summer institute on superconductivity held at McGill University in

1968. It is considerably less comprehensive than the Parks treatise, and marred by typographical errors, but some of the lectures are useful from a pedagogical point of view.

Applications:

FONER, S., and B. B. SCHWARTZ (eds.): "Superconducting Machines and Devices: Large Systems Applications," Plenum Press, Plenum Publishing Corporation, New York, 1974. Proceedings of a summer school. Broad coverage of international effort in transmission lines, rotating machinery, levitation, etc.

GREGORY, W. D., W. N. MATHEWS, and E. A. EDELSACK (eds.): "The Science and Technology of Superconductivity," in 2 volumes, Plenum Press, Plenum Publishing Corporation, New York, 1973. Proceedings of a summer school.

NEWHOUSE, V. L.: "Applied Superconductivity," John Wiley & Sons, Inc., New York, 1964. A well-written survey, but unfortunately predates the great developments of type II superconductivity and Josephson-effect devices.

Proc. IEEE, Special Issue on "Applications of Superconductivity," January, 1973 (vol. 61, no. 1).

RESEARCH ADVISORY INSTITUTE, "Device Applications of Cryogenics," 1971; publication AD 729697, available through the National Technical Information Service, U.S. Dept. of Commerce, Springfield, Va. 27151. A very complete survey of small-scale device applications, featuring Josephson effect, compiled by experts in the field. Coverage is complementary to that of Foner and Schwartz, above.

INDEXES

NAME INDEX

SUBJECT INDEX

SUFFOLK UNIVERSITY LIBRARY